The Cosmos

by Chris De Pree, PhD

A member of Penguin Group (USA) Inc.

For Papa, W. H. "Bud" Stuckey (1920–2013), who now knows all the secrets of the cosmos.

ALPHA BOOKS

Published by Penguin Group (USA) Inc.

Penguin Group (USA) Inc., 375 Hudson Street, New York, New York 10014, USA • Penguin Group (Canada), 90 Eglinton Avenue East, Suite 700, Toronto, Ontario M4P 2Y3, Canada (a division of Pearson Penguin Canada Inc.) • Penguin Books Ltd., 80 Strand, London WC2R 0RL, England • Penguin Ireland, 25 St. Stephen's Green, Dublin 2, Ireland (a division of Penguin Books Ltd.) • Penguin Group (Australia), 250 Camberwell Road, Camberwell, Victoria 3124, Australia (a division of Pearson Australia Group Pty. Ltd.) • Penguin Books India Pvt. Ltd., 11 Community Centre, Panchsheel Park, New Delhi—110 017, India • Penguin Group (NZ), 67 Apollo Drive, Rosedale, North Shore, Auckland 1311, New Zealand (a division of Pearson New Zealand Ltd.) • Penguin Books (South Africa) (Pty.) Ltd., 24 Sturdee Avenue, Rosebank, Johannesburg 2196, South Africa • Penguin Books Ltd., Registered Offices: 80 Strand, London WC2R 0RL, England

International Standard Book Number: 978-1-61564-603-6
Library of Congress Catalog Card Number: 2014935263

16 15 14 8 7 6 5 4 3 2 1

Interpretation of the printing code: The rightmost number of the first series of numbers is the year of the book's printing; the rightmost number of the second series of numbers is the number of the book's printing. For example, a printing code of 14-1 shows that the first printing occurred in 2014.

Printed in the United States of America

Publisher: Mike Sanders
Executive Managing Editor: Billy Fields
Senior Acquisitions Editor: Tom Stevens
Development Editor: Kayla Dugger
Senior Production Editor: Janette Lynn
Design Supervisor: William Thomas
Indexer: Johnna VanHoose Dinse
Layout: Brian Massey, Ayanna Lacey
Proofreader: Monica Stone
Illustrators: Toby Mikle, Paul Ikin

Contents

Part 1: The Sun, Earth, and Moon 3

Part 2: Other Planets and Moons in the Solar System 41

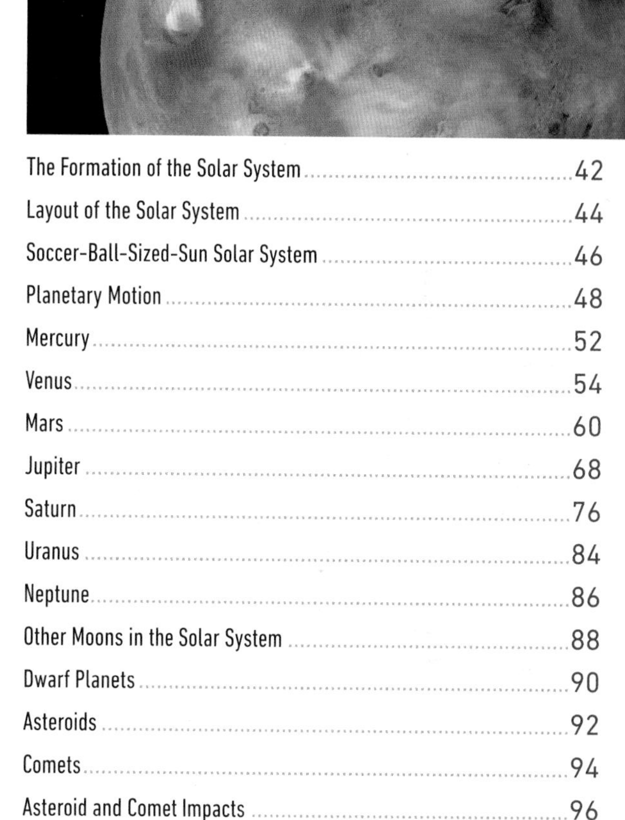

Part 3: The Births, Lives, and Deaths of Stars 99

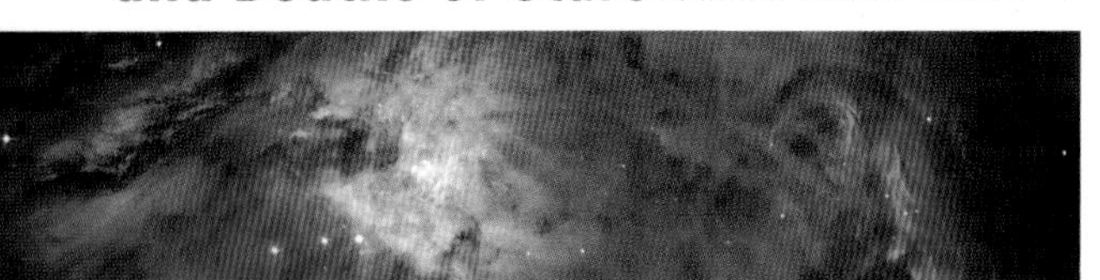

Part 4: The Milky Way Galaxy 141

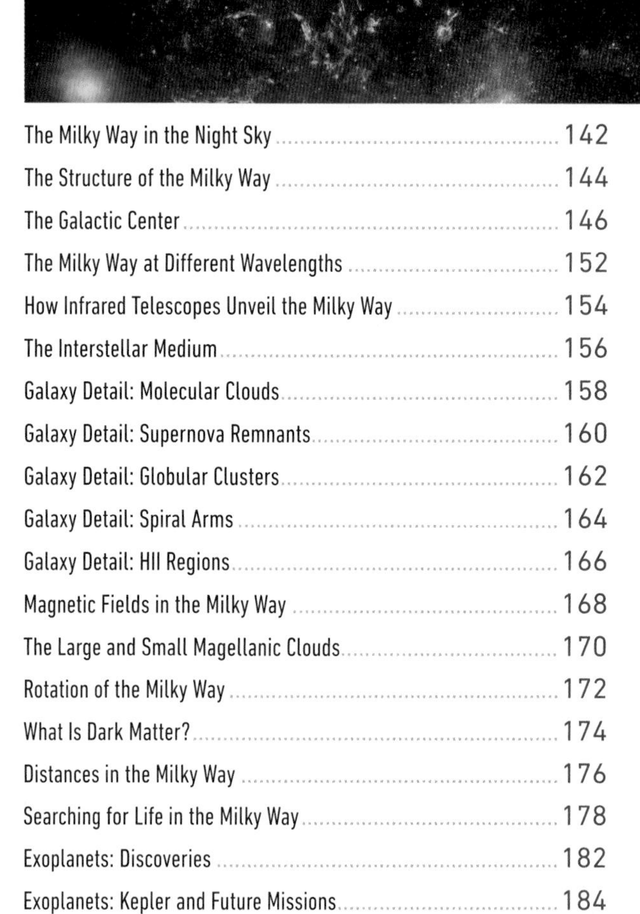

Part 5: Galaxies ... 187

Part 6: The Origin and Fate of the Universe ... 229

Introduction

The motions of the Sun, Moon, and planets visible without a telescope (Mercury, Venus, Mars, Jupiter, and Saturn) have been observed and recorded for thousands of years. The regular, predictable motions of these nearby objects have challenged humanity's ability to model the physical world. New observations often lead to new models, and as technology and observations improve, current models are challenged, modified, and (if needed) cast aside. That is the strength of the scientific method. Observations of the physical world are constantly improving, so what you're reading today could change. Still, that is the strength of the scientific method. As scientists—and as curious humans—it's our job to continue learning and refining our understanding of the cosmos.

In keeping with that idea, this book starts close and moves out in concentric circles from Earth—not because the universe is centered on this planet, but because people's understanding has moved outward over time.

In **Part 1, The Sun, Earth, and Moon,** I discuss the bodies most familiar to us: our planet, the Sun, and the Moon. In this part, I talk about how the motions of these shape how we tell time, as well as give you information on their surfaces and interiors.

In **Part 2, Other Planets and Moons in the Solar System,** I walk you through the other planets in the solar system, most of which are strange worlds that are nothing like Earth. The relative desolation of the other planets in the solar system gives reason to be grateful for the sanctuary of Earth.

Beyond the solar system are the other stars, and in **Part 3, The Births, Lives, and Deaths of Stars,** I help you explore the lives and deaths of stars. Throughout the universe are stars larger and smaller than the Sun, but improvements in telescopes and imaging technology have allowed astronomers to understand the lifetimes of stars, from their cold beginnings to their fiery ends.

In **Part 4, The Milky Way Galaxy,** I talk about the solar system's home galaxy, which is just one of a myriad of galaxies that fill the universe. Being inside the Milky Way presents both advantages and challenges to understanding its shape and composition.

Part 5, Galaxies, looks at galaxies in general and the rich variety of galaxies people are able to observe, both nearby (in the Local Group) and out to very great distances. As telescopes have gotten larger and more sensitive, astronomers have been able to observe more distant, faint galaxies and fit them into a picture of galaxy evolution.

In the final part of the book, **Part 6, The Origin and Fate of the Universe,** you arrive at the most-distant sources that are possible to observe. Because light travels at a finite speed, the most-distant sources observers can see are also the youngest, providing a peek of the early universe. I also talk about the potential end of the universe, as well as the unsolved questions about the cosmos.

Join me on a journey through the vastness and beauty of the cosmos, starting with the familiar, nearby objects of the solar system and ending with clues to the origin of everything.

Acknowledgments

Getting the chance to write a book like this is the result of many good people in my life, and I would like to thank them here. First, thanks to Alan Axelrod, my coauthor on *The Complete Idiot's Guide to Astronomy,* for taking a chance and contacting me after an Open House at Bradley Observatory over 15 years ago. You introduced me to the world of popular science writing, and I will be forever grateful to you for that. To John Kolena, who got me interested in astronomy and astrophysics as a senior at Duke University. To all my mentors in graduate school, especially Wayne Christiansen at UNC Chapel Hill and Miller Goss at the National Radio Astronomy Observatory—thank you for showing me just how cool radio astronomy could be. To my wonderful children, Dylan, Claire, Matilda and Maddie, for listening to my astronomy stories over the years, and for asking great questions. To my amazing wife and best friend Sheryl, for supporting me through many long days of writing. And finally, to Papa, W. H. "Bud" Stuckey, to whom this book is dedicated. "Bud" was a World War II Marine veteran and an avid amateur astronomer. He built his own telescopes for neighborhood kids to look through on his street in Macon, Georgia. Papa loved astronomy and couldn't wait to talk about the latest news. His kindness, humility, and enthusiasm for learning are an inspiration to everyone who knew him.

Special Thanks to the Technical Reviewer

Idiot's Guides: The Cosmos was reviewed by an expert who double-checked the accuracy of what you'll learn here, to help us ensure this book gives you everything you need to know about the cosmos. Special thanks are extended to Allison Smith.

Allison Smith received her MSc in physics from the University of Georgia and is an experienced instructor for introductory-level astronomy courses. She is currently pursuing a PhD in physics and astronomy, and her research involves using radio telescopes to study molecular gas in the interstellar medium.

Trademarks

How to View the Cosmos

Observing an object you have never seen before with your own eyes—from the surface of the Moon to the rings of Saturn—can be a thrilling experience.

Believe it or not, you can start getting familiar with the night sky with nothing more than your eyes and hands. In fact, astronomical observations were made for thousands of years before the telescope was invented in 1609, a little over 400 years ago. Telescopes have allowed people to explore and see more detail. However, no matter what tools you use to observe the cosmos, you are an astronomer.

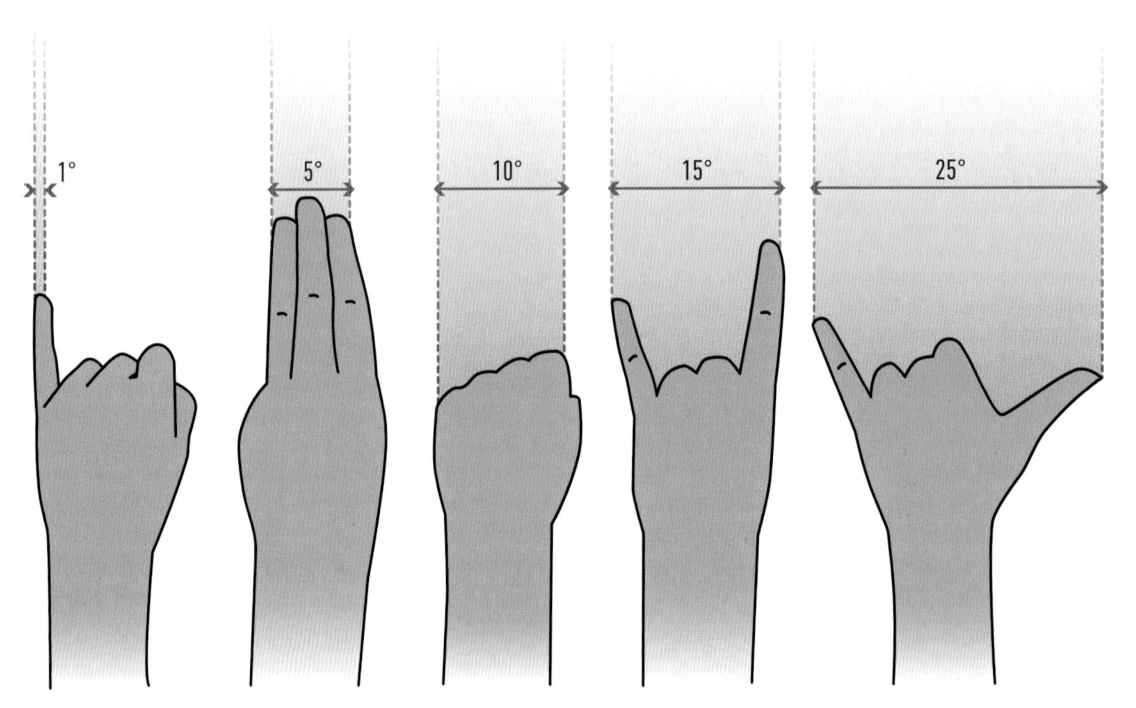

Using nothing more than your hand at arm's length, you can measure a number of useful angular sizes: 1, 5, 10, 15, and 25 degrees.

Measuring Angular Separation and Size with Your Hand

Angular separation is how far apart (in degrees) two objects are on the sky, while angular size is how large (in degrees) an object is on the sky. The whole sky from horizon to horizon is 180 degrees. From the horizon to directly overhead is 90 degrees.

Your hand held at arm's length provides a number of excellent, built-in ways to measure angular separation and angular size on the sky. To measure smaller angular sizes, just hold your hand out in front of you. The width of your pinky is about 1 degree; the Moon (and Sun) both have an angular size of about $^1/_2$ a degree. The observations that astronomers made for thousands of years basically used standardized and more accurate versions of this basic angular measuring device.

Observing with a Telescope

But what if you need to measure an object or separation smaller than 1 degree (which is divided into 60 arcminutes ′ or 3,600 arcseconds ″)? These smaller angular sizes are typically only important when you begin to observe with a telescope. For example, the human eye can see objects separated by about 4 arcminutes at best, while the 2.4-meter Hubble Space Telescope has a resolution of 0.05 arcseconds.

Optical telescopes greatly improve on the light-collecting ability and resolution of the human eye. Perfected in 1609 by the Italian astronomer Galileo Galilei, the first telescopes brought realms never before seen into human experience.

Telescopes do three things: collect light, resolve details, and magnify an image. Most people think that magnification is the most important thing telescopes do, but in fact the collection of light and the resolution of details are far more important to improving humans' understanding of the universe.

The basic structure of a telescope is the same, whether it is collecting optical, radio, or infrared radiation. These are all just different wavelengths of electromagnetic radiation, and in order to form an image at any wavelength, radiation has to be brought to a focus. Optical research telescopes use curved and coated glass surfaces (mirrors) to focus light, while radio telescopes use curved metal surfaces (dishes).

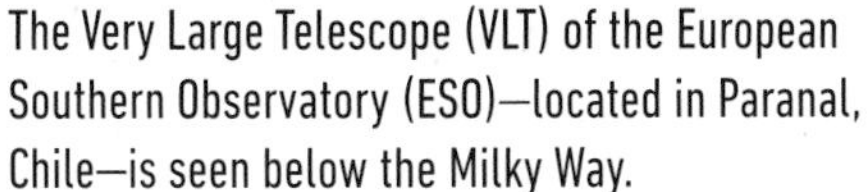

The Very Large Telescope (VLT) of the European Southern Observatory (ESO)—located in Paranal, Chile—is seen below the Milky Way.

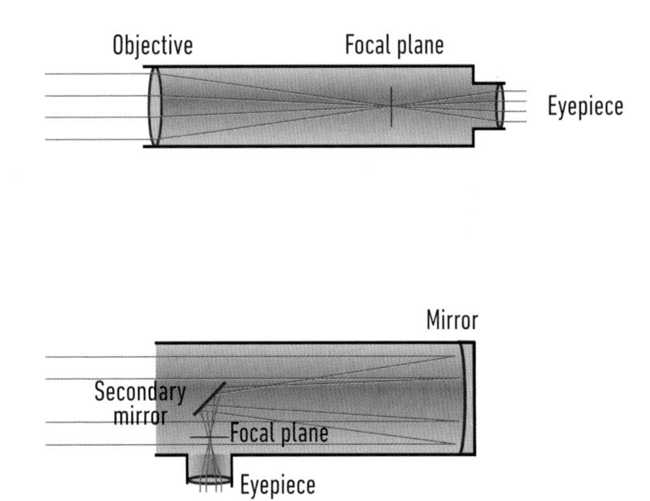

Optical telescopes focus light either by refracting, or bending, light (top) or reflecting light (bottom). The earliest telescopes were refractors and had glass lenses. Most modern research telescopes (like the Hubble Space Telescope) are reflectors.

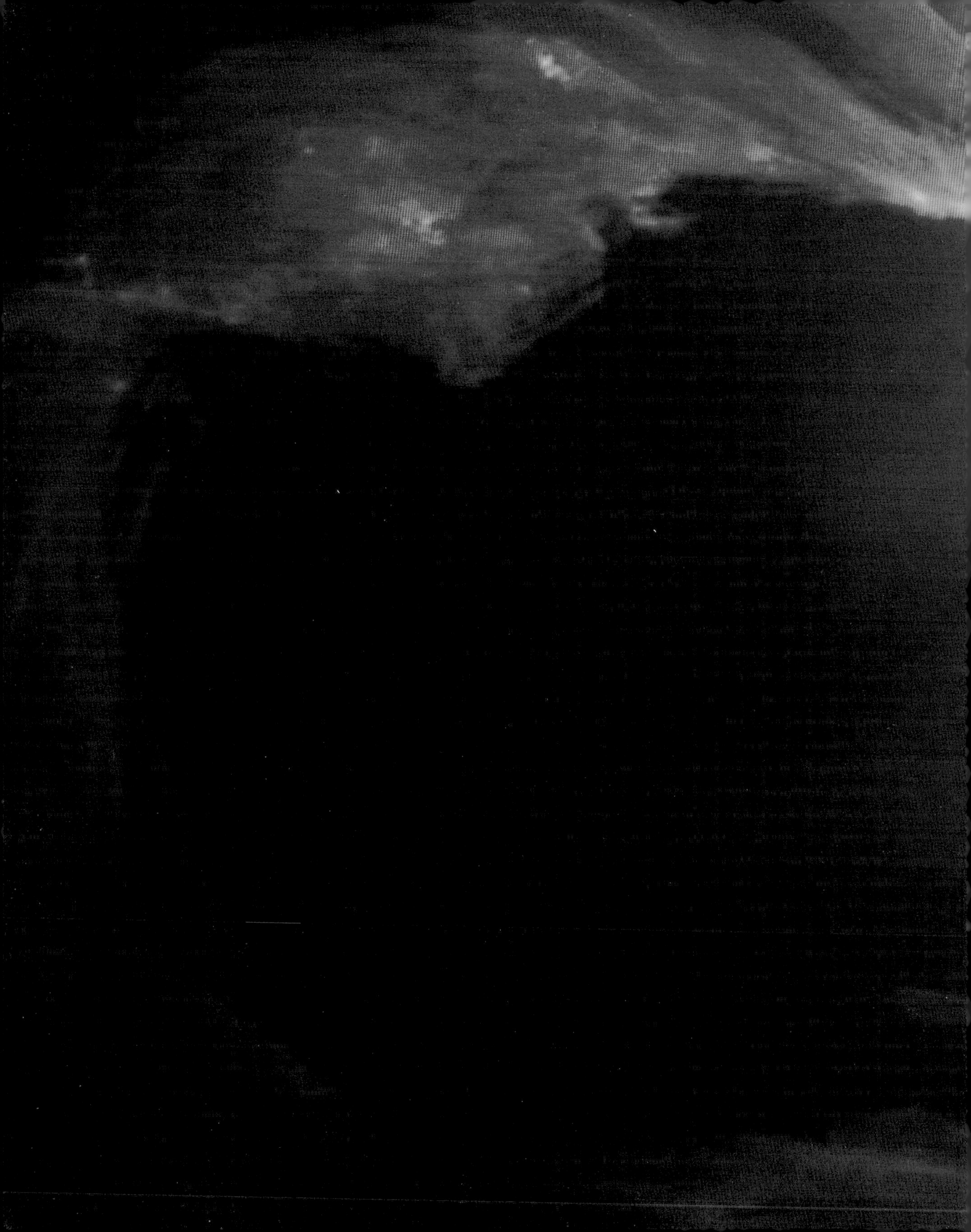

PART 1

The Sun, Earth, and Moon

The Sun, Earth, and Moon form a system that is familiar to everyone. The rising and setting of the Sun, the tilt of the Earth's axis, the slow orbit of the Moon—these motions and orientations determine the shape of your days, weeks, months, and seasons. In this part, I help you explore these three astronomical bodies: the Earth that you live on, the Moon that orbits the planet, and the life-giving Sun that Earth circles.

I begin with the ways in which the motions of the Sun, Earth, and Moon determine how you measure time. I then turn your attention to the details of the Sun's interior and surface, and the atmosphere and interior of planet Earth. Finally, I discuss Earth's nearest companion, the Moon, and how its violent formation changed the planet forever.

The Structure of the Night Sky

Mauna Kea in Hawaii is home to some of the world's greatest observatories. This image, captured over the course of several hours, reveals the rotation of Earth as seen in the trails of stars.

Since ancient times, humans have looked to the night sky and felt a connection. Ancient people noticed the sky's unchanging structures and subtle changes and employed careful observers to make sure that important decisions (such as planting crops, declarations of war, and marriages) happened at the right time. The relative positions of the fixed stars (which never seemed to change) and the variable positions of the Sun, Moon, and five planets visible without a telescope also gave ancient observers the ability to keep annual and monthly calendars. Even a casual observer today can notice the changing positions of the Sun and Moon, the phases of the Moon, and the presence and absence of wandering stars (planets).

One observation common to both ancients and people today is that the stars in the night sky appear to rise in the east and set in the west, like the Sun. In fact, stars move in arcs across the night sky, with all of the arcs centered on the North Celestial Pole, or the part of the night sky that is directly above Earth's North Pole. But while you might be tempted to say something in the night sky is "overhead" or "in the southwestern sky," these statements would not help another observer in another part of the world find your object. To deal with this problem, astronomers had to agree on a coordinate system. Two coordinate systems that are widely used are the altitude-azimuth (alt-az) system and the celestial coordinate system.

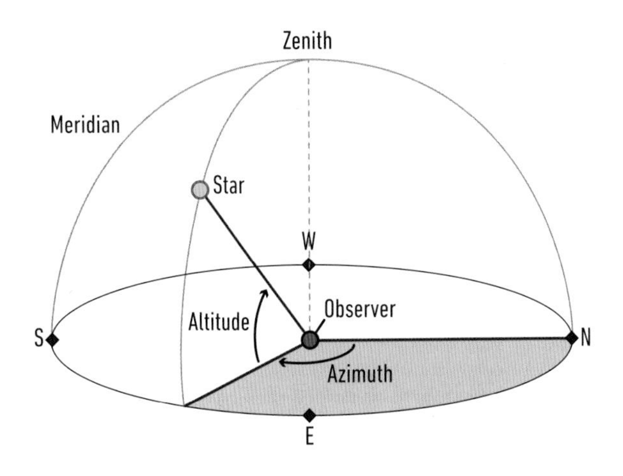

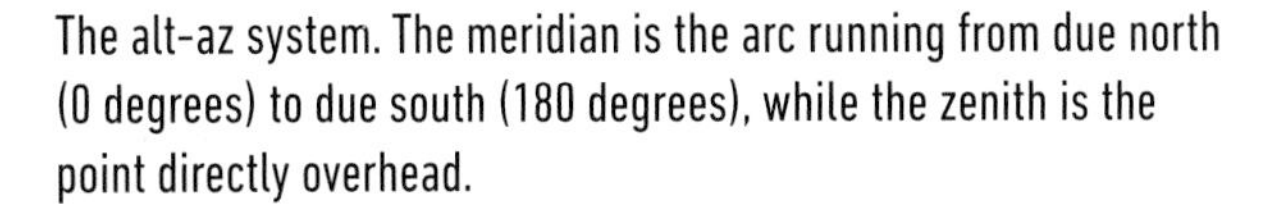
The alt-az system. The meridian is the arc running from due north (0 degrees) to due south (180 degrees), while the zenith is the point directly overhead.

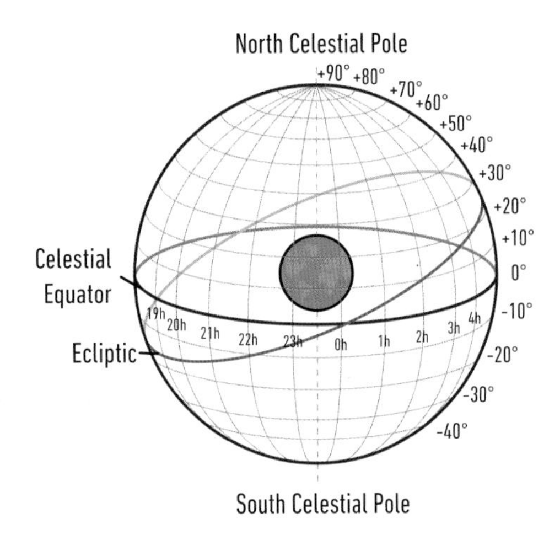

In order to use the celestial coordinate system, imagine that the entire universe outside of Earth is projected onto the inner surface of a sphere, and you are located at the center of that sphere.

The Altitude-Azimuth System

The altitude-azimuth (or alt-az) system tells you the degrees from the horizon to the object you're viewing in the sky, which is useful to give a sky position to a ground-based telescope. It is based on a system of 360 degrees of azimuth (0 degrees due north) and 90 degrees of altitude. So the alt-az position of an object due east and halfway between the horizon and the zenith (directly overhead) would have a coordinate of 90 degrees azimuth, 45 degrees altitude. The limitation of this system is that it is specific to one observing location on Earth.

The Celestial Coordinate System

A more general system of coordinates applicable to all observers is called the *celestial coordinate system.* The celestial coordinate system basically projects Earth's surface out onto the night sky. Earth's equator projected onto the sky forms the celestial equator, and Earth's North and South Poles projected onto the sky are the North and South Celestial Poles.

Using the celestial coordinate system, the position of an object in the night sky can be specified by two coordinates: right ascension (in hours, minutes, and seconds along the celestial equator) and declination (in degrees above or below the celestial equator). Right ascension can take on values between 0h 0m 0s and 23h 59m 59s, and declination can have values between -90 degrees (at the South Celestial Pole) and 90 degrees (at the North Celestial Pole).

The Sun's Motion: Telling Time

The daily, monthly, and annual motions of the Sun (in actuality the motion of Earth around it) in the daytime and nighttime sky provide a way to measure the passage of time.

Daily Motion

The Sun rises in the east and sets in the west due to the rotation of Earth. Depending on the time of year, the Sun passes from east to west in a low arc (winter) or a high arc (summer). As seen from Earth's North Pole, the planet rotates counterclockwise, causing this apparent motion on the part of the Sun.

Timekeeping takes advantage of this daily motion. For example, if you put a stick in the ground and point it toward the North Celestial Pole (Polaris is very close to this point in the night sky), you make a reasonable timekeeping device. The stick first will cast a shadow to the west when the Sun is in the east. At one moment, when the Sun is halfway across the sky (called *local noon*), the stick will cast a shadow due north. Late in the day, when the Sun is in the west, the stick will cast a longer and longer shadow to the east. If you imagine the motion of this shadow in the course of the day, you can see that the shadow moves "clockwise." The second, minute, and hour hands of analog watches today echo of the motion of that shadow.

Even digital clocks make a reference to the motions of the Sun. Imagine the sky divided in half by a huge arc running from north to south across the sky. This arc is called the *meridian.* When the Sun is in the eastern half of the sky (before noon), it is *antemeridianus* or A.M. When the Sun is directly on the meridian, the time is local noon. Once the Sun has crossed the meridian, it is *postmeridianus* or P.M.

This image, taken by the Mars exploration rover Spirit, shows the Sun setting below the rim of the Gusev Crater on Mars on May 19, 2006.

A Martian Sunset

Recently, humans have been allowed to see the Sun rise and set from a new vantage point—Mars. Scientists use images of the Sun's movements across the planet's sky to examine its atmosphere.

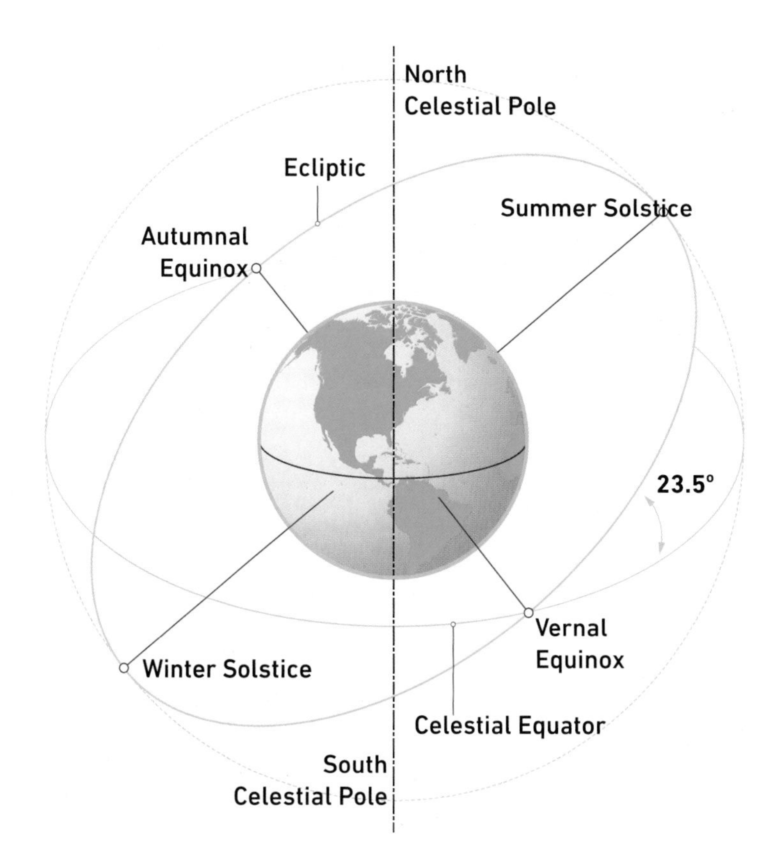

As the Sun moves along the ecliptic, it's in different parts of the sky at different times of year. In midwinter, the Sun is below the celestial equator at the winter solstice. In midsummer, the Sun is on the opposite side of the ecliptic at the summer solstice.

Monthly Motion

The ecliptic is the great circle on the sky along which the Sun moves a tiny amount each day. Because there are 360 degrees in a circle and 365.25 days in a sidereal year, that means the Sun moves a tiny bit less than a degree east along the ecliptic every day. While the Sun moves along the ecliptic, the stars remain essentially "fixed" on the celestial sphere, meaning that slightly different stars are visible every day of the year. For example, you may have noticed that different stars are visible at different times of year. The constellation Cygnus is high overhead in summer, and the familiar Orion constellation is high in the sky in winter. This seasonal change is the result of the Sun's small daily motion along the ecliptic.

Annual Motion

In the course of a year, the Sun will complete one full circuit of the ecliptic, passing through a set of particular constellations called the *zodiac*. A *sidereal year* (from the Latin word for "star") is the time it takes Earth to orbit the Sun with respect to the fixed stars, or the time required for the Sun to return to the same position with respect to the stars. A *tropical year* is the time it takes the Sun to return to the vernal equinox. Because Earth's axis (and thus the celestial equator) are rotating slowly (an effect called *precession*), the sidereal year and the tropical year are not exactly equal, with the tropical year being about 20 minutes shorter than a sidereal year. The passage of time is typically measured using the tropical year, since it's tied to Earth's precession.

The Sun's Motion: The Seasons

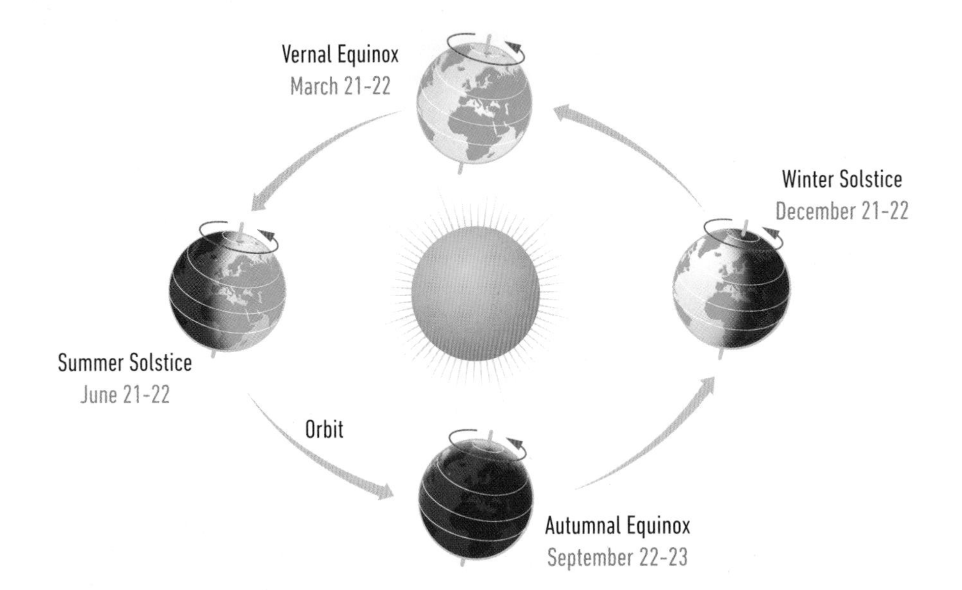

This diagram shows how the tilt of Earth's axis relative to its orbital path around the Sun causes the seasons.

There is a common misconception that the seasons are the result of how close the Sun is to Earth at different times of the year. While the distance to the Sun does change slightly during the year (since Earth's orbit is not exactly circular), the changes are too small to be the cause of the seasons. In fact, Earth and the Sun are closest (called *perihelion*) in January and most distant (called *aphelion*) in July, which fall in cooler and warmer months respectively in the Northern Hemisphere.

You can better understand the seasons you experience on Earth if you look more closely at the apparent daily motion of the Sun. The Sun rises in the east and sets in the west every day, but its daily rising and setting positions change very slightly. What causes the Sun to seemingly change position? The tilt of Earth's axis, which is thought to be the result of an impact that happened early in the history of its formation (see "The Formation and Structure of the Moon"). This tilt is the reason for the seasons you experience on Earth and why the seasons are opposite in the Northern and Southern Hemispheres.

Spring and Fall

On two days of the year, the Sun rises due east (at an azimuth of 90 degrees) on or about March 21 and sets due west (at an azimuth of 270 degrees) on or about September 22. These two days are known respectively as the vernal equinox and the autumnal equinox. As you can tell by the names, these refer to the spring and autumn seasons of the Northern Hemisphere, when the Sun's energy is equal in both hemispheres.

This time-lapse image shows the International Space Station view of the North Celestial Pole and Earth.

How Seasons Look from the North Pole

Because the Sun moves along the ecliptic during the course of the year, you can get an interesting vantage point on its movement from the North Pole. There, the star Polaris is found almost directly overhead, and the stars move around in circular orbits, never rising or setting. The celestial equator sits at the horizon of the North Pole, so once the Sun passes below the celestial equator on or about September 22, it does not rise until the vernal equinox six months later. The North Pole therefore experiences six months of just the stars and the Moon.

The Short, Cold Days of Winter

Because the Sun rises and sets farther to the south in winter, it spends less time above the horizon, giving people in the Northern Hemisphere the shorter days of winter between the autumnal equinox and the vernal equinox. In addition to the days being shorter, the 23.5-degree tilt of Earth's axis with respect to the plane of the solar system means that the Sun's rays strike Earth in winter time at an angle, spreading out their energy over a larger area. As a result, the Northern Hemisphere (tilted away from the Sun) experiences winter during the months between September and March, while the Southern Hemisphere (tilted toward the Sun) is in midsummer. The Sun reaches its southernmost setting point on or about December 21, which is known as the winter solstice. It then reverses direction and begins to set farther north every night.

The Long, Warm Days of Summer

After the vernal equinox, an observer in the Northern Hemisphere will see that the Sun continues to rise farther and farther to the north every day, meaning the Sun is both more directly overhead (concentrating its energy in a smaller area) and above the horizon longer. This combination leads to warmer temperatures in the Northern Hemisphere. The situation is reversed in the Southern Hemisphere, which is tilted away from the Sun during summer in the Northern Hemisphere. The Sun reaches its northernmost setting point on or about June 21, which is known as the summer solstice. After that, it sets farther south every night.

Lunar Motion and Phases

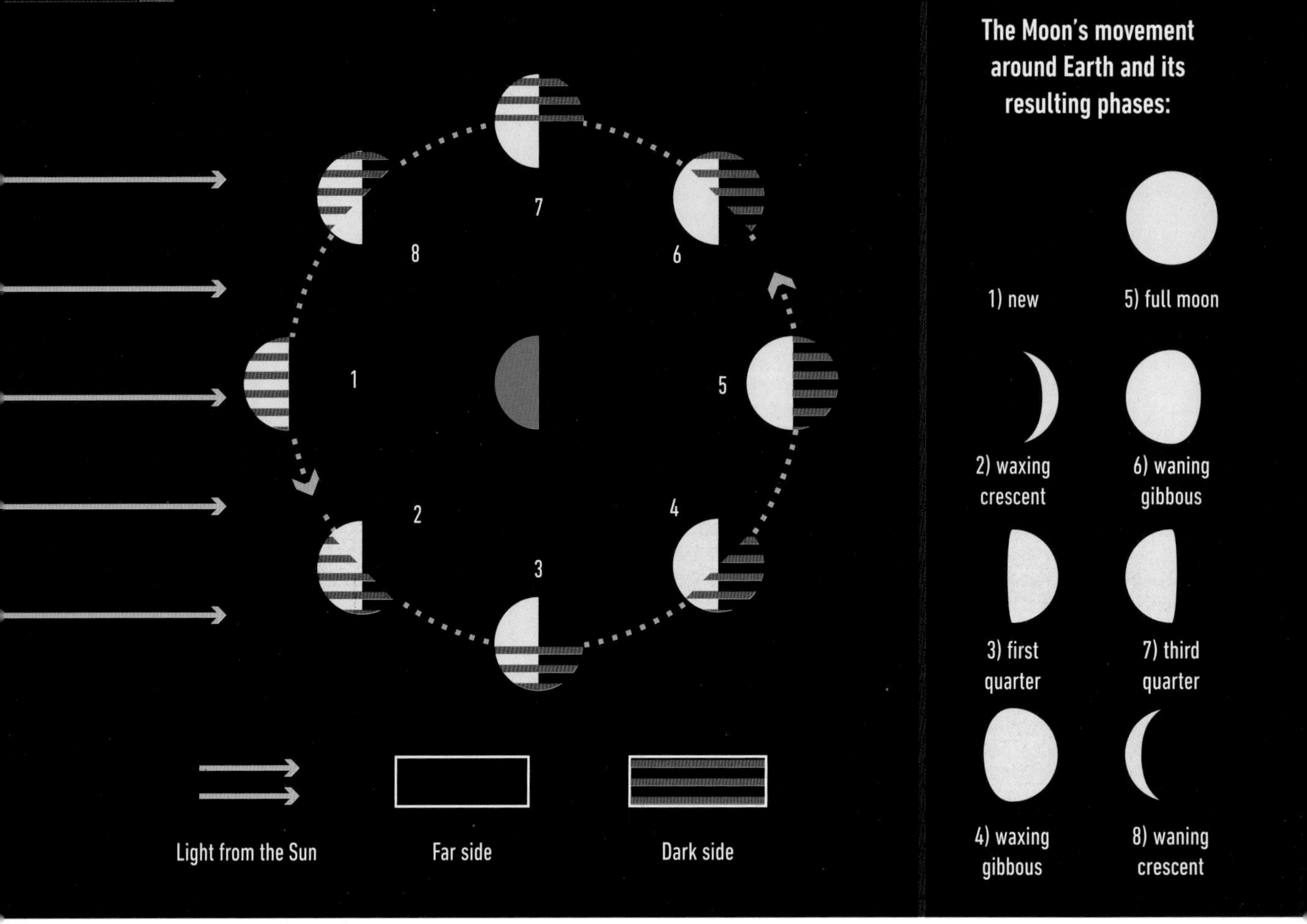

7

8

6

1

5

2

4

3

Light from the Sun

Far side

Dark side

The Moon's movement around Earth and its resulting phases:

1) new

5) full moon

2) waxing crescent

6) waning gibbous

3) first quarter

7) third quarter

4) waxing gibbous

8) waning crescent

The Moon is our closest astronomical companion whose motions and phases have been studied throughout recorded history. Like the Sun, the Moon moves across the background of the fixed stars each day. But the Moon makes its circuit against the stars much more quickly than the Sun, doing one full orbit of Earth (as measured against the stars) every 27.3 days. Because one full orbit is 360 degrees, that means the Moon moves about 13 degrees each day. Think about it like this: your hand's width with fingers extended is about 15 degrees on the night sky, so the Moon should move roughly that distance from night to night.

The full cycle of lunar phases, from waxing crescent (upper left) to new moon (lower right).

Lunar Phases and Human Culture

Many religions on Earth look to the Moon as their basic unit of measuring the passage of time. Islam, for example, uses a purely lunar calendar, in which a year consists of 12 synodic months. Because there are 354 days in 12 synodic months, the months in the Islamic calendar drift within the solar year (365.25 days) by 11 or 12 days. Other calendars try to account for this difference by injecting an additional month to make up for the difference in days between a solar year and 12 synodic months.

Islam, Christianity, and Judaism also use lunar cycles to determine the dates of important festivals. In fact, it was the steady drift of Easter away from spring that motivated the change in the West from the Julian calendar to the Gregorian calendar in 1582.

Phases of the Moon

One of the most familiar aspects of the night sky is the apparent nightly change of the shape of the Moon. Many highly educated people think the phases of the Moon are the result of Earth's shadow falling on the Moon; in fact, the phases of the Moon are the result of the relative position of the Moon, Earth, and the Sun (see "Lunar and Solar Eclipses" for how Earth's shadow can affect how you see the Moon and Sun). The half of the Moon's surface that faces the Sun is always illuminated. As the Moon orbits the planet (counterclockwise as seen from the North Pole of Earth), you on Earth can see different portions of that lit-up surface.

When Earth, the Moon, and the Sun are in a straight line, the Moon is new, meaning you can't see any of its bright half. As it orbits Earth, you start to see more and more of its lit surface, going from waxing crescent, to first quarter, to waxing gibbous, to full moon. As the Moon continues its orbit, you then start seeing less and less of the surface that faces the Sun each day, going from waning gibbous, to third quarter, to waning crescent, to new moon.

Sidereal and Synodic Months

A *sidereal month,* or a month as reckoned by the Moon's position with respect to the background stars, is how long it takes the Moon to orbit Earth one time. A sidereal month is 27.3 days. However, months are generally measured in *synodic months.* A synodic month measures the passage of a full cycle of the Moon's phases—from new moon to new moon or full moon to full moon. A synodic month is 29.5 days. The reason that a synodic month lasts slightly longer than a sidereal month is that Earth is orbiting the Sun at the same time and in the same direction that the Moon is orbiting Earth.

Lunar and Solar Eclipses

During a solar eclipse, the Sun, the Moon, and Earth line up with the Moon in the middle.

During a lunar eclipse, the Sun, the Moon, and Earth line up with Earth in the middle.

Because the Sun's bright, visible atmosphere (the photosphere) is blocked by the Moon during a solar eclipse, it is possible to see the Sun's hot, tenuous outer atmosphere (corona).

The Moon orbits Earth, and Earth orbits the Sun. In their orbits, occasionally these three objects line up so the shadow of one falls on the other. When Earth falls in the shadow of the Moon, the result is a *solar eclipse*. When the Moon falls in the shadow of Earth, the result is a *lunar eclipse*.

Viewing a Solar vs. a Lunar Eclipse

A solar eclipse is an effect that is localized on Earth—that is, not everyone gets to see a solar eclipse. Think of a solar eclipse as a small, circular shadow that moves across Earth as it rotates. Depending on the exact orientation of the Sun, Earth, and the Moon, a solar eclipse will be visible at different places on the surface of Earth.

In comparison, a lunar eclipse can be seen from all observers on Earth. As the Moon orbits Earth, it will occasionally fall into Earth's shadow; when it does, the Moon (near its full phase) will darken and finally be visible in the reddish light refracted through Earth's atmosphere.

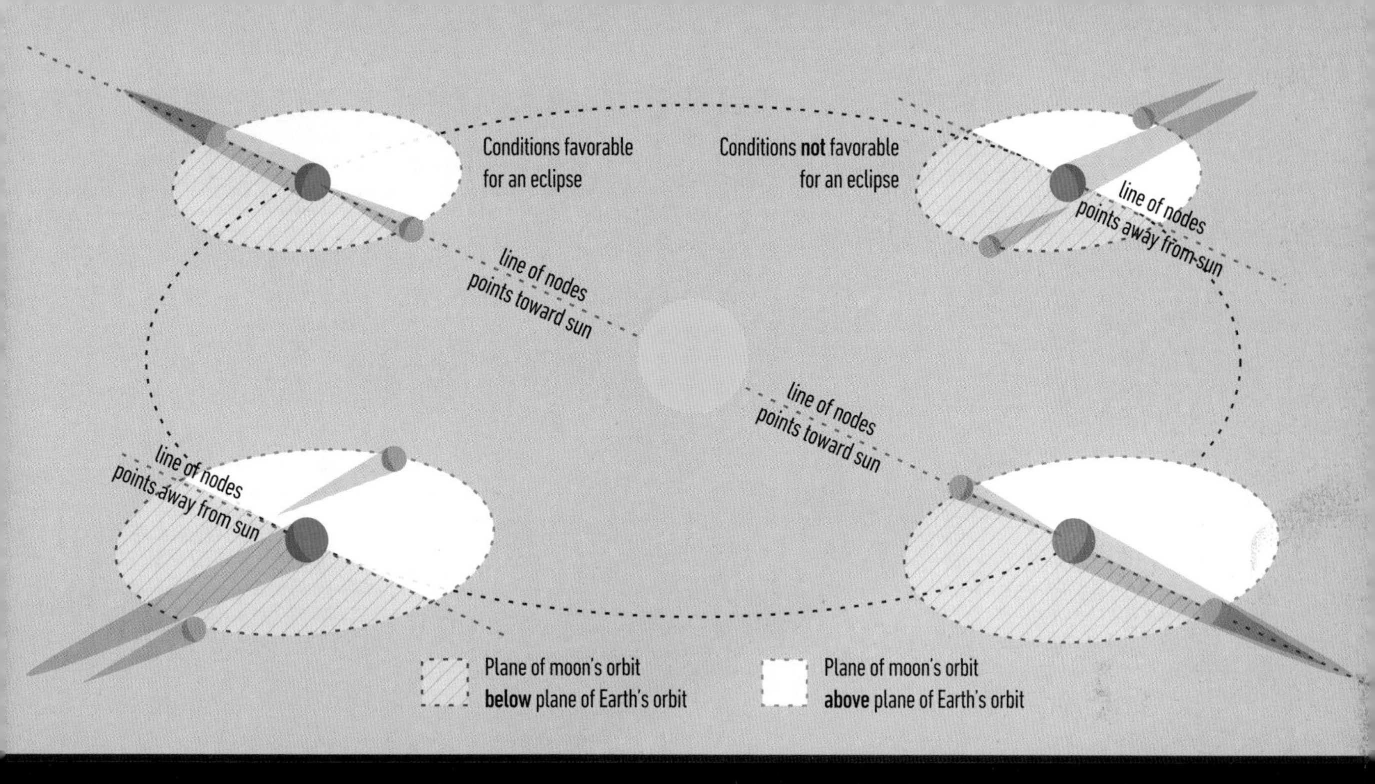

This shows how the line of nodes made by the intersection of the orbits of the Moon and Earth can lead to favorable and unfavorable conditions for eclipses.

Why Are Eclipses Rare?

Why do solar and lunar eclipses not happen every time the Moon orbits Earth? After all, a solar eclipse happens when the Moon is on the Sun side of Earth (new moon), and a lunar eclipse happens when the Moon is on the opposite side of Earth (full moon). However, these eclipses are rare because the plane of Earth's orbit around the Sun is tilted with respect to the plane of the Moon's orbit around Earth.

The intersection between two planes is always a line; in the case of the orbits of the Sun, Earth, and the Moon, this line is called the *line of nodes*. If the line of nodes is pointing toward the Sun, the orientation is favorable for eclipses. When the line of nodes points anywhere else (which is generally the case), then the shadows do not exactly line up.

How Angular Size Contributes to Eclipses

The angular size of an object is the result of two measurements: its physical size and its distance. In a coincidence of chronology, the Moon and the Sun currently have the same angular size in the sky: about ½ degree. As a result, when the Moon comes between Earth and the Sun (a solar eclipse), it almost completely blocks the Sun's bright photosphere.

Because the Moon is slowly moving away from Earth, its angular size is becoming smaller. Therefore, in the future, the Moon's angular size will be too small to cover the Sun, meaning it will pass in front of the Sun without fully blocking its disk, creating only a partial eclipse.

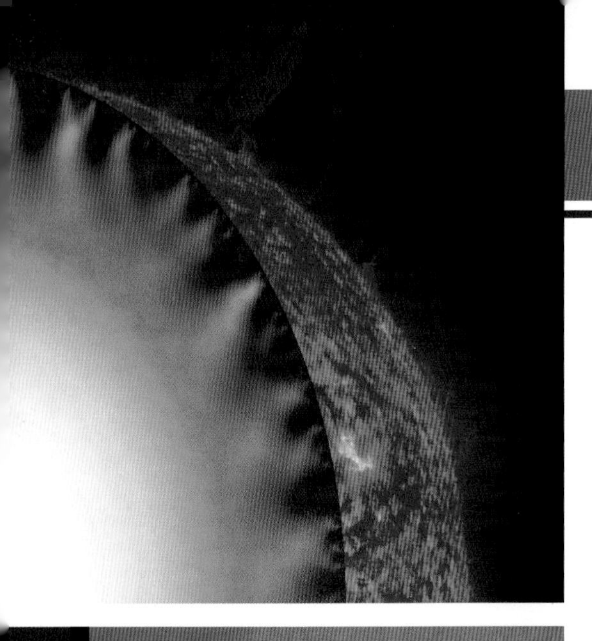

The Solar Interior

The Sun is composed of mostly hydrogen and helium (over 98 percent by mass), but because it formed from the exploded remnants of previous generations of stars, it also contains all of the elements of the periodic table in smaller amounts. The interior of the Sun is divided into three major regions: the core, the radiative zone, and the convective zone.

Statistics

EQUATORIAL RADIUS
696,342 km (109 Earths)

VOLUME
1.412×10^{18} km^3 (1.3 million Earths)

MASS
1.989×10^{30} kg (333,000 Earths)

DENSITY
Center: 1.6×105 kg/m^3
Photosphere: $2 \times 10\text{-}4$ kg/m^3
Corona: $1 \times 10\text{-}12$ kg/m^3

COMPOSITION OF PHOTOSPHERE (BY MASS)
73.46% hydrogen, 24.85% helium, 0.77% oxygen, 0.29% carbon, and 0.63% other elements

ROTATIONAL PERIOD (AT EQUATOR)
25.05 days

TEMPERATURE
Center: 1.6×10^7 K
Photosphere: 5,778 K
Corona: 5×10^6 K

The Three Regions of the Sun

At the Sun's center is the high-density, high-temperature *core* that has the extreme conditions necessary to fuse hydrogen into helium (see "Fusion in the Sun"). In models of the solar interior, the core extends out to about 30 percent of the Sun's radius.

Outside the core is the *radiative zone,* which continues out to about 70 percent of the Sun's radius. The Sun's temperature drops from about 7 million to about 2 million Kelvin from the bottom to the top of the radiative zone. Atoms in this region transfer energy by radiation—a photon of light is emitted by one atom, and that photon is quickly absorbed by another nearby atom. This region is so dense that the transfer of radiation proceeds very slowly. In fact, photons emitted in the core can take 100,000 years to find their way out of the radiative zone.

Above the radiative zone is the *convective zone,* which is located from 0.7 solar radii out to the Sun's visible surface (known as the photosphere). In the convective zone, energy is transferred out to the Sun's surface by the motions of material. Thermal cells of hot material at the base of the convective zone rise to the surface, where they cool, increase in density, and then sink back to the bottom of the convective zone.

Neutrinos in the Sun

One of the products of the thermonuclear fusion occurring in the Sun's core is neutrinos. Measurements of the flux of neutrinos detected on Earth are how people estimate the rate of fusion that is going on in the Sun's interior. These particles are thought to be massless and electrically neutral, so they do not interact with ordinary matter in any of the ways that other particles (or photons) would. As a result, the neutrinos produced in the Sun's core come streaming out of the Sun unimpeded. So while photons produced in the Sun's core may take 100,000 years to make their way to the surface of Earth, neutrinos make the journey at the speed of light in about 8.5 minutes. However, neutrinos are like the "greased pigs" of the particle world—they are extremely hard to catch. Neutrino detectors on Earth (like the Sudbury Neutrino Observatory in Canada) must be located deep in Earth so astronomers can be sure that the only objects interacting with the detectors are neutrinos that are streaming through Earth's interior.

This image from the Hinode Solar Optical Telescope clearly shows granulation at the solar surface, the result of energy transport in the convective zone. The Hinode mission is led by the Japan Aerospace Exploration Agency (JAXA) and is a collaboration between Japan, the United States, the United Kingdom, and Europe.

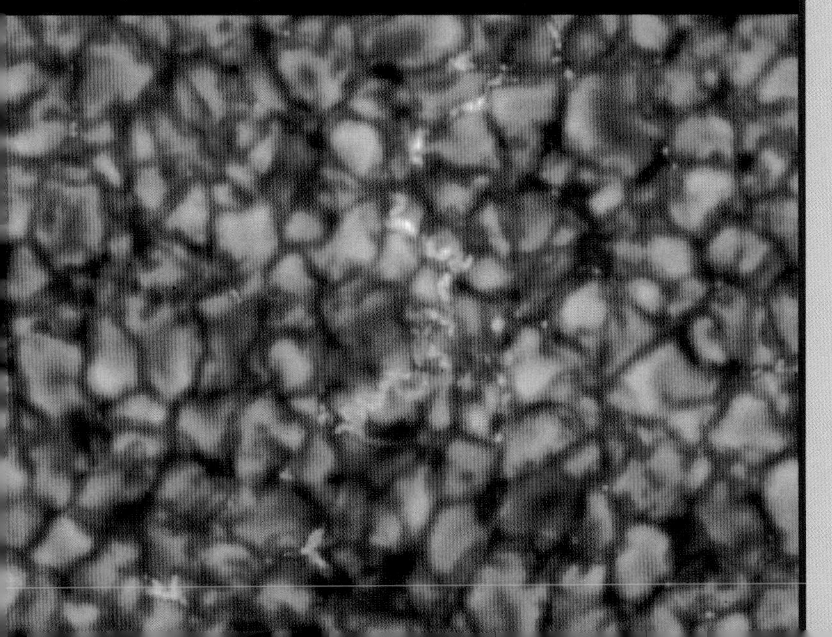

Convection Near the Surface

You can observe the result of convection at the Sun's surface in the form of granulation. Convective cells become smaller and smaller as they approach the solar surface. Like a boiling pot of water, the Sun's surface has regions where hotter (brighter) material rises and cooler (darker) material sinks. Typical granules have diameters that are thousands of kilometers across.

Fusion in the Sun

One puzzle that confounded scientists in the nineteenth and early twentieth centuries was how the Sun was able to produce enormous amounts of energy for so long. None of the traditional explanations made sense. If the Sun were burning like an enormous fire with chemical energy, its current energy generation rate would exhaust its mass in just a few thousand years. The Sun also generates some energy by its gravitational collapse (which it is still doing very slowly), but this was also insufficient. It was time for a new idea.

The Structure of the Atom

The new idea came in the form of the smallest indivisible particle: the atom. While ancient Greek philosophers came up with the concept of the atom, chemists and physicists in the nineteenth and early twentieth centuries made a lot of progress in figuring out its structure, including the fact that most of the mass of an atom is concentrated in its center or *nucleus*. Other experiments indicated that the massive nucleus carried a positive charge, and that there was a less-massive particle (the electron) that carried a negative charge. This negative charge somehow surrounded the positively charged nucleus.

The simplest atom is hydrogen, which consists of nothing more than one positively charged particle (a proton) and one negatively charged particle (an electron). But the universe is filled with a periodic table full of other elements. Where did these other elements come from? It turns out that the origin of all of those elements and the source of the Sun's energy output have a common explanation: thermonuclear fusion. Thermonuclear fusion in the cores of stars makes heavy elements from lighter ones and produces large amounts of energy in the process.

Making Elements and the Proton-Proton Chain

For a brief moment early in its history, the entire universe had the temperature and density of the center of a star, making it hot and dense enough to fuse some lithium, helium, and beryllium (see "Matter and Energy"). But once the expanding and cooling universe was only 20 minutes old, this early fusion was no longer possible. More elements couldn't be synthesized until the gravity of the universe had assembled stars.

Fusion can only occur when high temperatures and densities force protons into very close collisions, overcoming the tendency of two positively charged objects to repel one another. When the solar system formed, most of the mass was collected into the center of the solar system: the protosun. Gravity's attractive force continued to increase the temperature and density of the center of the protosun until it was hot enough and dense enough to once again make elements.

In 1920, British astrophysicist Arthur Eddington first suggested that the Sun's energy came from the fusion of hydrogen into helium. In stars like the Sun, the most important process is called the *proton-proton chain*. In this process, four protons are converted into a nucleus of the helium atom, which contains two protons and two neutrons (particles with mass very close to protons but no charge). A small amount of the initial mass of the original protons is converted into energy. This energy in the form of photons of light and some neutral particles called *neutrinos* are all that escape the Sun's interior. The other intermediate products of the reaction—positrons and isotopes of hydrogen and helium—are used up in the Sun's interior. The net result is that hydrogen is converted into helium and energy.

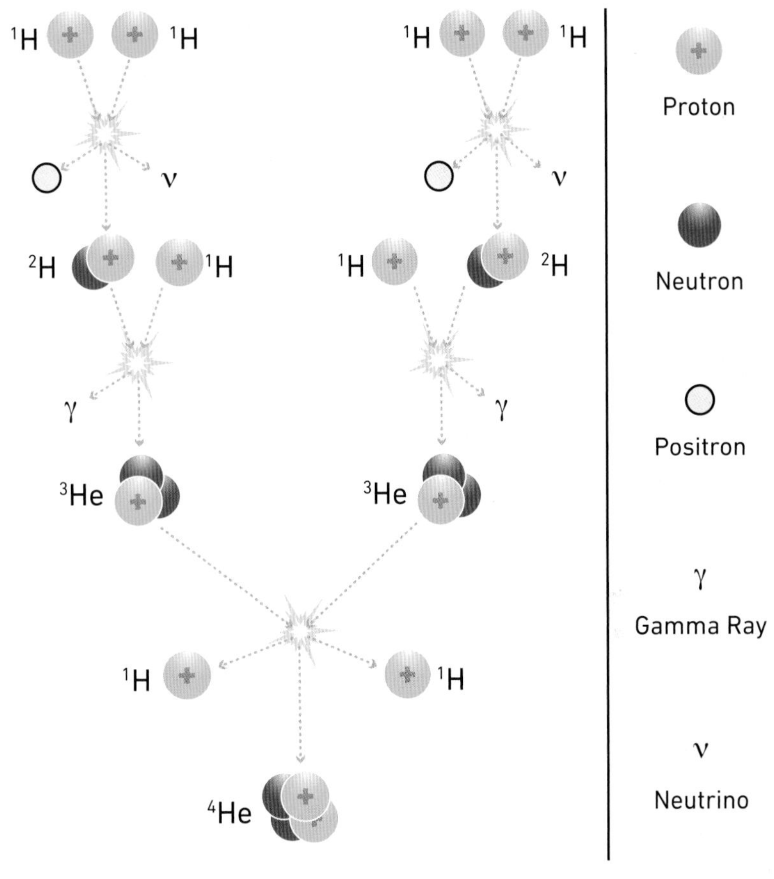

The proton-proton chain is the main way that stars like the Sun turn hydrogen (protons) into helium (two protons and two neutrons).

Deuterium and tritium are short-lived, radioactive isotopes of hydrogen. Deuterium is briefly created in the proton-proton chain.

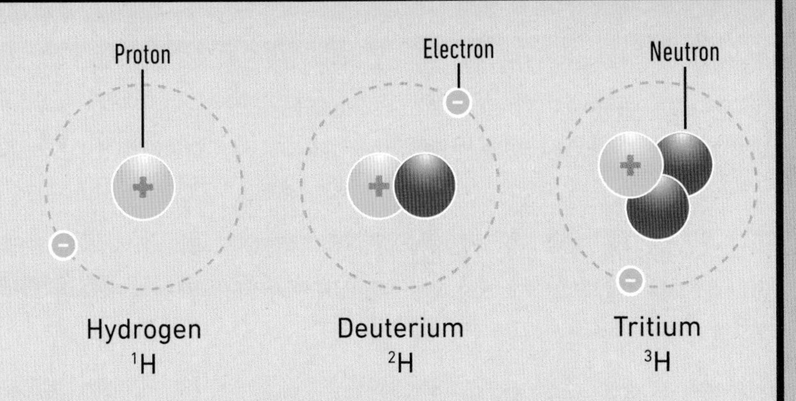

What's in an Isotope?

A hydrogen atom consists of just a proton in the nucleus and an electron. That is true of the most common isotope of hydrogen. But the nucleus of the hydrogen atom can also accommodate one or two extra neutrons. If you add in one neutron to the nucleus of hydrogen, you have *deuterium* (written ^{2}H). If you add two neutrons, you have *tritium* (^{3}H).

Sunspots and the Solar Surface

While we all know not to stare at the Sun, the Sun's surface has been an object of fascination since the first telescope was invented in 1609 by Galileo Galilei. The Sun's atmosphere is divided into three layers: the photosphere, the chromosphere, and the corona. The photosphere is the layer of the Sun's atmosphere you observe as its surface, while the chromosphere and corona are only visible under particular conditions (I will discuss the latter two in "The Sun's Upper Atmosphere and the Solar Wind").

The Sun's disk is relatively small, taking up only a 1/2 degree on the sky, so it is very difficult to make out details in the photosphere with the naked eye. However, under the magnification of a telescope, the Sun's photosphere reveals a wealth of information.

Sunspots, shown in this image, are darker because they are cooler. They are thought to be related to the presence of large magnetic fields at the surface of the Sun.

What Is a Sunspot?

One of the first observations of the Sun's surface noted by Galileo and others was that it was not perfectly smooth and unblemished (as previously thought). In fact, the Sun's surface was seen to be covered from time to time in small dark spots called *sunspots*.

Sunspots appear as small, irregular dark areas on the surface of the Sun. They vary in size but are typically a few tens of thousands of kilometers across, or about the size of a terrestrial planet. Galileo noticed early on that sunspots were observed to move across the face of the Sun and quickly deduced that this motion was due to the Sun's rotation. Galileo then used sunspot rotation to determine the rotation rate of the Sun (near its equator) to be about 25 days.

Sunspots have been observed since the middle of the nineteenth century to go through periods of maxima and minima—that is, the number of sunspots and their location on the Sun's surface change in a regular way. Sunspots appear at northern latitudes (about 30 degrees north and south of the Sun's equator) and increase in number as they move toward the Sun's equator over the course of 11 years. They then disappear and reappear at northern and southern latitudes again, with their magnetic poles reversed.

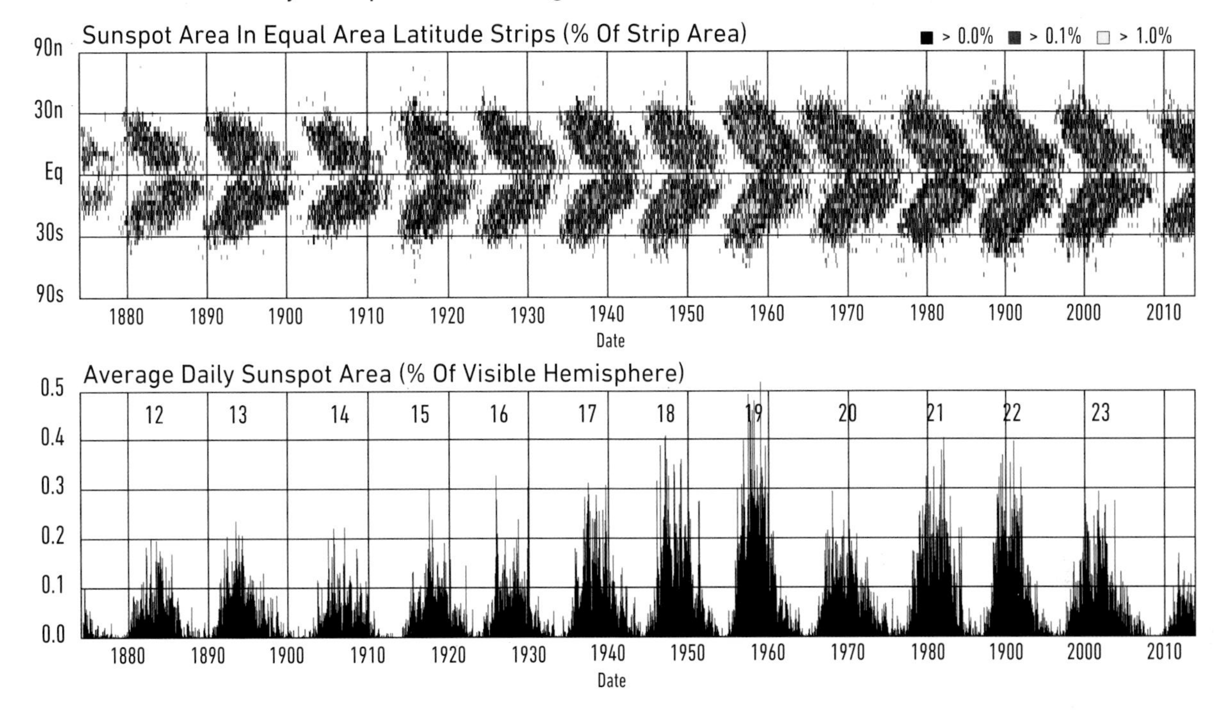

Data going back to the nineteenth century shows the location of sunspots (upper part of graph) and their number (lower part of graph) go through repeated 11-year cycles.

In the early twentieth century, American solar astronomer George Ellery Hale found that the location of sunspots are associated with high-intensity magnetic fields. Magnetic fields are detected at the Sun's surface, and sunspots are observed to mark the positions of magnetic field lines that run through the photosphere. The Sun's photosphere contains many atoms that are ionized (have had electrons stripped away) and thus have a positive charge. This hot, charged gas (called *plasma*) is pushed away by the presence of the magnetic fields at the Sun's surface. With the hot plasma pushed aside, these regions are relatively cool and thus look darker to observers.

Limb Darkening

Another observation of the Sun's surface shows a phenomenon known as *limb darkening,* which refers to the center of a star (in this case, the Sun) being brighter than its edge. This effect is related to the fact that the Sun's photosphere (into which astronomers can only observe 400 km) is hotter at its base and cooler at its top. So when you observe the Sun, the edges are dark because you are looking into the cooler (darker) upper layers of the Sun's photosphere. On the other hand, the Sun's center appears bright because you are looking into the hotter (brighter) lower layers.

The Sun's Upper Atmosphere and the Solar Wind

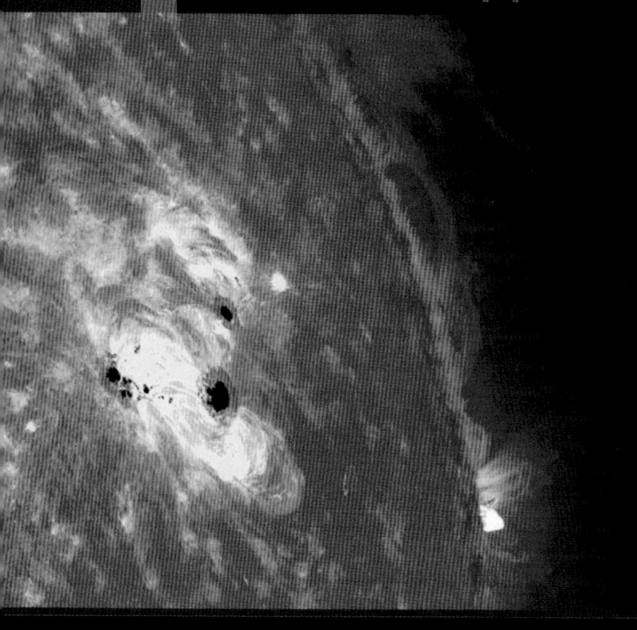

An H-alpha image of the Sun's chromosphere.

Emission Lines

Atoms can be detected by the photons of light that they emit. One type of radiation that atoms emit are emission lines. These lines arise as the result of electrons that are bound to the atom's nucleus but change from a higher to a lower energy level. When the electron moves between levels (from high to low), the difference in energy is emitted as a photon of light found at a particular wavelength. For example, the hydrogen atom is associated with an emission line that is found at 656.3 nm. This line is red in color and is easily found in gases that have temperatures of thousands of Kelvin. The Sun's chromosphere is hotter than the photosphere and is often observed in this line of hydrogen, called the *Hydrogen alpha* (or Hα) line.

The Sun's upper atmosphere is a dynamic place, and most of the action is driven by the presence of intense magnetic fields.

Chromosphere

A thin layer of the Sun's atmosphere located above the photosphere is called the *chromosphere*. Unlike the photosphere, which is seen mostly through absorption lines, the chromosphere is observed most frequently in emission lines, which are seen toward regions of hot, low-density gas. The chromosphere is often observed in the emission lines of hydrogen that occurs at 656.3 nanometers (nm). The chromosphere is both less dense and hotter than the photosphere below it and is the location of regions called *spicules,* which are jets of rising gas that surge upward into the Sun's outer atmosphere. There are hundreds of thousands of spicules covering the Sun's surface at any given time. The spicules are lofted upward by the Sun's surface magnetic fields.

Corona

The *corona,* or crown, is the outermost layer of the Sun's atmosphere. Despite its very high temperatures, the corona is exceptionally faint, because it is so low density—that is, there are hot atoms and molecules out there, just not very many of them. As a result, the corona is difficult to observe unless the bright photosphere is blocked. This happens naturally during a solar eclipse, when the Moon almost perfectly blocks the Sun's bright disk. During this time, the corona becomes visible even to the human eye (see "Lunar and Solar Eclipses"). Like the chromosphere, the corona has emission lines that are visible. One emission line of iron indicates that the temperature in the corona is at least 2 million Kelvin.

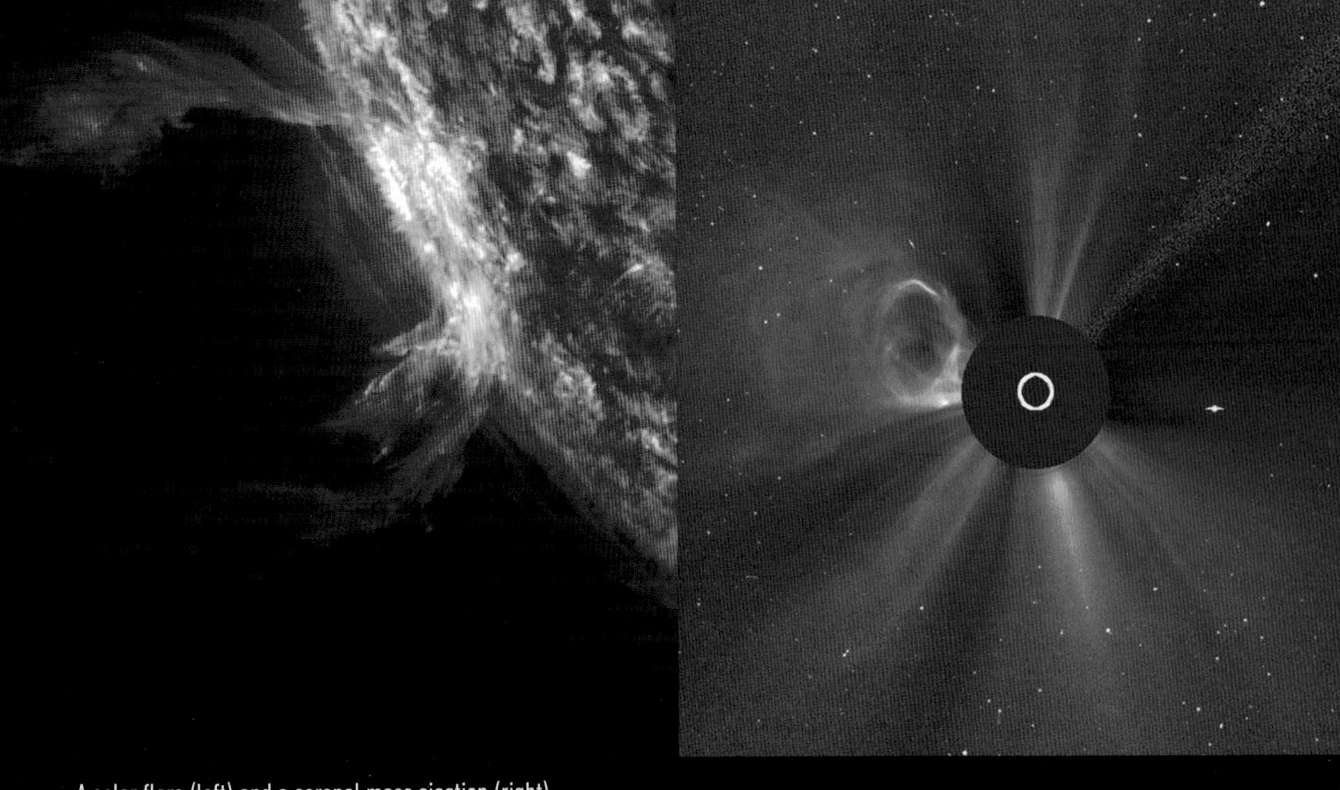

A solar flare (left) and a coronal mass ejection (right).

Solar Flares vs. Coronal Mass Ejections

Solar flares are frequent releases of energy from the Sun's surface that are associated with sunspot groups and the magnetic fields that formed them. Coronal mass ejections are rare events that release large amounts of matter and electromagnetic radiation.

Solar Wind

The material in the Sun's corona moves at a high velocity, and some of the particles have sufficient velocity to escape the Sun's gravitational field. These atoms and ions flow away from the Sun and form the *solar wind.* The solar wind is made mostly of high-energy electrons and hydrogen and helium nuclei. It is estimated that that the Sun loses a billion (10^9) kg of material each second into the solar wind. It is a testament to the enormous mass of the Sun that even losing this amount of mass each second will result in a tiny loss of its total mass during its 10-billion-year lifetime (perhaps a tenth of a percent).

When these high-energy ions and electrons get to Earth, they are accelerated by Earth's magnetic field into Earth's atmosphere. Interactions between these high-energy particles and Earth's upper atmosphere produce the effect known as the *Northern and Southern Lights* (*Aurora Borealis* and *Aurora Australis*).

Earth's Interior

Earth has a number of layers below its surface. However, unlike most observations of the universe, you can't observe Earth's interior through telescopes. How, then, do people know with confidence Earth's interior structure?

Finding Out the Layers

People have only probed Earth's interior to a few kilometers at best with the deepest drills, yet there are models of Earth's interior that talk about changes thousands of kilometers below the surface. It turns out that Earth itself provides a good probe of its own interior in the form of earthquakes.

Earthquakes produce three kinds of waves: surface waves, P (primary) waves, and S (secondary) waves. Surface waves travel through the crust and are generally what people feel when experiencing an earthquake. P waves are longitudinal waves (vibrations in the direction of motion), while S waves are transverse waves (vibrations perpendicular to the direction of motion). Both travel through Earth's interior and serve as excellent probes for the location of different layers and boundaries that are impossible to probe any other way.

In the same way that light bends (or refracts) through glass, P waves and S waves bend as they move through the varying densities of material in Earth's interior. Seismologists use detectors located at positions all over Earth to trace the movement of these waves during an earthquake.

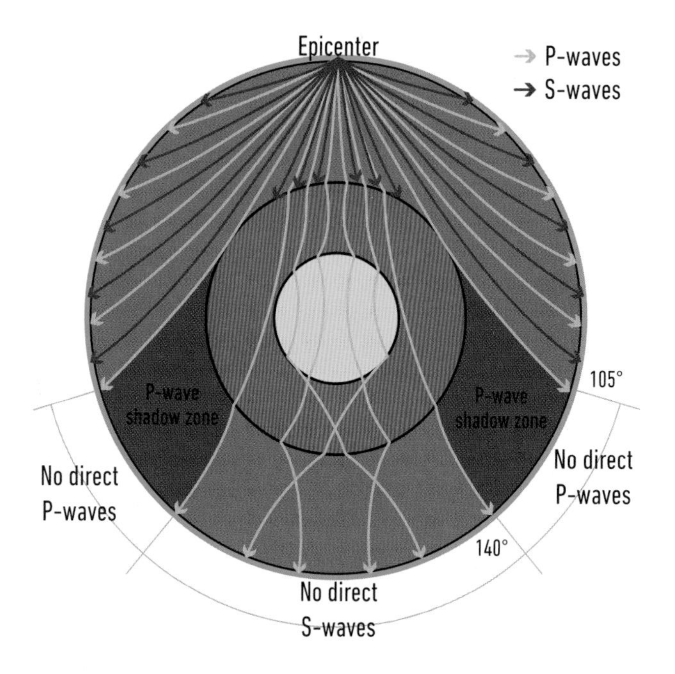

This image shows how an earthquake, located at the epicenter at the top, generates waves that are refracted by the different layers in Earth's interior. The size estimate of Earth's molten core is the result of the fact that S waves can't travel well through liquids, but P waves can.

The layers of Earth's interior.

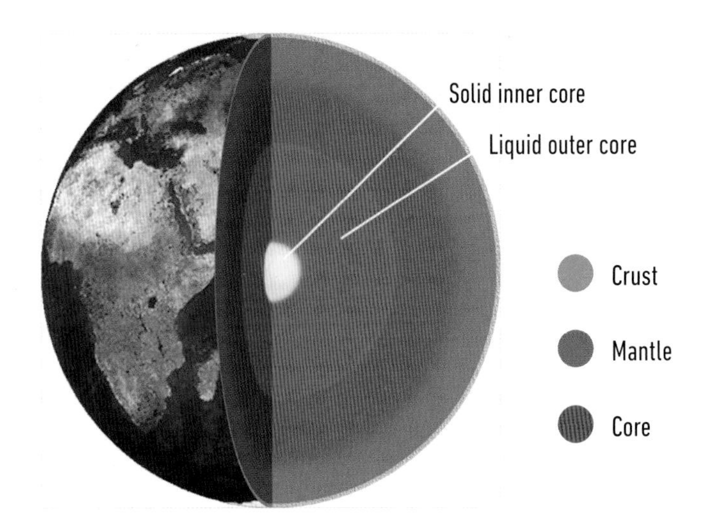

Interior Layers

Based on these readings, the interior layers and their sizes can be mapped. The upper layer, called the *crust,* is very thin, varying between 5 and 35 km thick. It is thickest beneath Earth's continents and thinnest beneath the oceans. Between the crust and the mantle is a thin layer of silicon-rich rock called the *asthenosphere.* This layer is plastic (able to flow slowly) because its temperature is slightly greater than the melting temperature. The crust is separated into plates called *tectonic plates,* which float on the asthenosphere (see "Earth's Surface: Plate Tectonics").

The *mantle* is much thicker than the crust, at almost 3,000 km. It is made of silicate rocks, which contain large amounts of magnesium and iron. These elements were the heavy ones left in the inner solar system when the planets were first forming; the lighter elements were mostly pushed into the outer solar system to form the gas giant planets. The mantle experiences convection, with hot material rising up, cooling below the crust, and then sinking down again. This convection is thought to drive the motion of the tectonic plates.

Earth's *core* has a solid inner part with a radius of about 1,300 km and a very high density of 13,000 kg/m^3. Outside this solid inner core is a liquid outer core that is over 2,000 km thick and ends at the bottom of the mantle at a radius of about 3,500 km. The liquid outer core is composed of mostly iron with some nickel and is thought to be the source of Earth's magnetic field. The inner core is kept solid by the enormous pressures found there.

Differentiation

How did Earth's interior get to be the way it is? When it was forming, the entire Earth was molten, so materials were able to move about freely. During this period, called *differentiation,* the heavier elements (for example, iron and nickel) sank to Earth's interior, and the lighter elements (for example, silicon and magnesium) rose to the surface.

Earth's Surface: Oceans

This image shows the long-term average ocean temperature (in degrees Celsius) in a part of the Pacific Ocean for the month of February.

The planet is unique in the solar system in having most of its surface covered in liquid water. Earth's oceans cover almost three quarters of its surface and are a major player in the planet's stable environment. Earth has four major oceans—the Pacific, Atlantic, Indian, and Arctic oceans—with the ocean surrounding Antarctica sometimes called the Southern Ocean.

The oceans absorb carbon dioxide from Earth's atmosphere and are home to the algae that produce about half of the planet's oxygen. In fact, early bacteria floating in Earth's oceans and absorbing sunlight are the likely cause of the transformation of Earth's atmosphere from one rich in carbon dioxide to one rich in oxygen.

The Transformation of the Oceans

While Mars and Venus may have had water oceans in the past, only Earth's oceans have survived to this point in the history of the solar system. Early in its molten history, Earth was too hot to have a liquid ocean, and there have been times in Earth's history when it was so cold that its surface was almost completely covered in ice, even at the equator. Life probably survived this "Snowball Earth" (about 700 million years ago) only because of one of the unique properties of the water molecule: its density actually decreases in solid form, causing bodies of water to ice over. So while the surfaces of lakes, rivers, and even oceans freeze over, life can continue in the warmer depths below the surface.

Temperature Scale

Most of the planet uses the Celsius temperature scale. In this scale, 0° Celsius is the freezing temperature of water, and 100°C is the boiling point (at atmospheric pressure). However, astronomers most often use the Kelvin scale, because it does not allow negative temperatures. The coldest temperature possible—absolute zero, at which all atomic and molecular motion would cease—is 0 Kelvin, which corresponds to -273°C.

The Role of Earth's Oceans

The oceans play a big role in Earth's climate by transferring heat from the warm regions (tropics) to the colder regions (poles). The oceans also provide an energy source for hurricanes.

The ocean is divided by oceanographers into different depth regimes. The photic zone is the upper 200 m of Earth's oceans in which photosynthesis can occur. In regions below 200 m, the primary sources of energy are hydrothermal vents. Earth's oceans get colder with depth dropping to 277 K at a depth of 2 to 4 km below sea level.

About half of the oxygen in Earth's atmosphere is created by phytoplankton, or microscopic organisms that live in the oceans and convert carbon dioxide into oxygen and organic compounds using energy from the Sun.

This image shows a source of energy in Earth's oceans: hydrothermal vents. These deep-sea fissures support their own local ecosystems.

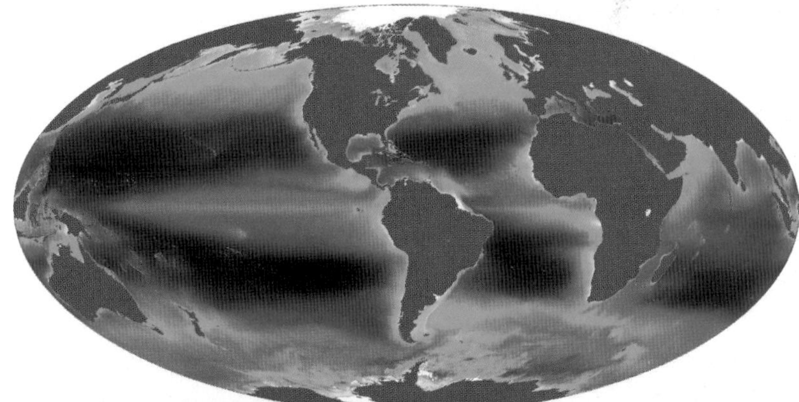

This image is a nine-year average from 1997 to 2006 of chlorophyll levels due to phytoplankton in the oceans. The highest concentrations appear yellow, and the lowest appear blue.

Earth's Surface: Plate Tectonics

This image shows the entire surface of Earth, including the sea floor. The Mid-Atlantic Ridge is clearly visible between the continents of North and South America and Africa.

Earth's crust is a thin layer that floats on the upper layer of the mantle called the *asthenosphere* (see "Earth's Interior"). This layer of crust is not a solid surface, but in fact is broken up into a series of plates that are observed to move as a result of convective motion in the material below the crust. The motion of these plates was once referred to as continental drift but is now called *plate tectonics,* since it is the plates and not the continents specifically that are drifting. The plates' motions are responsible for the observed geological activity of our planet: earthquakes, volcanoes, and the formation of mountains and deep ocean trenches.

What Plates Have Formed

In regions where plates have come together, Earth has formed some of its highest mountain ranges, like the Himalayas and the Tibetan Plateau. When plates come together under the ocean, they can create deep trenches. Regions between separating plates are called *rifts.* The Mid-Atlantic Ridge is just one such rift, hidden by the vast Atlantic Ocean.

Geology and biology both support the idea that in the relatively recent past (240 million years ago), the continents of South America and Africa were connected. The slow drift of these two plates have carried the continents away from one another. The continents as you see them today are a recent creation, having taken shape in just the last 100 million years.

How Tectonic Plates Come Together

Depending on their location, there are three ways in which plates can come together: oceanic-continental convergence, oceanic-oceanic convergence, and continental-continental convergence.

An *oceanic-continental convergence* happens when an oceanic plate meets a continental plate. A good example of an oceanic-continental convergence is the west coast of South America, where the Nazca Plate is being pushed below the South American Plate in a region called a *subduction zone*. This subduction has created a deep trench off the coast of South America and has also lifted up the Andes Mountains (which sit on top of the western edge of the South American Plate).

An *oceanic-oceanic convergence* is the result of two oceanic plates meeting. For example, in the Pacific Ocean, the Marianas Trench is the location of an oceanic-oceanic convergence, where the Philippine Plate is moving beneath the Pacific Plate. Oceanic-oceanic convergences like this result in volcanism and the creation of island chains or island arcs above the subduction zone. The Aleutian Islands are an example of this type of an island arc.

Finally, when two continental plates meet, you have a *continental-continental convergence.* While one of the plates is still subducted, the crustal material from both plates is pushed upward, creating enormous mountain ranges. For example, the collision of the Indian Plate into the Eurasian Plate has created the Himalayas and the Tibetan Plateau, with most of this growth in this mountain range occurring in just the past 10 million years.

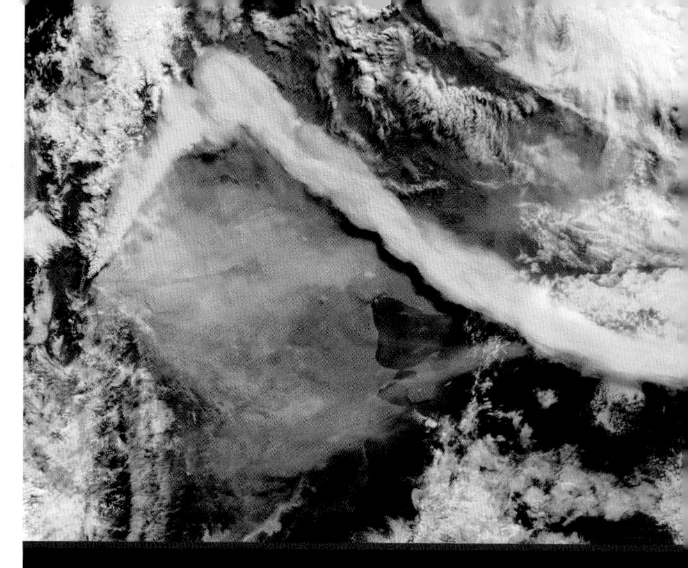

The ash plume from the eruption of southern Chile's Puyehue-Cordon Volcano. The eruption of volcanoes like this one are caused by tectonic plate motion.

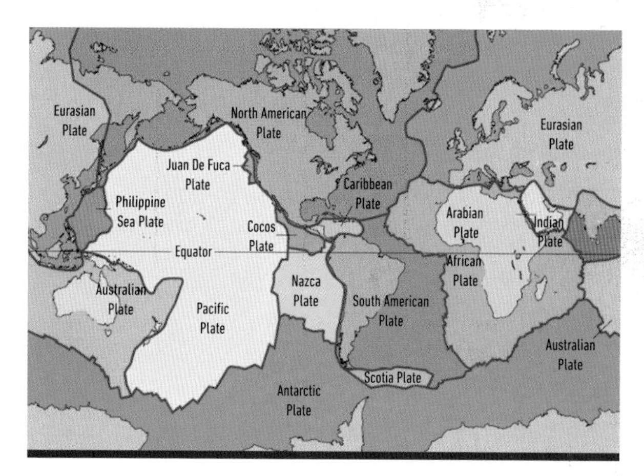

A map of Earth overlaid with the major known plates. The boundaries between these plates are known to be regions of active geological activity, including earthquakes and volcanoes.

How Fast Are They Moving?

Earth's plates move at different speeds—ranging from 1 to 8 cm per year—but average to less than 5 cm per year. That might not seem like much, but after 100 million years, that amounts to a motion of 5,000 km, or about the width of the continental United States.

Earth's Magnetic Field

Like the Sun and many of the larger planets in the solar system, Earth has a strong magnetic field. But Earth is unique in having the strongest magnetic field of any of the terrestrial planets, with geological records indicating this protective field has been in place throughout the evolution of life on Earth.

Earth's magnetic field is thought to be the result of motions in the liquid iron found at high temperatures and pressures in Earth's outer core. While the field is generated by motions in its interior, the effects of the field stretch out into the solar system to many times Earth's diameter.

While in detail Earth's magnetic field is much more complex, it is often approximated to be a magnetic dipole, with the South Magnetic Pole located near to Earth's geographic North Pole. A dipole magnetic field is most familiar as a bar magnet; in the case of Earth, that "magnet" is tilted by about 10 degrees with respect to the rotational axis of Earth.

The Magnetosphere

Earth's magnetic field is responsible for generating its magnetosphere, which is the region around Earth that is formed and shaped by its magnetic field. The magnetosphere protects Earth from high-energy charged particles that would otherwise strip away the ozone layer. So things like the charged particles in the solar wind would be much more damaging to Earth's atmosphere and surface if it were not for the presence of the magnetosphere.

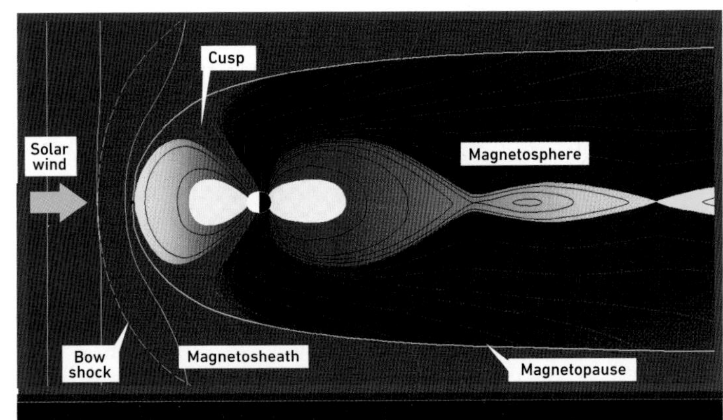

Earth's magnetosphere takes on a "teardrop" shape as a result of its interaction with material streaming from the Sun (the solar wind). Magnetic field lines originate near Earth's North and South Poles.

The shape of Earth's magnetosphere is not spherical; rather, it's shaped by its interactions with the solar wind. On the Sun-facing side of Earth, the magnetic field is compressed and is located about seven Earth diameters (90,000 km) from Earth.

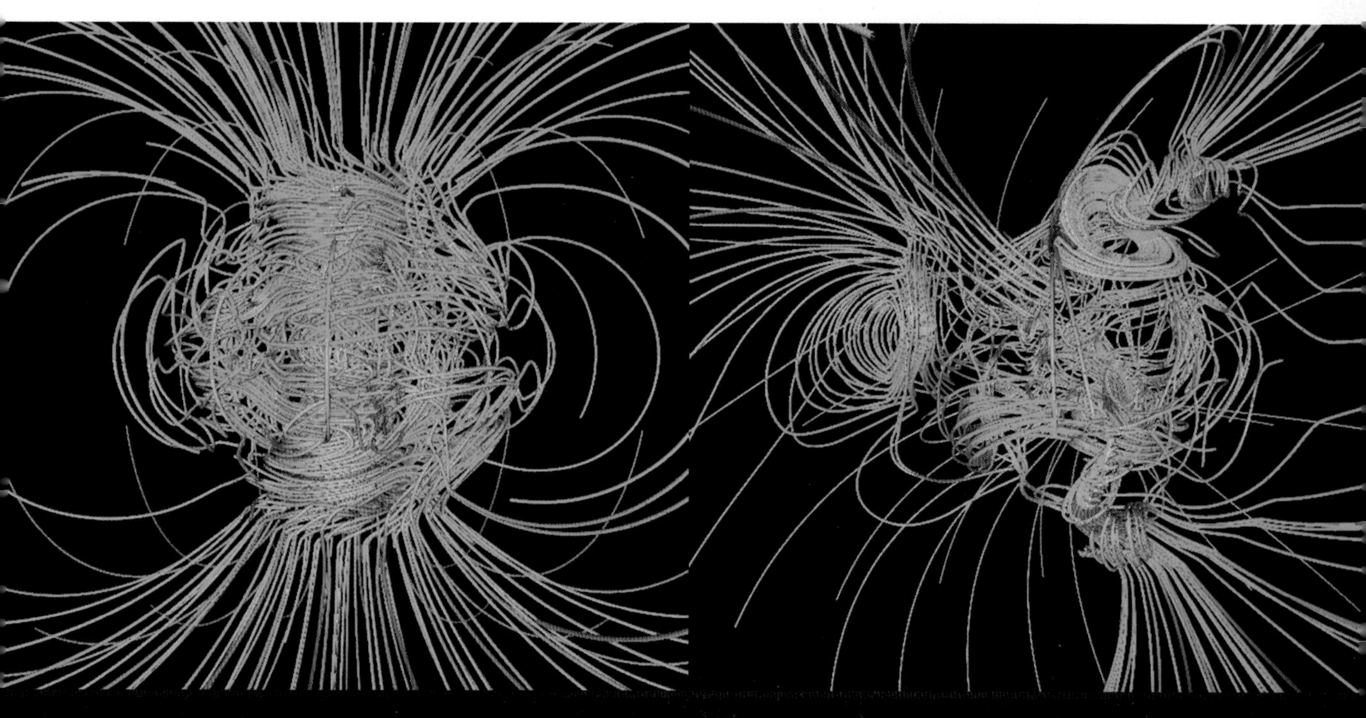

Computer models of Earth's magnetic field at the time of constant polarity (left) and during a magnetic field pole flip (right). The chaotic patterns seen in the right image are thought to correspond to a period of weak magnetic field strength on Earth that could potentially allow many more high-energy—and therefore damaging—charged particles reach Earth's atmosphere and surface.

Earth's Changing Magnetic Field

Earth's rocks tell us that the geomagnetic field has existed for at least 3 billion years. And the field would have long since decayed away were it not being generated by an actively moving, charged core.

Earth's magnetic field is known to have changed over time, undergoing a complete pole reversal every few hundred thousand years, known as a "pole flip." Based on geological records, the last pole flip is thought to have happened about 750,000 years ago. Models indicate the magnetic field would change polarity more frequently but that Earth's solid iron core keeps the polarity changes from happening more often.

The direction of Earth's magnetic field is known to change slowly, and current observations show that Earth's magnetic field is decreasing in strength. One casualty of a weak magnetic field is likely to be Earth's ozone layer. This layer is highly effective at absorbing ultraviolet radiation, and if it were stripped away, it is likely that skin cancer rates among surface-dwelling animals would increase.

Earth can clearly survive these pole flip events, since they have happened hundreds of times since it formed. However, the impact on living things will likely be significant.

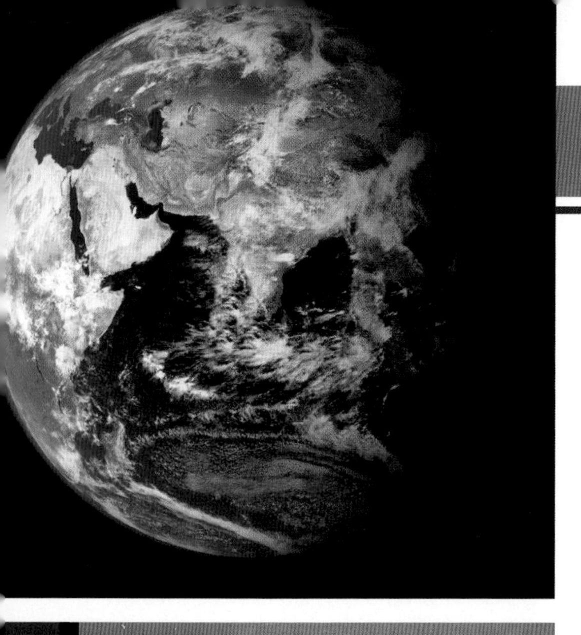

Earth's Atmosphere

Earth is a unique jewel in the solar system. There may be billions of Earthlike planets in the Milky Way Galaxy (see "Searching for Life in the Milky Way"), but there is only one Earth-like planet in the solar system that is accessible to humans for the foreseeable future. And while a few other planets and moons in the solar system have atmospheres, none (as far as people know) have an atmosphere that has been so fully shaped by life.

Earth's atmosphere protects life in a variety of ways: it moderates temperature changes as the planet rotates, its upper layers contain ozone that shields life from harmful ultraviolet radiation, and its very presence keeps large numbers of meteoroids from ever striking the surface.

Statistics

RADIUS
6,378 km

SIDEREAL PERIOD
365.26 days

SIDEREAL ROTATIONAVL PERIOD
23h 56m 4.1s

VOLUME
1.1×10^{12} km^3

SATELLITES
The Moon

MEAN DISTANCE FROM THE SUN
1.0 AU (1.5×10^8 km)

MASS
6.0×10^{24} kg

ATMOSPHERIC COMPOSITION (BY VOLUME)
78% nitrogen, 21% oxygen, 0.93% argon, 0.039% carbon dioxide, and trace amounts of water vapor and other molecules

Composition

Earth's current atmosphere is composed mostly of molecular nitrogen (~80 percent), with the remainder mostly molecular oxygen (~20 percent) and a tiny amount of argon. The remaining elements in Earth's atmosphere are found in trace amounts, including carbon dioxide. These ratios are a big change from Earth's original atmosphere, which was heavy with carbon dioxide. Starting about 2 billion years ago—thanks to photosynthesis by cyanobacteria in the oceans—Earth's atmosphere started to have more and more oxygen, rising to its current levels only about 200 to 300 million years ago. The presence and current abundance of oxygen in the atmosphere is the result of life on Earth. The abundance of nitrogen in the atmosphere is also related to bacteria that remove oxygen from nitrates and release nitrogen as a waste product.

Layers in Earth's Atmosphere

The lowest layer of Earth's atmosphere is called the *troposphere,* which extends from the surface up to an average of 12 km. The temperature drops with altitude in this layer, as you experience when you see ice forming on the window of our airplane as it reaches cruising altitude. This is the layer of Earth's atmosphere that contains almost all of the planet's weather patterns. The troposphere is heated from below by Earth's surface, and the temperature difference between top and bottom drives convective cells.

Above this layer is the *stratosphere,* which stretches from 12 to 50 km. The stratosphere contains triatomic oxygen (O_3) called *ozone*. The ozone layer efficiently absorbs ultraviolet radiation, effectively shielding Earth's surface from these harmful rays. Because the stratosphere is absorbing radiation, its temperature increases with altitude, meaning there is no convection in it.

Above the stratosphere is the *mesosphere,* which extends from 50 km to about 85 km above Earth's surface. There is little ozone found in the mesosphere, and less radiation is absorbed, so the temperature again drops in this layer.

The *thermosphere* is the layer above the mesosphere that extends out to hundreds of kilometers. This layer warms again because it contains molecules that absorb high-energy ultraviolet photons from the Sun.

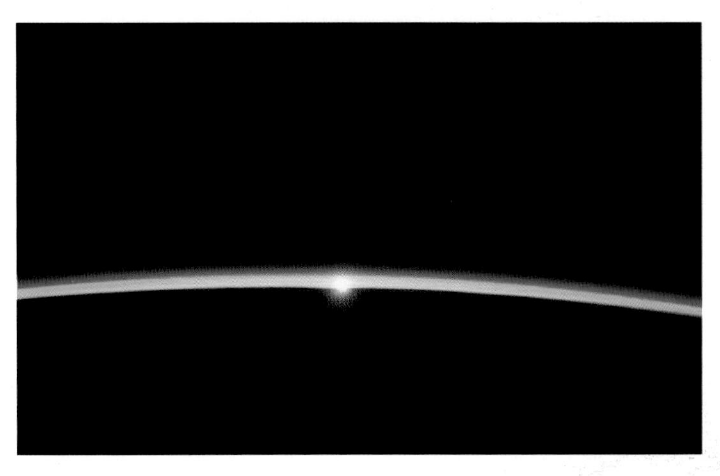

As this image shows, Earth's atmosphere is a very thin protective layer, comprising only about 1 percent of its radius (out to the top of the mesosphere).

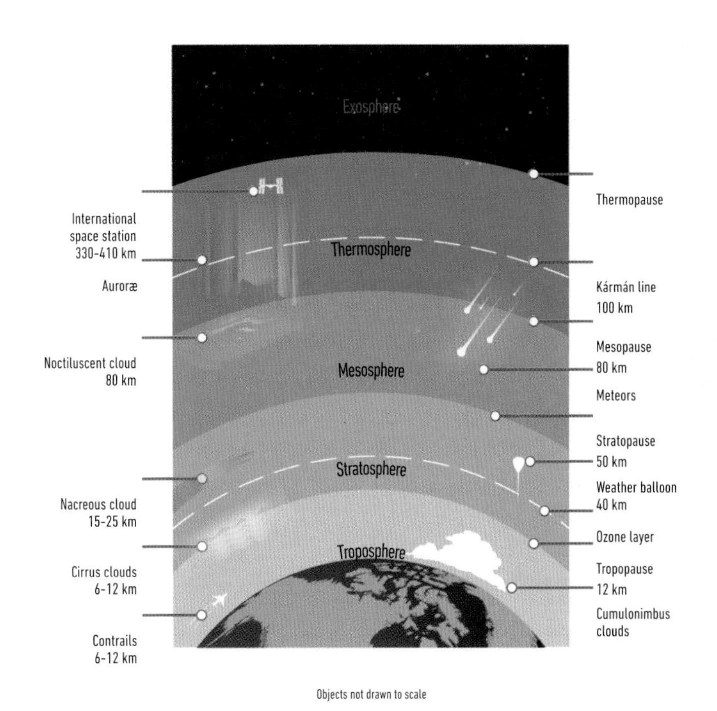

The layers of Earth's atmosphere.

Global Climate Change

As you learned earlier, the early atmosphere of Earth consisted of mostly carbon dioxide (CO_2) before bacteria transformed it into nitrogen and oxygen. Even after stabilizing into its current composition of mostly nitrogen and oxygen a few hundred million years ago, Earth's climate has experienced great changes in temperature. Geologists recently confirmed that almost 700 million years ago, Earth was almost entirely covered in ice (sometimes referred to as "Snowball Earth"). On the other end of the temperature spectrum, during the reign of the dinosaurs, there were periods of no surface ice, with swamps at the North Pole.

Earth's ice has preserved a wonderful record: the levels of CO_2 in Earth's atmosphere in parts per million back 800,000 years. What this archive shows is that never in the past 800,000 years has the level of CO_2 in the atmosphere risen above about 300 parts per million (ppm). In 2013, however, the level passed 400 ppm, due potentially to the burning of fossil fuels and deforestation.

And in recent years, large amounts of data are indicating that Earth's oceans and atmosphere are warming. The National Oceanic and Atmospheric Administration (NOAA) recently published a number of different studies that all point to a warming globe. There is no denying that Earth has been warmer in the past than now, but the danger in allowing Earth to continue to warm is that the secondary effects of global warming will likely cause large disruptions to Earth's large and growing human population, 20 to 25 percent of whom live in coastal regions.

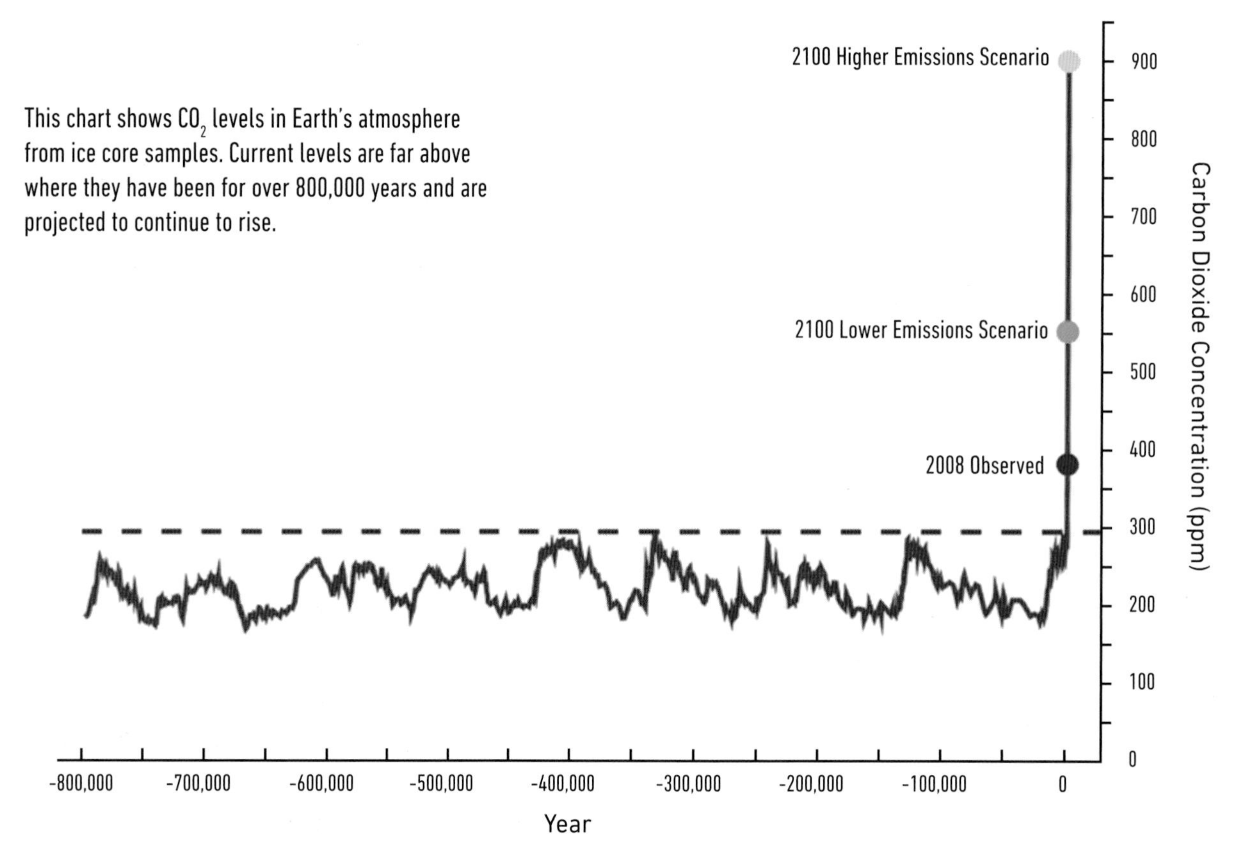

This chart shows CO_2 levels in Earth's atmosphere from ice core samples. Current levels are far above where they have been for over 800,000 years and are projected to continue to rise.

This satellite image from the National Snow & Ice Data Center shows floating pieces of ice from the 2008 Wilkins Ice Shelf Collapse. Rising sea and air temperatures are causing a rapid retreat in the planet's ice cover.

One of the more controversial claims is that the rise in CO_2 and the rise in Earth's sea and air temperatures are related, since CO_2 is a known "greenhouse gas." Others have proposed that the changes in Earth's temperature are part of a natural cycle, and that the source of temperature variations is the Sun's variable output. However, the measured variations in the Sun's energy output do not correlate with the rise in Earth's average surface temperature.

While the governments of various countries have had a hard time coming to a consensus as to what to do about rising levels of CO_2 in the atmosphere and rising global temperatures, humans have shown the ability to act as a planet in the recent past. For example, when scientists discovered that chlorofluorocarbons were acting as catalysts to destroy Earth's ozone layer, an international agreement—The Montreal Protocol in 1987—called for drastic reductions in the production of these chemicals.

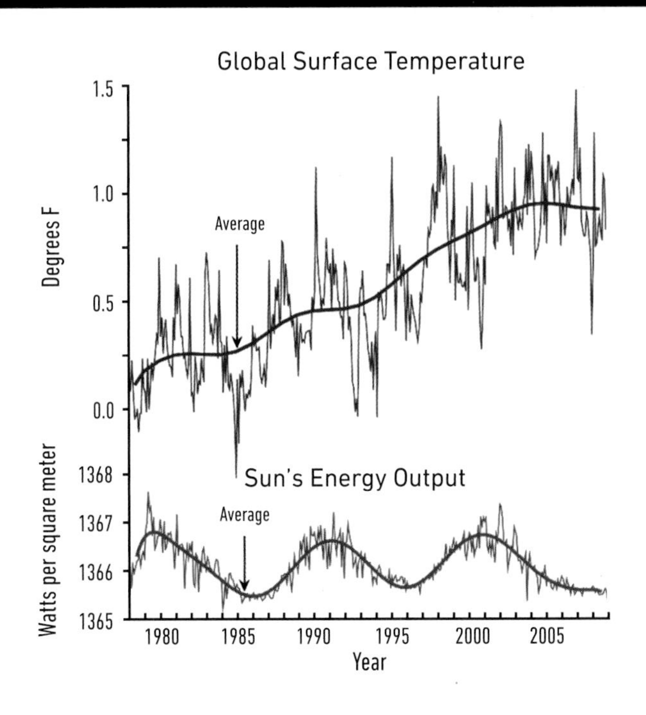

This image shows that the normal variations in the Sun's energy output in the past 30 years do not seem to correlate with the rise in the global surface temperature of Earth during the same period of time.

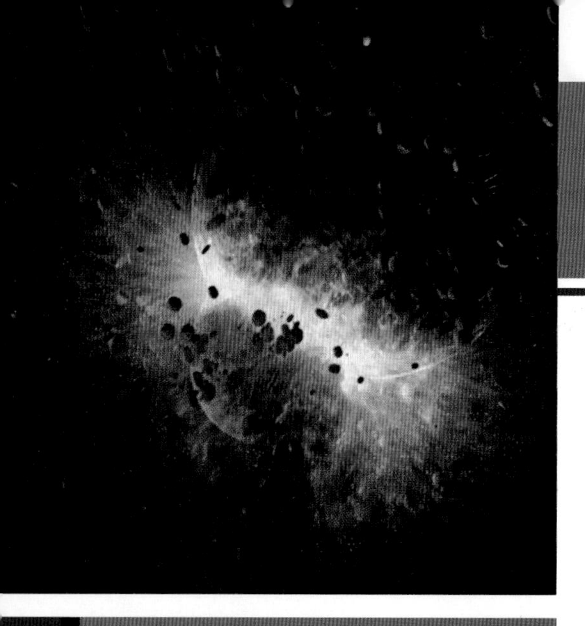

The Formation and Structure of the Moon

Statistics

MEAN RADIUS
1,737 km (0.27 Earth radii)

ORBITAL PERIOD
27.3 days

SYNODIC PERIOD (ONE FULL PHASE CYCLE)
29.5 days

SIDEREAL ROTATION PERIOD
27.3 days

VOLUME
2.2×10^{10} km^3 (0.02 Earths)

SATELLITES
None

MEAN DISTANCE FROM THE EARTH
384,400 km

MASS
7.3×10^{22} kg (~0.012 Earth masses)

ATMOSPHERIC COMPOSITION (BY VOLUME)
Almost nonexistent

The Moon is a natural satellite of Earth and is the fifth-largest moon in the solar system. The Moon has been geologically dead for a very long time, its only changes being small surface impacts and its slow drift away from Earth (the latter confirmed when Apollo astronauts placed reflectors on the surface of the Moon in 1969 and thereafter to measure its motion precisely).

How the Moon Formed

Early Moon formation theories in the late nineteenth century suggested that a rotating proto-Earth spun off a molten Moon, giving rise to both the orbit and the Moon's outward motion. However, theorists were unable to show how such a "fission" theory would work in detail, so the idea was rejected in favor of a more catastrophic beginning.

In the middle of the twentieth century, it was suggested that the Moon's origin might have been a collision between the proto-Earth and a Mars-sized object almost 4.5 billion years ago. The debris of this collision, containing material from the impactor and Earth, orbited the wounded Earth until gravity collected it into the Moon. Once this object was assembled, gravity pulled it into the shape of a sphere, and the low-density material began to "float" on the surface of an ocean of magma. Since that time, the Moon has suffered periods of bombardment that have pocked its surface with large and small craters.

In an interesting modification to this theory, astronomers have recently shown that the creation of other "moonlets" probably accompanied the formation of the Moon. One of the smaller objects that orbited Earth in the same orbit as the Moon eventually was shaken from its orbit and struck the far side of the Moon at low velocity, creating a Moon with the thicker lunar highlands on the far side where the "pancake" collision occurred.

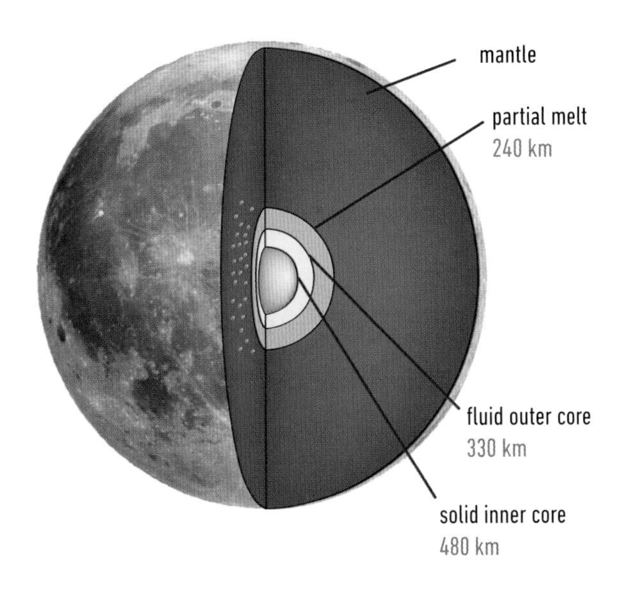

The internal structure of the Moon.

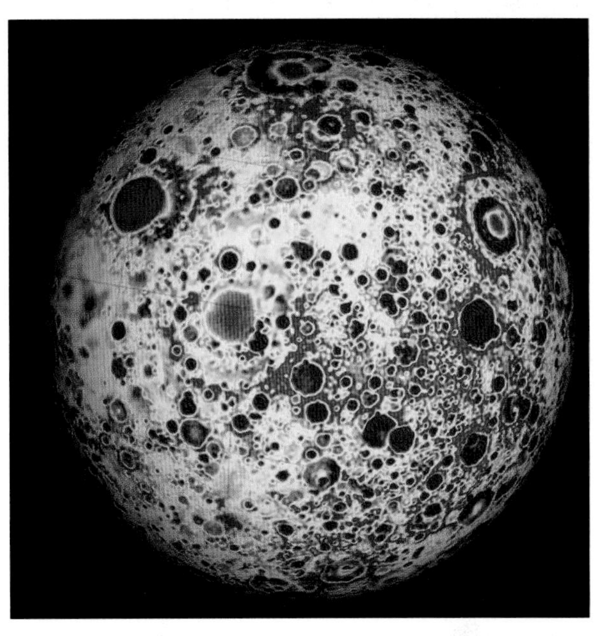

An image of the Moon's gravitational field from the GRAIL-A and GRAIL-B satellites. The colors indicate variations in the strength of the Moon's gravitational field, from red (stronger) to blue (weaker).

The Moon's Structure

The Moon's inner structure is relatively simple compared to that of many other solar system objects. It is thought to have a very thick mantle, with its only remaining geologically active material confined to the central 500 km of its radius. Like Earth, it has a solid inner core and a fluid outer core.

The recent GRAIL (Gravity Recovery and Interior Laboratory) mission to the Moon measured its gravitational field carefully using two spacecraft. The surface gravity of the Moon is related to the bulk density of its crust, so the measurements from GRAIL gave a new estimate of the average of the thickness of the Moon's crust, which ranges between 34 and 43 km.

Debunking the Capture Hypothesis

Computer simulations have been essential to explore different models of the Moon's origin. These simulations have been able to show that the Moon could not have been simply "captured" by Earth's gravity, which would have only deflected it. The only way for the Moon to stay around was for it to hit.

The Lunar Surface

The colors of features on the Moon's surface are related to their smoothness and elevation. The dark regions are lower and smoother, while the bright regions are higher and more heavily cratered.

The Moon is a familiar companion, and even the unaided eye can make out regions of light and dark on its grey surface. Because the Moon orbits Earth at the same rate that it rotates on its axis (an effect called *tidal locking*), you always see the same side of the Moon. The side of the Moon that faces away—the far side—was not observed until the Soviet Union and later the United States orbited the Moon with cameras. The side of the Moon you do not see is sometimes referred to as the "dark side of the Moon"; however, this is not an accurate description. All sides of the Moon are illuminated as it orbits Earth—you are just not privy to the surface features on its far side (either when they are dark or when they are lit up).

The Moon has no appreciable atmosphere, and early thoughts that it might be inhabited in the seventeenth and eighteenth centuries were untenable once telescopes gave people a clear view of its harsh surface. The main elements of the Moon's surface visible are its *craters,* its dark *maria,* and its light-colored regions called *lunar highlands*.

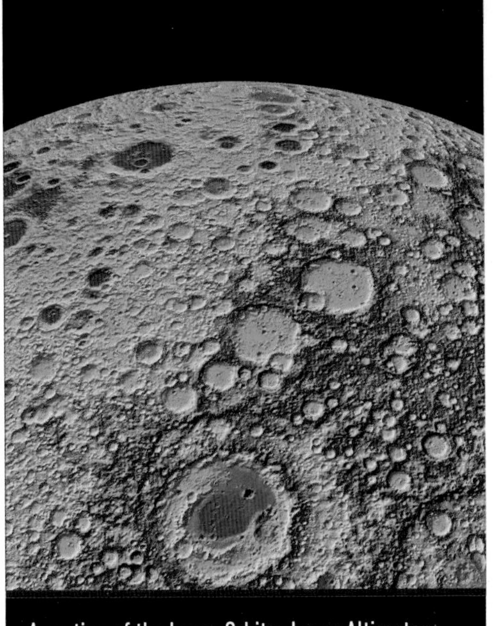

A portion of the Lunar Orbiter Laser Altimeter (LOLA) data from the far side of the Moon showing the elevations of the surface of the Moon is shown in this false-color image. The blue regions are lower and red regions are higher.

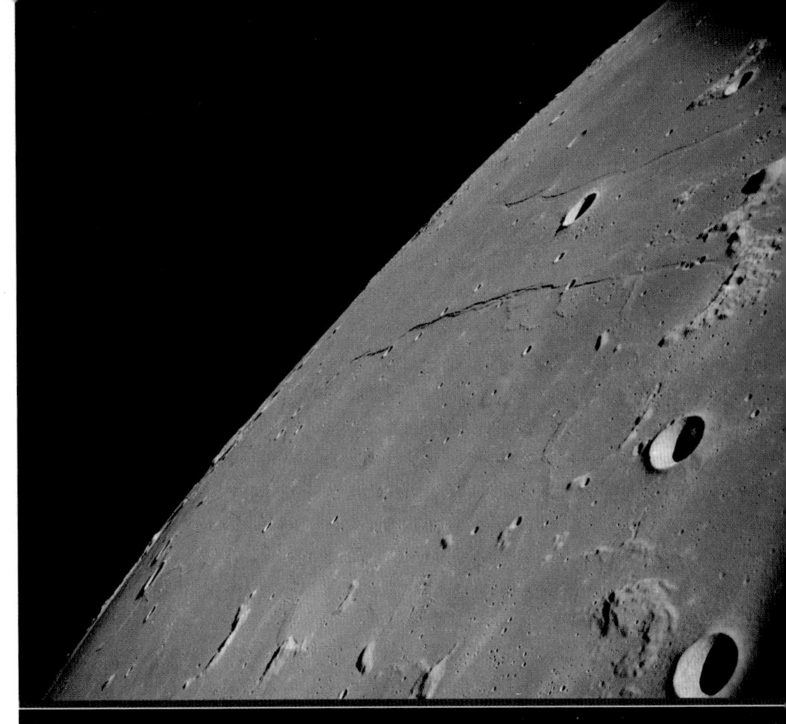

This image of the Sea of Tranquility was taken by the Apollo 8 mission in 1968. The large craters visible are 10 to 15 km across.

Craters

The craters on the surface of the Moon are believed to have been formed by collisions early in the history of the solar system. In the typical scenario, a meteoroid approaches the lunar surface and ejects material out radially from the impact site. The result is a crater wall, a central peak of rubble, and an ejecta blanket (often visible as rays of lighter material around the crater itself). Craters are often surrounded by smaller secondary craters from ejecta, or material thrown out by the collision. If the impact was particularly large, it might have also allowed lava to flow in from the Moon's younger interior, forming a *mare* (the singular of *maria*).

Maria

Early observers thought these large, grey-colored regions may have been bodies of water, leading to their poetic names (for example, The Sea of Tranquility). However, later observations and visits to the surface of the Moon have made it clear the maria may have once been seas of lava, not water. They are observed to have far fewer craters than the rest of the lunar surface, indicating the maria are regions of the surface that are relatively young. They were most likely formed after lava filled in the basins created by large early impacts.

Lunar Highlands

The regions around the maria are called *lunar highlands,* because on average they are a few kilometers above the average lunar elevation. In addition to being higher than the maria, the highlands are heavily cratered, indicating they are also older. One of the big surprises that came from observations of the far side of the Moon is that lunar highlands cover almost the entire surface of that side.

Visiting the Moon

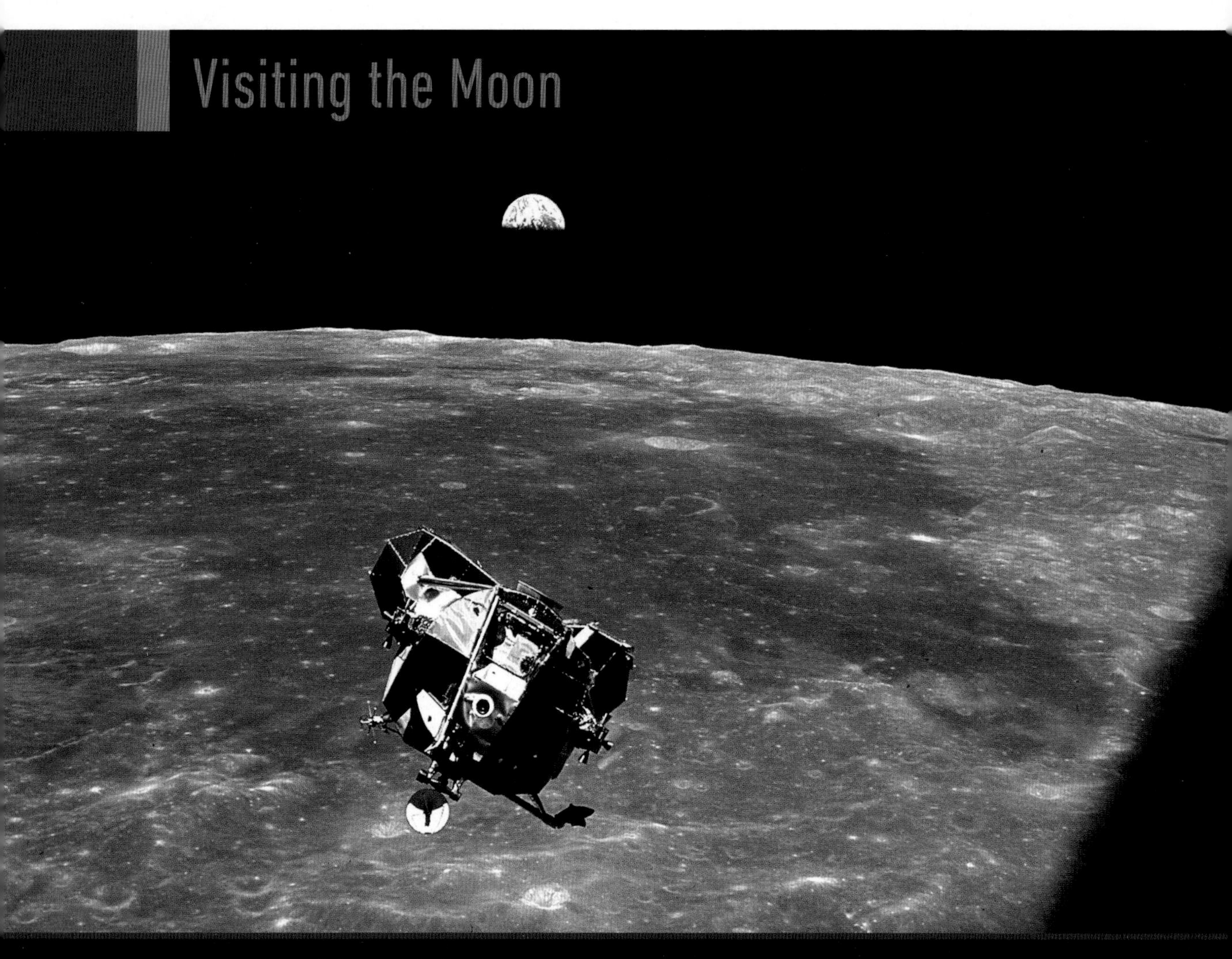

This image—taken by Michael Collins on July 21, 1969—shows the Lunar Module "Eagle" returning to rendezvous with the Command/Service Module. The lunar surface and Earth in first quarter phase are visible in the background.

In the solar system, humans have physically visited only the Moon. Within NASA, there were years of debate about the best way to get people to the surface of the Moon and back. All of the possibilities involved significant risks. The way that it finally happened was the result of a plan called *Lunar Orbit Rendezvous (LOR)*. This plan—which was one of three possibilities—was championed by Dr. John Houbolt, who was part of the Lunar Mission Steering Group. Using LOR, three astronauts would launch from the surface of Earth in a two-piece spacecraft (the Command/Service Module and the Lunar Module). The two-piece spacecraft would then go into lunar orbit. Only two of the astronauts would descend to the Moon's surface in the Lunar Module, with the third astronaut remaining in orbit around the Moon. This mode allowed the fuel needed to return to Earth to stay with the Command/Service Module.

LOR turned out to be highly successful, with six complete lunar roundtrip missions. From 1969 to 1972, the Apollo 11, 12, 14, 15, 16, and 17 missions launched from Earth and successfully delivered a crew of two to the Moon's surface.

After the successful landing of Apollo 11 in July 1969, the Apollo missions carried out more and more sophisticated experiments. One mission (Apollo 13) famously experienced an explosion in a fuel tank on route to the Moon but still returned its crew home safely. These landings were such a monumental undertaking that they have not been repeated since.

Other Missions to the Moon

There was a long gap in lunar exploration from the 1970s until the mid-1990s, when a number of missions were launched. The Clementine mission (1994) was an orbiter that carried a number of instruments, including a LIDAR (Light Detection and Ranging) imaging camera, which made detailed elevation maps of the lunar surface.

Most recently, in December 2013, The China National Space Administration (CNSA) landed a spacecraft (Chang'e 3) and a rover (Yutu) on the surface of the Moon. This mission is notable as the first soft landing on the surface of the Moon since 1976. The Chang'e 3 lander is equipped to sample lunar soil to a depth of 30 m. This mission will also test technology to be used in Chang'e 5, a future mission that plans to return lunar material to Earth.

The Lunar Roving Vehicle (LRV) of Apollo 15 allowed astronauts to venture farther from the landing site than previously. The final three Apollo missions (15, 16, and 17) brought LRVs to the surface of the Moon, where they remain today.

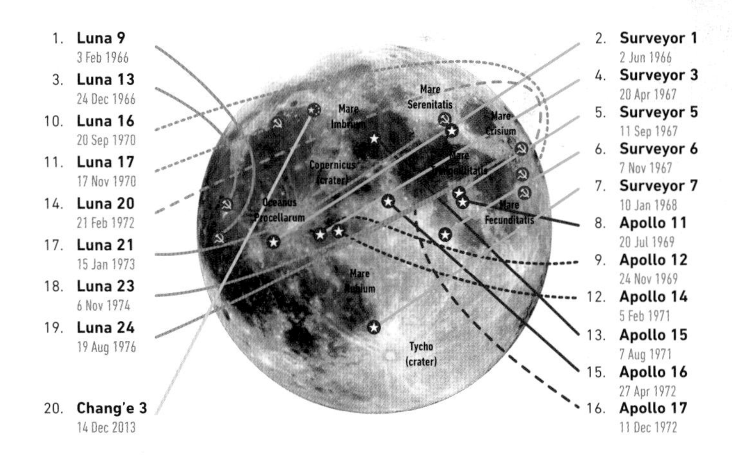

Missions to the Moon throughout the years. On the right are the U.S. missions, both unmanned and manned. On the left are the unmanned missions by Russia and China.

The solar system is filled with a rich variety of planets and moons, from the inner terrestrial planets to the outer gas giants. In this part, I give you a rundown of Earth's companions and their moons, as well as information on asteroids and comets found in the solar system.

First, you learn about the current theories of how the solar system formed. Afterward, you visit each of the planets, learning their basic properties and how each of them compares to planet Earth. I then show you other moons; some, such as Jupiter's moon Europa and Saturn's moon Titan, are places where life may have found a foothold in the solar system. Finally, you learn about some of the smallest and oldest objects in the solar system—asteroids and comets—which have had big impacts on the planets and moons in the solar system.

The Formation of the Solar System

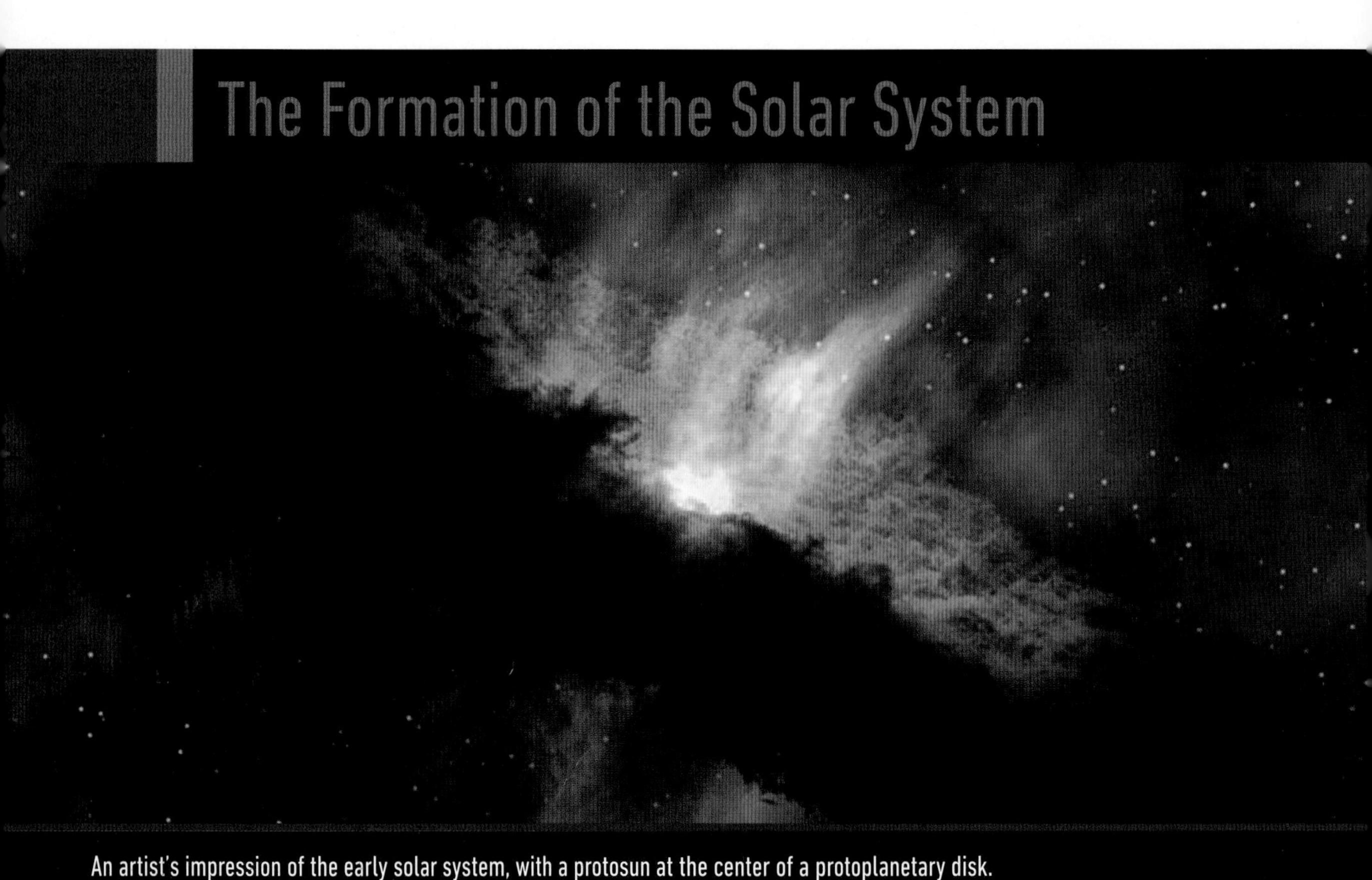

An artist's impression of the early solar system, with a protosun at the center of a protoplanetary disk.

Observations of the solar system today provide a lot of clues about how the planetary system might have formed. What do astronomers know about the solar system?

- Most of its mass is in the center of the solar system, located in the star that the planets orbit (the Sun).
- The planets orbit in the same direction—counterclockwise as viewed from the North Pole side of the solar system.
- Most of the planets rotate on their axes counterclockwise as viewed from the North Pole side of the solar system. The exceptions to this rule are Venus (which rotates clockwise) and Uranus (which is tilted on its side).
- All of the planets are located in approximately the same plane, known as the plane of the solar system or *ecliptic* (see "The Sun's Motion: Telling Time").
- The inner four planets (Mercury, Venus, Earth, and Mars) have rocky surfaces, are relatively high density (mass per unit volume), and are relatively small. They are called the *terrestrial planets.*
- The outer four planets (Jupiter, Saturn, Uranus, and Neptune) are much more widely spaced, have relatively lower density, and are both larger and more massive than the inner planets. They are called the *Jovian planets* and are composed mostly of hydrogen and helium.

Early Hypotheses on the Formation

Based on the information you have, you might speculate, for example, that the Sun formed and its enormous gravity captured the planets that now orbit it. This hypothesis has a number of problems. If the planets were randomly captured, they are unlikely to have orbits on the same plane, with orbits and rotations in the same direction. Random capture also has a hard time explaining the segregation of the planets into inner, rocky, small planets and outer, gassy, large planets.

Another hypothesis suggests a nearby encounter of the Sun with another star pulled out a "tail" of material, and that this tail (which would naturally orbit in one direction) broke up into the planets. Again, this idea does not naturally explain the locations of the terrestrial and Jovian planets. Other tests of this model have shown that any tidal forces strong enough to pull out enough material to form the planets would not allow the planets to stay in orbit around the Sun.

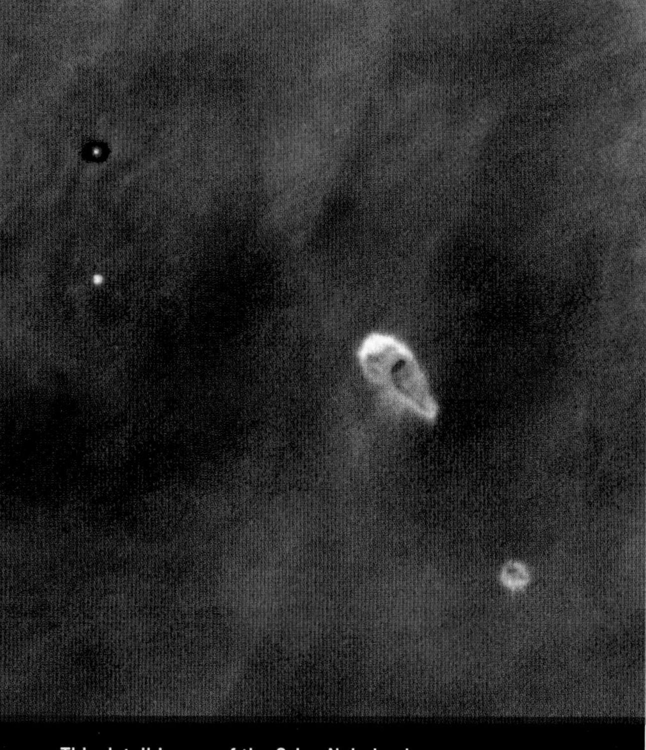

This detail image of the Orion Nebula shows some young planetary systems as they are just beginning to form. The gas and dust in the protoplanetary disk that surrounds the star absorbs background light, so the disk in the upper-left corner looks dark.

The Nebular Hypothesis

An early version of the best current model of the formation of the solar system was suggested independently in the 1700s by German philosopher Immanuel Kant and French mathematician and astronomer Pierre-Simon LaPlace. They both proposed that the Sun and all the planets began as a large, slowly rotating cloud of gas and dust. Gravity caused this cloud to contract in size and rotate faster and faster. That little bit of rotation is important, because the gravitational forces in a rotating cloud of gas and dust will concentrate most of its mass at the center (this is the protosun), and the remaining mass will fall into a flat disk that orbits it. This scenario, called the *nebular hypothesis,* describes what many today think happened in the solar system.

In fact, proplyds (see "Protoplanetary Disks [Proplyds]") can be seen in star-forming regions like the Orion Nebula. Similar to these objects in the Orion Nebula, the planets that orbit the Sun are thought to have started in a disk orbiting the protosun.

Layout of the Solar System

Before I get into the layout, let's first talk about the units of measurement used for bodies in the solar system. Astronomers tend to use units that "fit" the problem at hand. For example, a meter is a perfectly good unit when you are talking about things that are the scale of humans, while kilometers (or 1,000 meters) are a good unit when discussing the distance between cities. However, on the scale of the solar system, meters and even kilometers are a bit small; therefore, astronomers use a unit called, appropriately, an *Astronomical Unit (AU)*. One AU is the equivalent of about 1.5×10^{11} meters, or the mean distance between Earth and the Sun. In AU, the distances between the planets and the distances from the Sun to each planet become much more manageable, as you can see in the following table.

For objects that are even farther away, an even bigger unit is needed. Astronomers define the term *light-year* to be the distance that light with a velocity of 300,000 km/s travels in a year. A light-year is 9.5×10^{15} m.

Solar System Geography

Our solar system contains eight planets and a large number of dwarf planets (see "Dwarf Planets") and other smaller objects that orbit the Sun. The planets are the most familiar of these bodies, and they orbit in a well-known pattern. Close to the Sun are the terrestrial planets, all located within about 1.5 AU of the Sun. The four Jovian planets are located between 5 and 30 AU from the Sun.

The solar system is also populated with a number of other less well-known objects: asteroids, dwarf planets, and comets. You tend to hear about these objects only when they come close to Earth and light up the night sky, either with the long tail of a comet or the streak of an asteroid burning up in Earth's upper atmosphere, but these objects have a home in the solar system. Most asteroids are contained in a region between the inner and outer planets (between 2 AU and 3 AU) known as the *asteroid belt*. The asteroid belt does not contain a large amount of mass. In fact, about half of the belt's mass is found in just four asteroids: Ceres, Vesta, Pallas, and Hygiea.

Planet	Mean Distance from the Sun in AU
Mercury	0.39
Venus	0.72
Earth	1.00
Mars	1.52
Jupiter	5.20
Saturn	9.55
Uranus	19.19
Neptune	30.07

Ceres (shown here to scale with Earth and the Moon) is the only known asteroid belt object to be considered a dwarf planet, with a diameter of 950 km. For comparison, the diameter of Earth is 12,700 km, or over 10 times the diameter of Ceres.

Far-Flung Objects

The solar system does not end with the orbit of Neptune. The region from the orbit of Neptune at 30 AU out to about 50 AU is called the *Kuiper belt.* Because they are so far from the Sun, many Kuiper belt objects consist of frozen ices of water, methane, and ammonia.

Much farther from the Sun (perhaps 50,000 AU away) is a region known as the *Oort cloud.* This region, almost a light-year away from the Sun, consists of icy objects that were most likely scattered out to these large distances as the solar system formed. The Oort cloud is thought to be the primary source of long-period comets.

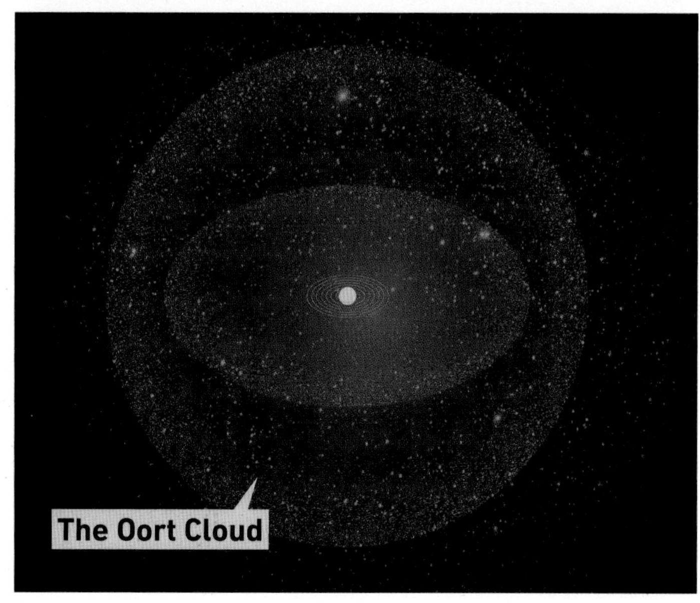

This image shows the location and orbit of objects in the Kuiper belt and Oort cloud.

Soccer-Ball-Sized-Sun Solar System

Sun	Mercury	Venus	Earth	Mars
1,391,000	4,879	12,104	12,742	6,779
	57,909,100	108,208,000	149,598,261	227,939,100
	0.0035	0.0087	0.0092	0.0049
	9.37	17.50	24.20	36.87

The relative distance of each planet from the Sun is shown to scale. On the scale of a soccer ball-sized Sun, the distances to the planets are given in meters. For example, if the Sun is the size of a soccer ball, the Earth is the size shown above, and orbits at a distance of almost 9.5 meters.

The relative sizes of the planets in the solar system are shown compared to a Sun the size of a soccer ball. Note that Saturn has been tipped to fit.

▲ Distance from the Sun to each planet on a 366-mm-line scale

Planetary Motion

For thousands of years, astronomers assumed that the apparent motion of the Sun, Moon, and planets represented their actual motions—that is, Earth was a fixed object and the Sun, Moon, planets, and stars all moved around it. Many ancient cultures adhered to this geocentric or Earth-centered model of planetary motions. While the Sun, Moon, and planets generally moved east against the background of fixed stars, ancient astronomers knew that from time to time, certain planets would stop in their eastward motion, move in the opposite direction, and begin their eastward motion again. This "retrograde motion" was regular and—in more developed geocentric models—predictable.

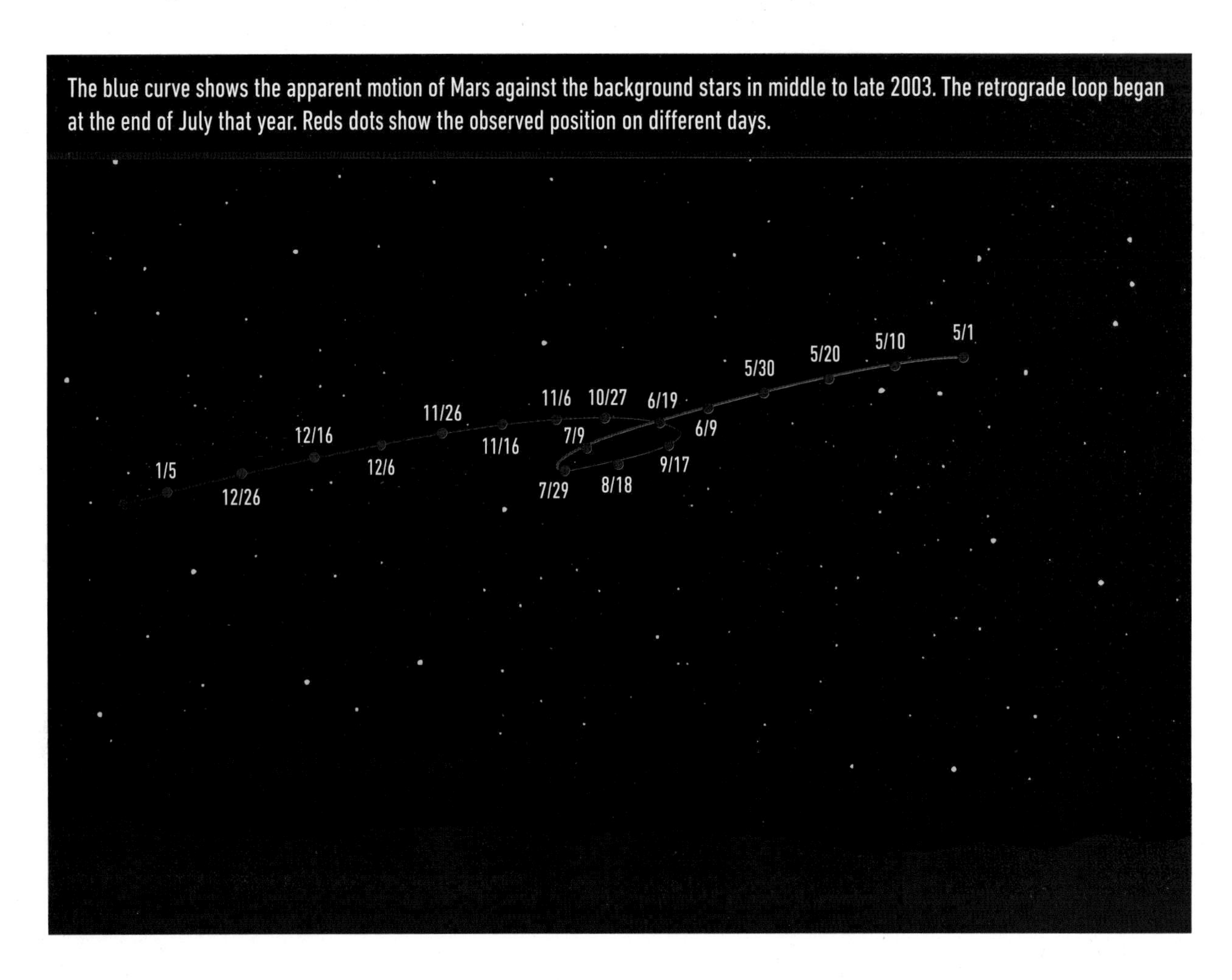

The blue curve shows the apparent motion of Mars against the background stars in middle to late 2003. The retrograde loop began at the end of July that year. Reds dots show the observed position on different days.

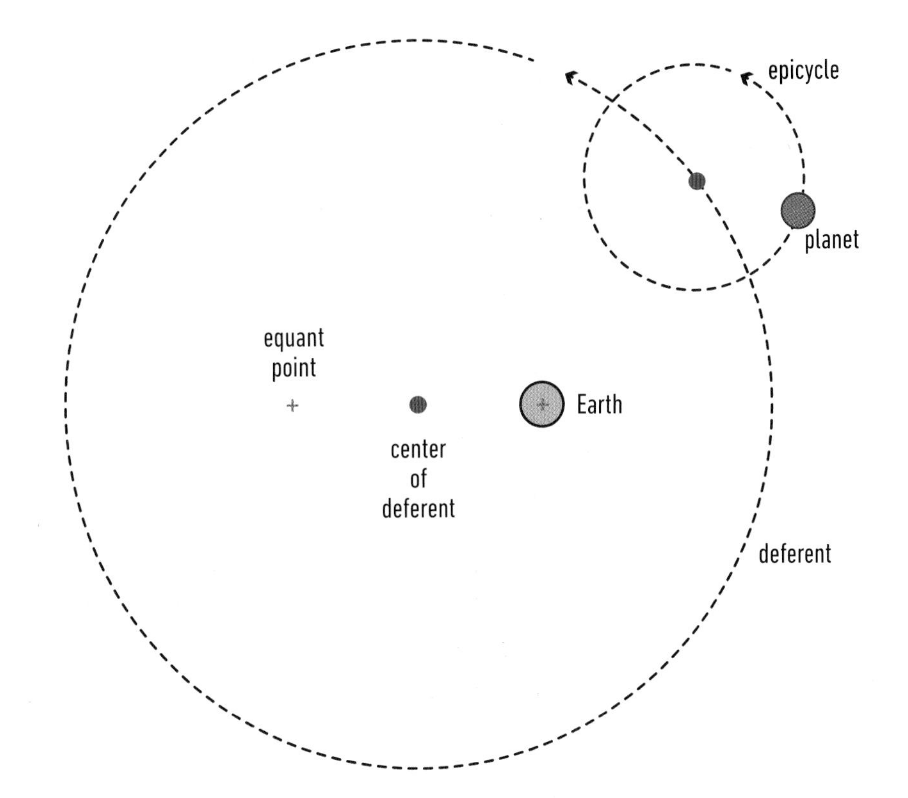

The Ptolemaic model of the solar system. The center of each planetary orbit was slightly offset from Earth's position in order to better match observations of planetary positions.

Ptolemaic Model

One of the most detailed models of the solar system, called the *Ptolemaic model* after Greek astronomer and mathematician Ptolemy, held sway from the second century B.C.E. until the Renaissance.

In the Ptolemaic model, the celestial sphere (see "The Structure of the Night Sky") was thought to represent the actual structure of the heavens. Earth was located at the center of the celestial sphere and was orbited by the Moon, Mercury, Venus, the Sun, Mars, Jupiter, and Saturn (the rest of the planets were not discovered until the invention of the telescope). Beyond the orbit of Saturn were the fixed stars.

Each planet's orbit was the result of two circles: the *deferent* (centered near the position of Earth) and the *epicycle* (centered on the deferent). As a planet moved along the epicycle centered on the deferent, it would appear to move east against the background stars until it was on the near side of the epicycle, when it would appear to move west. The Ptolemaic system provided a way to predict the positions of the Sun, Moon, and planets for more than 1,000 years.

This seventeenth-century illustration of the Copernican system indicates that Earth is just one of the planets orbiting the center of the solar system, the Sun.

The Copernican System

While Aristarchus, a Greek astronomer, was the first to suggest the idea of a Sun-centered or heliocentric model of the solar system in the third century B.C.E, it was sixteenth-century Polish mathematician Nicolaus Copernicus whose work first challenged the Ptolemaic model.

In a heliocentric model, the Sun is located at the center of the solar system, and the planets orbit the Sun. The Copernican system, published shortly before his death in 1542, has Earth and all the planets orbiting the Sun in circular orbits. The phenomenon of retrograde motion is explained by the fact that Earth orbits the Sun more rapidly than the outer planets, meaning as Earth passes them (like a runner lapping another on an inside track), they appear to move backward against the fixed stars.

The main limitation of the Copernican system was that the orbits of the planets are circles. As a result, while conceptually correct, the Copernican system was not an accurate way to predict the positions of the planets.

Kepler's Laws of Planetary Motion

Tycho Brahe, a sixteenth-century Danish astronomer, amassed an enviable dataset of planetary motions by carefully noting the positions of the visible planets with respect to background stars. These planetary positions were accurate to 1 arcminute, making them the best dataset yet assembled.

In the seventeenth century, German mathematician Johannes Kepler analyzed Brahe's data and had an epiphany: the planets did orbit the Sun, but not in circles. They orbited in ellipses (which look like slightly squashed circles).

Kepler described the motions of the planets with three simple laws that bear his name:

1. The planets orbit the Sun in elliptical paths, with the Sun at one focus of the ellipse.
2. The planets sweep out equal areas of the ellipse in equal times, effectively meaning the planets move faster when they are closer to the Sun.
3. There is a relationship between the period of each planet's orbit in (P) and the planet's mean distance from the Sun in (A) : $P^2 = A^3$.

Kepler's laws are useful to describe any orbit driven by gravitational forces, from the solar system out to galaxy clusters.

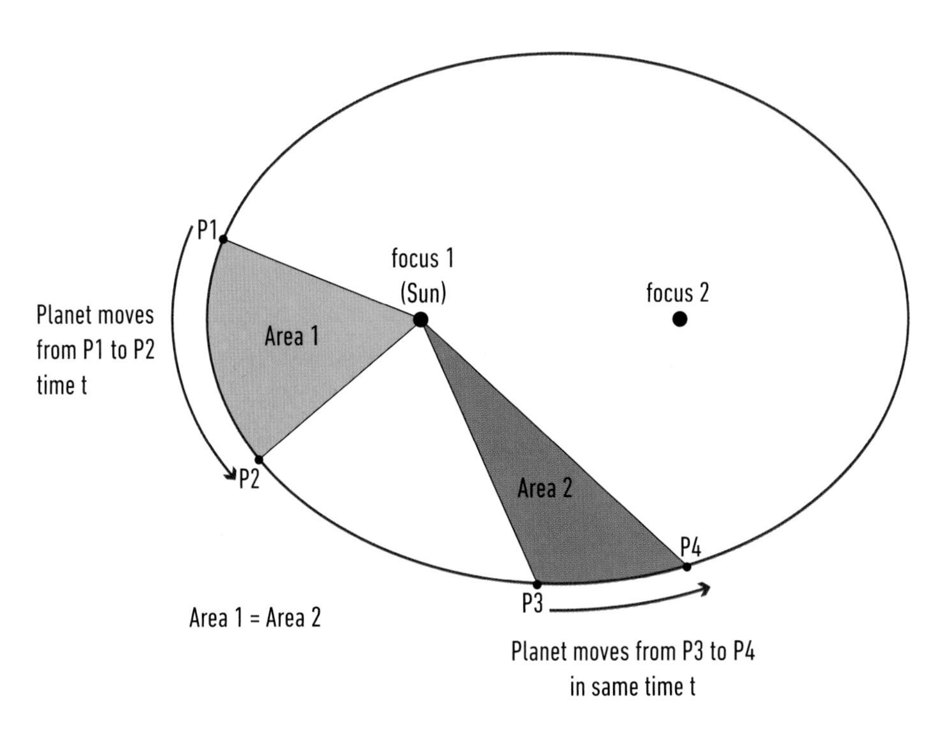

This image shows the calculations used in Kepler's Laws of Planetary Motion.

Mercury

Mercury is the smallest planet in the solar system and the closest planet to the Sun. It is one of four terrestrial (or Earthlike) planets, but unlike the other three terrestrial planets, Mercury has no appreciable atmosphere and a heavily cratered surface, much like that of the Moon. Mercury has not been geologically active for billions of years and has the largest day-night changes in surface temperature of the planets, varying between 100 K and 700 K.

Statistics

RADIUS
2,440 km (0.329 Earth radii)

SIDEREAL PERIOD
87.97 days

ROTATIONAL PERIOD
58.7 days

VOLUME
6.1×10^{10} km^3 (0.056 Earths)

SATELLITES
None

MEAN DISTANCE FROM THE SUN
0.387 AU (5.8^{10} m)

MASS
3.3×10^{23} kg (~0.055 Earth masses)

ATMOSPHERIC COMPOSITION (BY VOLUME)
Almost nonexistent, but consists of 42% molecular oxygen, 29% sodium, and 22% hydrogen

Formation of Mercury

One of the most unusual aspects of Mercury is the high level of metals found in its interior. That is, compared to its fellow terrestrial planets Venus, Earth, and Mars, Mercury has a much higher ratio of metal to silicates (rock). For many years, the leading explanation for the low ratio of silicates was that Mercury suffered an enormous impact early in its history that vaporized and removed most of the surface (silicate) materials, leaving behind a massive iron core. Other theories involved Mercury forming from different materials than other solar system objects.

These theories have recently been tested by the Gamma-Ray Spectrometer aboard the MESSENGER spacecraft currently orbiting Mercury. Early results indicate that the abundance of potassium, thorium, and uranium on the surface of Mercury are not consistent with a giant impact or with the early Sun vaporizing the rocky surface of the planet. Instead, it appears that Mercury formed from very primitive material in the solar system with a high metal abundance.

The presence of a magnetic field indicates that Mercury must still have a partially liquid iron core under a 600-km-thick rocky mantle.

Atmosphere and Magnetic Field

One might think, because it's so close to the Sun, Mercury would be the hottest planet; however, its lack of atmosphere means it has a hard time retaining heat. Because Mercury has a weak gravitational field, it can't hold on to its atmosphere for any significant time. The hydrogen and helium found in its atmosphere most likely come from the solar wind (see "The Sun's Upper Atmosphere and the Solar Wind"), and heavier elements come from the radioactive decay of elements in its interior.

Mercury does have a magnetic field, but it is very weak, only about 1 percent as strong as Earth's field. In order to have a magnetic field at all, Mercury must have moving conductive liquid material in its core. Mercury's weak magnetic field creates a magnetosphere that protects it from some of the solar wind.

Orbit and Rotation

Because of its proximity to the Sun (about 40 percent of the Earth-Sun distance), Mercury orbits the Sun faster than any other planet, completing each orbit in just 88 days. This closeness to the Sun also makes Mercury experience strange effects on its rotation and orbit. Mercury goes through only three rotations for every two orbits it makes around the Sun. By comparison, Earth goes through over 730 rotations every two orbits it makes around the Sun. This effect is the result of the Sun's strong gravitational pull.

Mercury also has the most elliptical orbit of any of the planets, its distance from the Sun varying between 0.31 and 0.47 AU.

Venus

Venus is the second planet from the Sun. It is a strange sister planet to Earth, with a mass and volume quite similar to it. Venus is located about 30 percent closer to the Sun than Earth and has a rocky surface that, until revealed in the 1990s with the radar images from the Magellan spacecraft, was literally shrouded from view by its atmosphere. These images revealed that the surface of Venus is relatively young due to being resurfaced by widespread volcanic activity.

Statistics

RADIUS
6,050 km (0.95 Earth radii)

SIDEREAL PERIOD
224.7 days

ROTATIONAL PERIOD
243 days (retrograde rotation)

VOLUME
9.3×10^{11} km^3 (0.866 Earths)

SATELLITES
None

MEAN DISTANCE FROM THE SUN
0.723 AU (1.08×10^8 km)

MASS
4.9×10^{24} kg (~0.82 Earth masses)

ATMOSPHERIC COMPOSITION (BY VOLUME)
96.5% carbon dioxide, 3.5% nitrogen, and small amounts of sulfur dioxide and other elements and molecules

The Phases of Venus

Long before the age of space exploration, in the seventeenth century, Italian physicist Galileo Galilei made some of the first detailed observations of Venus. One peculiar effect he noticed was that the size and shape of the planet changed with time. Venus appeared to be a full circle when it was smallest in diameter and in the same part of the sky as the Sun. However, as it got larger, the disk of Venus (now a sunset object) shrank until it was just a thin crescent in the same part of the sky as the Sun, illuminated from the right side. It then reappeared as a sunrise crescent object, with its disk illuminated on the left side.

In a geocentric view of the solar system, with Earth at the center and all the planets orbiting Earth, there would be no reason for Venus's size and phase to change; it should have appeared as a fully illuminated disk whenever it was observed. The changing phase and size of the planet was strong evidence supporting a heliocentric solar system.

In a heliocentric model of the solar system, these changes occur as Venus goes from the far side of the Sun (small and "full") to the near side (large and "crescent").

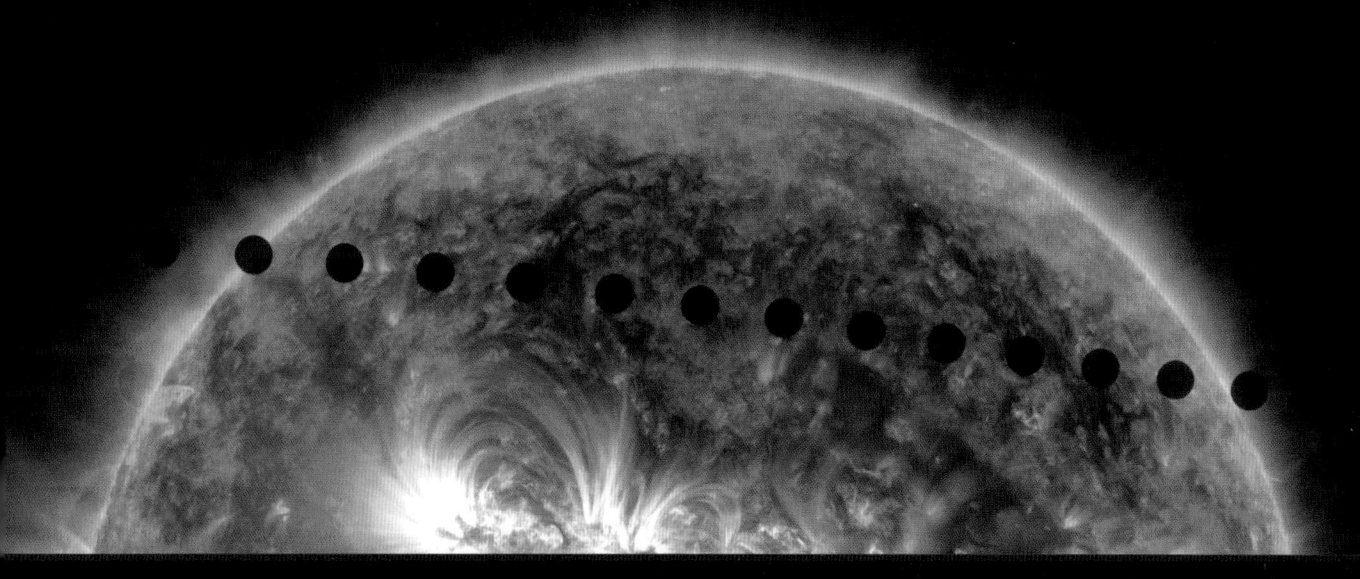

Because Venus is located between Earth and the Sun at a mean distance of 0.7 AU, it occasionally passes in front of the Sun's disk. The last transit of Venus happened in June 2012, as recorded here in a time lapse of several images made at 171 nanometers by the Solar Dynamic Observatory (SDO).

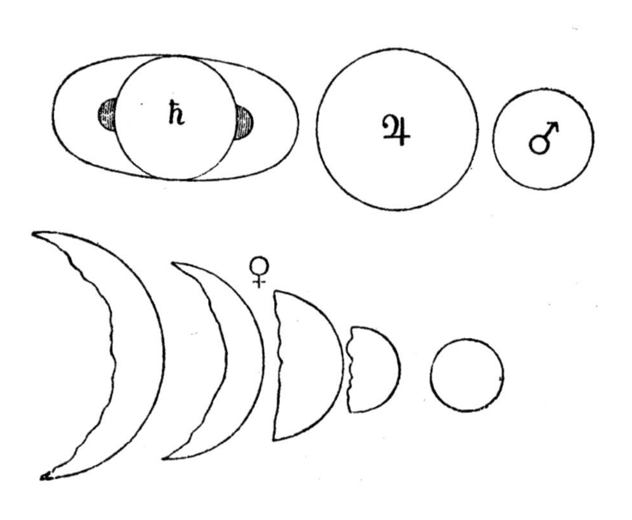

Clockwise from the top, Galileo's observations of Saturn, Jupiter, Mars, and Venus. Unlike the other planets, Venus goes through phases and changes significantly in size.

This is an ultraviolet, false-color image of Venus taken by the Hubble Space Telescope. Unlike the clouds on Earth, which consist of water vapor, the clouds of Venus consist of sulfuric acid. The bands of light and dark show how the atmosphere of Venus circulates. Here, Venus is in a crescent phase as it swings around the near side of the Sun.

Atmosphere

Early in the twentieth century, there was popular speculation that Venus, shrouded in clouds, might be a lush tropical paradise. But observations of Venus's atmosphere by the 1940s had made it clear the planet's surface temperatures were very high. By the 1960s, American and Soviet missions had established the atmosphere consisted mostly of carbon dioxide, with no rain or surface water whatsoever. Spectroscopy also established that the visible cloud layers consist mostly of sulfuric acid. Below the cloud deck, at an altitude of 50 km above the surface, is a layer of sulfuric acid haze.

The atmosphere of Venus at its surface is hot and very dense, with surface pressure that is approximately 100 times that of Earth. Under these conditions, normal materials do not fare well. As a result, surface exploration of Venus has lagged well behind the robotic exploration of the surface of Mars. However, a current mission of the European Space Agency, Venus Express, is exploring the dynamics and chemistry of Venus's strange atmosphere to determine whether there are cycles in the abundance of water, carbon dioxide, or sulfuric acid that would indicate ongoing volcanism on the planet.

Global volcanism is one of the main suspects in Venus's transformation from a planet very much like the Earth to a planet with a runaway greenhouse effect. Volcanism can release greenhouse gases like carbon dioxide into a planet's atmosphere, raising its average surface temperature above the boiling point of water.

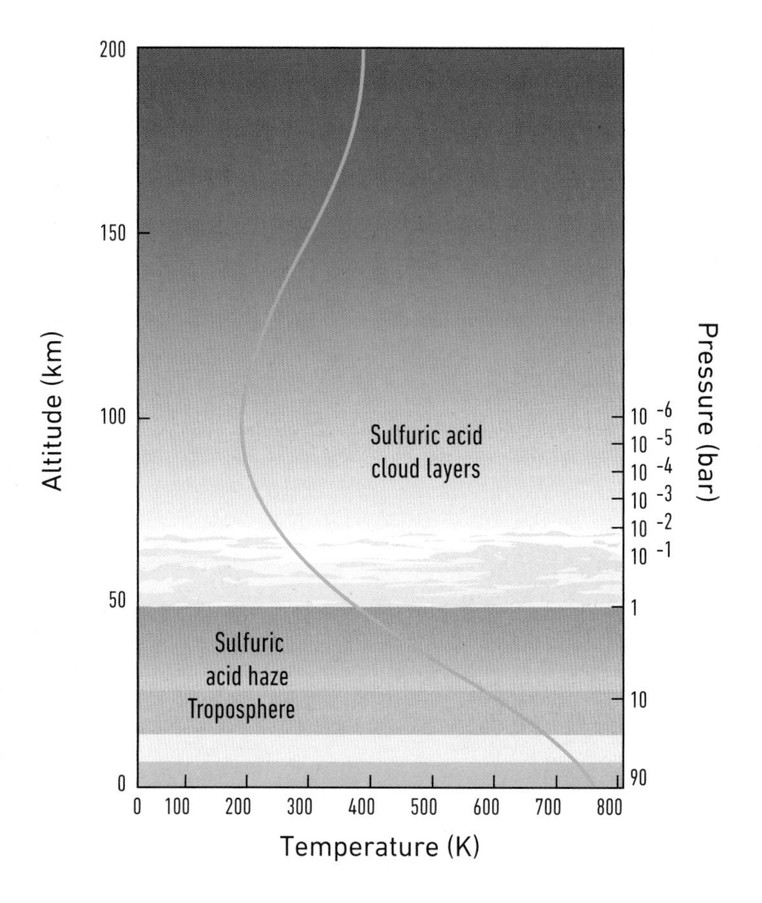

This is a graph of the temperatures of the surface and atmosphere of Venus. The temperature increases closer to the surface, rising from Earthlike temperatures in the upper cloud layers to an incredible 700 K at the surface.

This artist's impression, looking for all the world like Tolkien's Mount Doom, shows an active volcano on the surface of the solar system's hottest planet. The fluctuations in levels of sulfur dioxide (SO_2) detected by Venus Express could have a number of sources, one of which is volcanic activity.

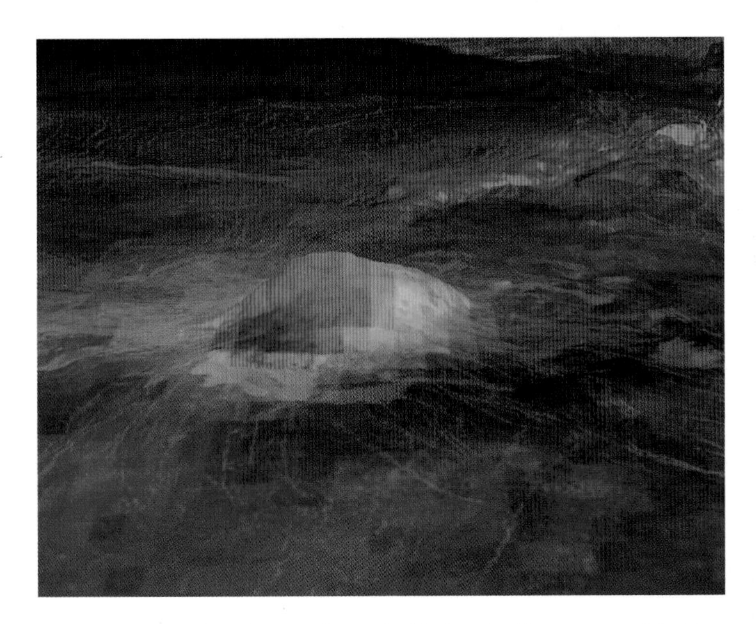

This shows the surface features imaged with the Magellan spacecraft in the late 1990s overlaid with the Infrared Thermal Imaging Spectrometer data from the Venus Express spacecraft, which gives an estimation of surface temperature. As you can see, the regions near the volcanic peak Idunn Mons are highest (red-orange) in temperature.

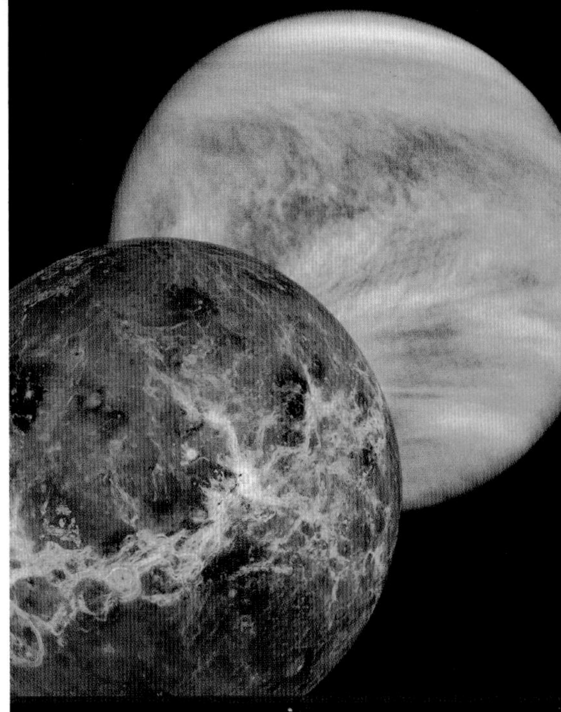

The surface of Venus as imaged in radar (left) by the Magellan spacecraft and the cloud layers of Venus as imaged in visible light (right). The image of Venus's clouds was taken by the Pioneer spacecraft that visited Venus in 1978. A thick cloud layer consisting mostly of sulfuric acid shields the surface from the view of optical telescopes.

A Foggy View

Astronomers often have to make observations at a different wavelength to see through intervening material. For example, if you have been outside on a foggy day, you know that you can't see very far because the water droplets in fog are opaque to visible light—that is also why you can't see the sky behind clouds. The atmosphere of Venus is opaque to visible light, so radio and infrared waves are used to observe its surface.

This color 170-degree panorama taken by Venera 13 shows the surface of Venus near the surface feature known as Phoebe Regio near Venus's equator.

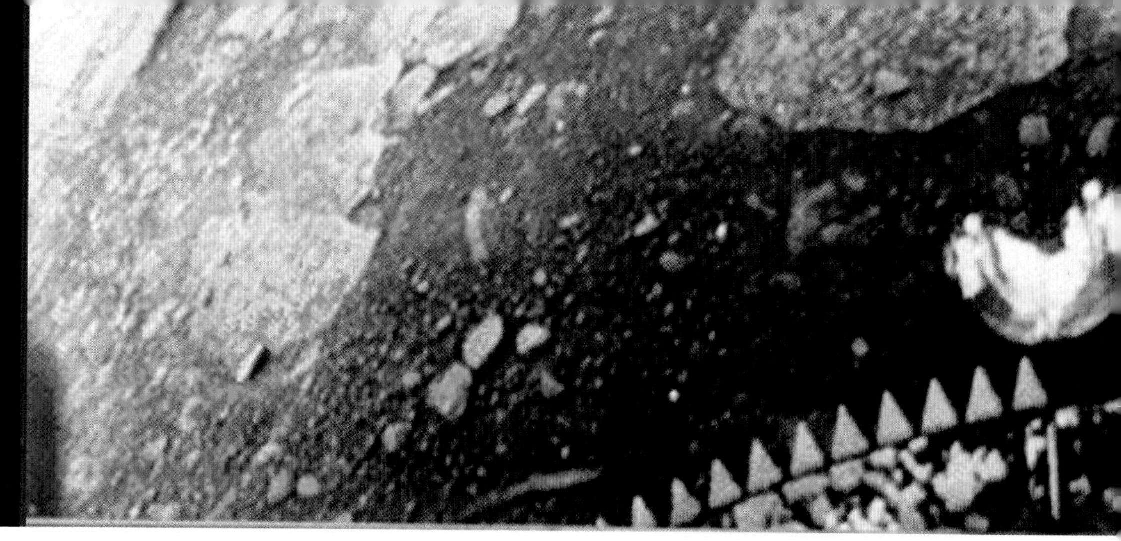

Surface

Early observations of Venus indicated the surface was extremely hot, with a mean surface temperature of 735 K and a surface pressure close to 100 times that on Earth at sea level.

In an early attempt to study the surface in detail, the Soviet Union's Venera 13 landed on the surface in 1982. It sent back a few grainy images of its surroundings before its metal structures succumbed to the temperature and pressure of the harsh environment.

The entire surface of Venus was first revealed when the Magellan spacecraft went into orbit around the planet in the 1990s and started beaming back scans of radar range finding. A global map of the surface of Venus based on the scans revealed that most of the planet is covered by flat volcanic plains, with a few areas significantly above the planet's average radius (known as highlands). While its carbon dioxide atmosphere makes it too hot now, the planet's distance from the Sun and location in the solar system make it likely that Venus had oceans in the past.

This false-color image of one hemisphere of the surface of Venus is the result of a decade of observations and analysis. The Magellan spacecraft determined the elevation and reflectivity of the entire Venusian surface using microwave radar between 1990 and 1994. Low elevations on the surface are blue, and higher elevations are reddish-brown. Volcanoes and volcanic flows are widespread.

A region of Venus's surface dubbed "The Crater Farm" shows three impact craters in the Lavinia Planitia. Venus has very few craters, due in part to its thick atmosphere burning up small meteorites before they hit the surface.

The small number of craters per unit area on the surface of Venus indicates the surface is only about 500 million years old and that the entire planet was resurfaced at the same time. The Magellan images made it clear that Venus does not have plate tectonics like Earth; rather, its entire surface is a single plate. This can explain how the planet was resurfaced. Because it has no tectonic plates, it's thought the interior of the mantle heated up until material there was hot enough to force its way to the surface and cover the planet.

There is widespread evidence that Venus has experienced surface volcanism in the past few million years, and bright areas in the Magellan images of the surface indicate regions of lava flows. Between 1,000 and 2,000 volcanoes and volcanic features have been identified on the planet's surface, many of which are shield volcanoes with gently sloping sides. Without tectonic motion, the volcanic material piles up in a single location on the surface. Volcanic activity is also thought to replenish the highly reactive sulfur compounds in Venus's atmosphere.

The 5-km-high shield volcano Maat Mons, the highest on the planet's surface, towers above the surrounding plane, covered in lava flows. The vertical scale in this 3D Magellan image is amplified by a factor of 10.

Mars

Mars is the fourth planet from the Sun. It has a surface that is visible even through a modest optical telescope from the surface of Earth. Perhaps as a result, Mars has served as a canvas onto which humans have projected many ideas about life in the solar system.

Statistics

RADIUS
3,386 km (0.53 Earth radii)

SIDEREAL PERIOD
686.97 days

ROTATIONAL PERIOD
1.03 days

VOLUME
1.63×10^{11} km^3 (0.151 Earths)

SATELLITES
Two small moons, Phobos and Deimos (see "Other Moons in the Solar System")

MEAN DISTANCE FROM THE SUN
1.52 AU (2.27×10^8 km)

MASS
6.4×10^{23} kg (~0.11 Earth masses)

ATMOSPHERIC COMPOSITION (BY VOLUME)
93.5% carbon dioxide; 2.7% nitrogen; 1.6% argon; and smaller amounts of other molecules and atoms, including oxygen and water

Water on Mars?

At the beginning of the twentieth century, American astronomer Percival Lowell described a complex pattern of dark lanes on the surface of Mars as *canals,* or waterways. He proposed a Martian civilization that was in its death throes, desperately trying to bring water from its visible white polar regions to the dry regions at the equator.

However, better observations made it clear that the surface of Mars held neither life nor water, and that the canals that had been proposed were merely an optical illusion caused by the tendency of the human brain to connect dark areas into patterns.

Early images from the Mariner and Viking missions seemed to indicate the presence of water in Mars' past, but its current conditions—cold with a very thin atmosphere—would not allow for any water at its surface (though that doesn't stop science fiction authors from writing about an inhabited Mars).

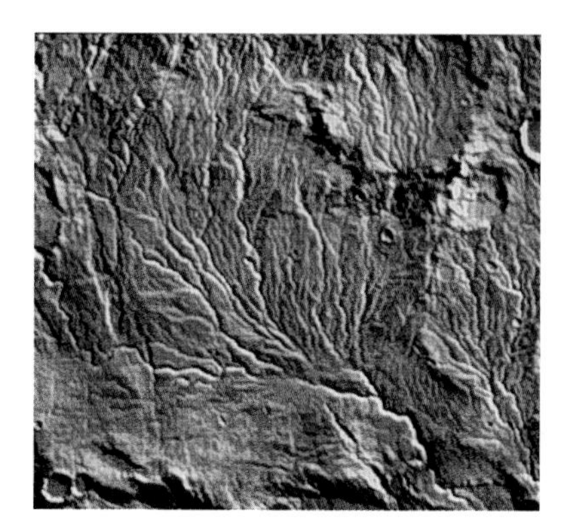

One long-standing question has been whether Mars ever had surface water. Images like this from the Viking orbiter in the 1970s revealed the presence of what appeared to be water-carved channels.

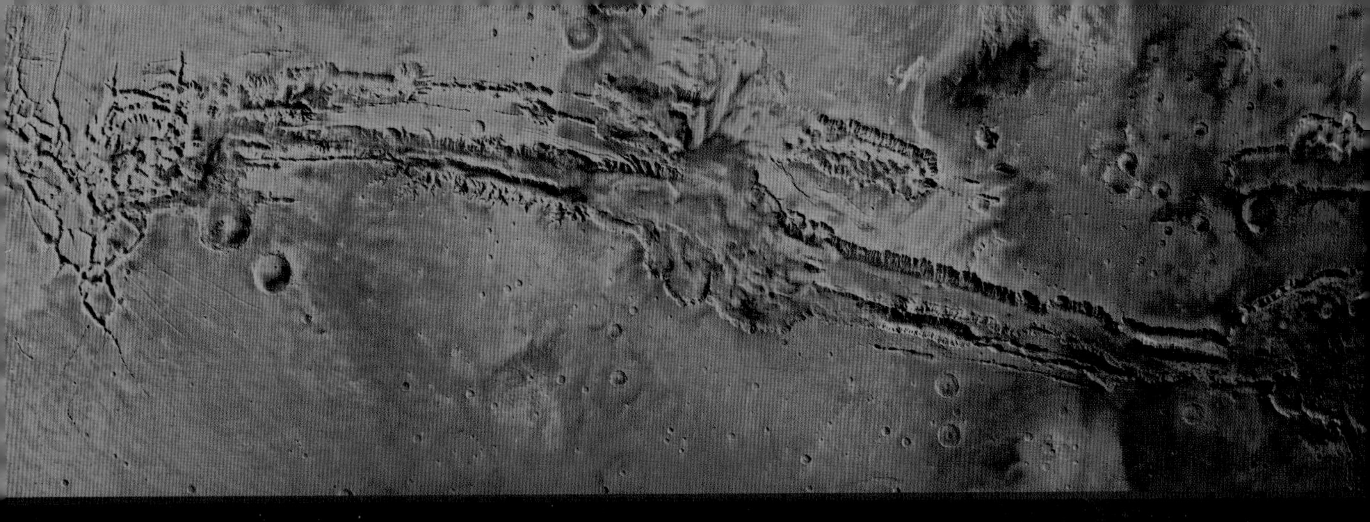

This color Viking orbiter image of the Valles Marineris shows the entire canyon system, which stretches for over 3,000 km across the surface of Mars. Unlike Earth's Grand Canyon, the Valles Marineris was not formed primarily by water erosion, but there are channels from the canyon into nearby lower regions.

General Observations

In naked-eye observations, Mars has a reddish appearance, which is due to the iron oxide common on the surface of Mars. Observations of Mars through a small telescope when it is closest to Earth reveal its white polar caps, made of a combination of water and carbon dioxide ice. Like the ice caps on Earth, Mars' polar ice caps are observed to expand and recede with the seasons.

The planet is also shown to be tilted on its axis by about 25 degrees, close to Earth's 22.5 degrees. As a result, Mars experiences seasons like Earth; however, its 687-day trip around the Sun means the seasons last almost twice as long.

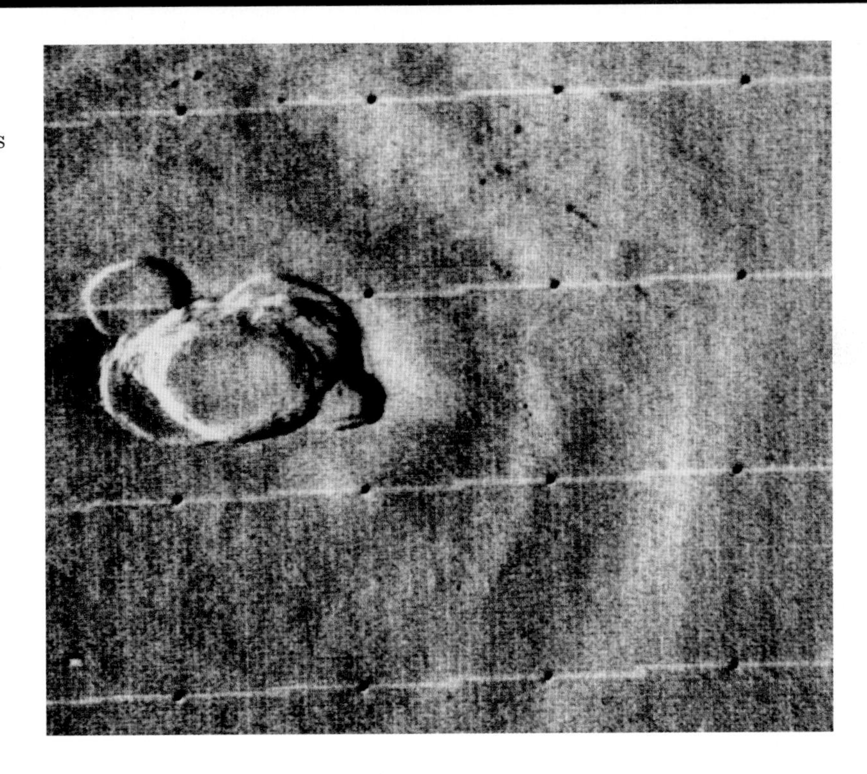

Understanding of Mars has been greatly enhanced by orbiters and landers. The Mariner 9 spacecraft was the first mission to orbit another planet, and its orbit allowed it to send back detailed images showing that Mars had surface volcanoes (such as Olympus Mons, shown here), a vast canyon system now called Valles Marineris, and ancient riverbeds.

Wispy clouds are often observed in the Tharsis region of Mars. The water vapor in the thin Martian air condenses into ice crystals as warm currents of air move up the slopes of the volcanoes and cool.

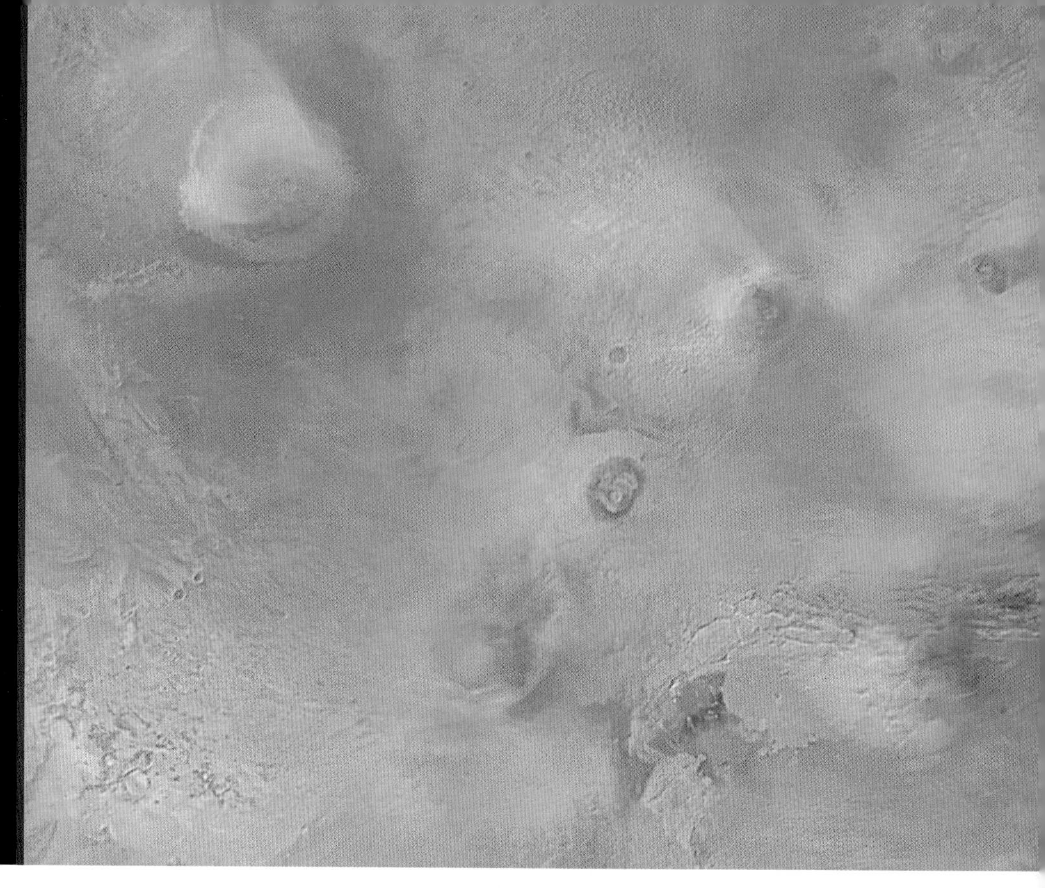

Atmosphere

Mars has a thin atmosphere that consists of mostly carbon dioxide, nitrogen, and argon, with tiny amounts of other components (including oxygen, water vapor, and methane). The surface pressure on the planet is very low, less than 1 percent of Earth.

The atmosphere of Mars has two layers of clouds—a lower layer of water ice clouds and a higher bank of carbon dioxide ice clouds—that form within 50 km of the surface. The carbon dioxide that makes up the latter type of cloud is frozen at the poles in the winter and released into the atmosphere during the warmer temperatures of summer.

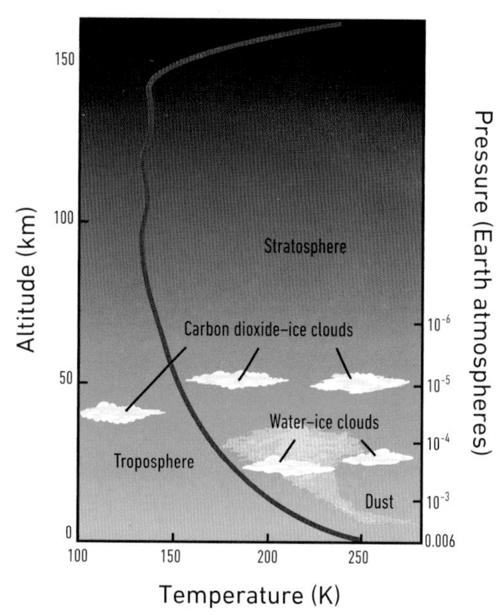

Mars' temperature (indicated by the red line in the diagram) reaches its maximum at the planet's surface at about 250 K.

Dust Storms

While thin, the atmosphere has enough density to kick up dust storms that sometimes blanket the entire planet.

Dust storms are a common occurrence on the surface of Mars, and several Mars rovers have detected small dust storms ("dust devils") moving near them across the cold Martian desert. Larger events—like the global dust storm that engulfed the planet in 2001—are rarer and can warm the upper layers of Mars' atmosphere while cooling the surface.

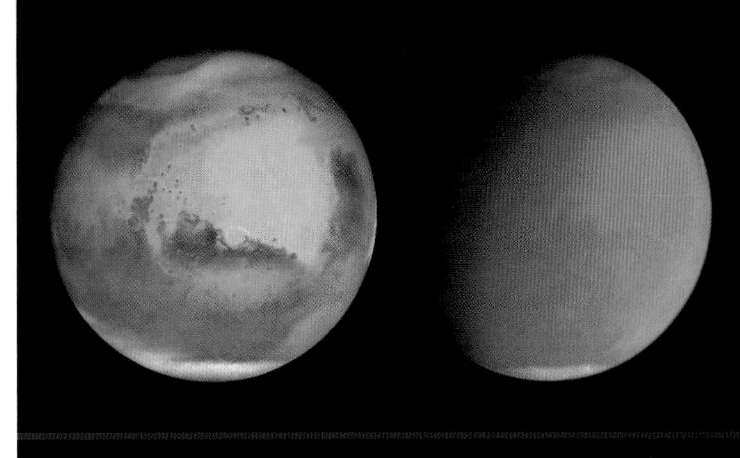

Dust storms are a well-known feature of Martian weather. In this pair of images from the Hubble Space Telescope, Mars is seen on June 26 and September 4 of 2001. In the September image (right), most of the surface of the planet is shrouded in a global dust storm.

Sunset on Mars as seen by the Mars Pathfinder Lander in this true-color image. This panorama view shows about 60 degrees across the horizon.

A Changing Atmosphere

Like Earth, Mars has an atmosphere that has evolved with time. Several lines of evidence point to a Mars in the distant past that had oceans and a more substantial atmosphere. Recent missions to Mars, including the Curiosity rover, have measured isotope ratios for evidence that Mars lost a substantial fraction of its original atmosphere. The enrichment of heavier isotopes of carbon and oxygen in Martian carbon dioxide indicates that lighter isotopes present in the upper atmosphere were preferentially lost.

Mars' small size has worked against the longevity of its atmosphere in several ways. Its relatively weak gravitational field made it harder for the planet to hold on to atoms and molecules. Also, the small volume of the planet cooled more quickly than Earth, resulting in the loss of the magnetic field that protected it from the solar wind.

Surface

The surface of Mars has been mapped in exquisite detail by a number of orbiting missions, including the Mars Global Surveyor and the Mars Reconnaissance Orbiter. These maps of the planet show the northern hemisphere of Mars is fairly smooth and low, while regions south of the equator are higher, more heavily cratered, and older.

The main large features of the Martian surface are the cratered highlands south of the equator, the Tharsis region containing three shield volcanoes, Olympus Mons to the west of Tharsis, and the Valles Marineris to the east of Tharsis.

The crust of the planet Mars is very different in the two hemispheres, with a potential thickness of about 80 km in the heavily cratered southern hemisphere and only half that in the smooth northern hemisphere. The cause of the difference is still debated by planetary geologists—some possibilities are an ancient northern ocean or a large meteorite impact relatively late in the life of the planet.

The Mars Global Surveyor mission detected magnetization in the Martian crust, which reveals that Mars had a global magnetic field like Earth. However, its interior has cooled enough that its core can't support a magnetic field. As you learned in the previous section, a magnetic field protects a planet from solar radiation, so the weakening of its magnetic field may have been related to the loss of surface water and atmosphere on Mars.

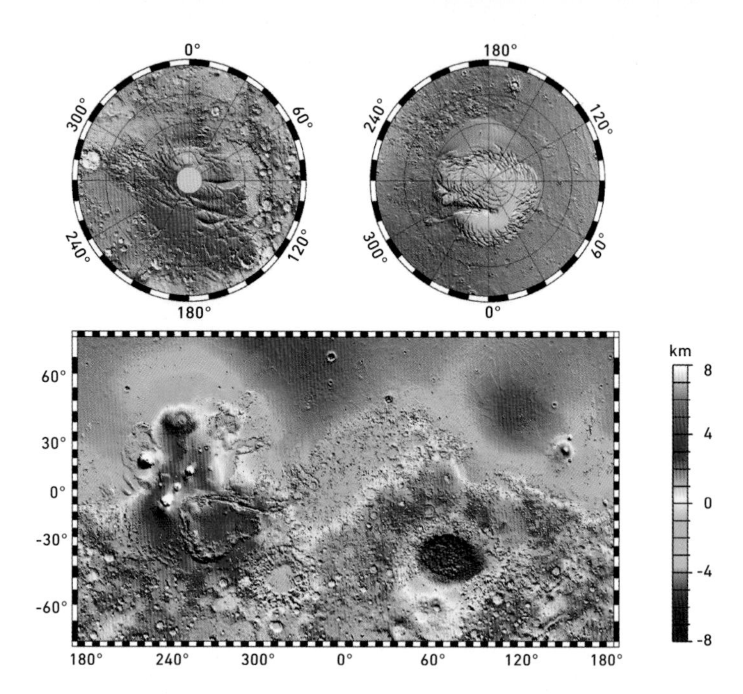

These images show the topography of the entire planet. The lower image shows the planet to 70 degrees N and S latitude, with the poles of the planet shown separately to avoid distortion. The scale on the right shows the range of elevations, from the depths of the Hellas basin (45 degrees S, 70 degrees E) to the peaks of Olympus Mons (18 degrees N, 225 degrees E).

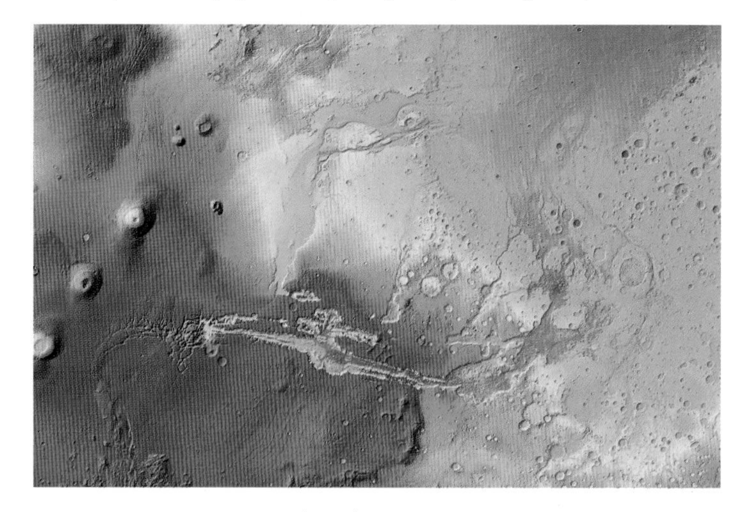

Mars Orbiting Laser Altimeter (MOLA) data showing a topographic map of the Valles Marineris region of Mars, with blue representing low elevations and white representing high elevations. Outflow channels run from the canyon toward the northern plains. Notice the lack of heavy cratering near the Tharsis Ridge volcanoes to the left and the more heavily cratered regions on the lower right.

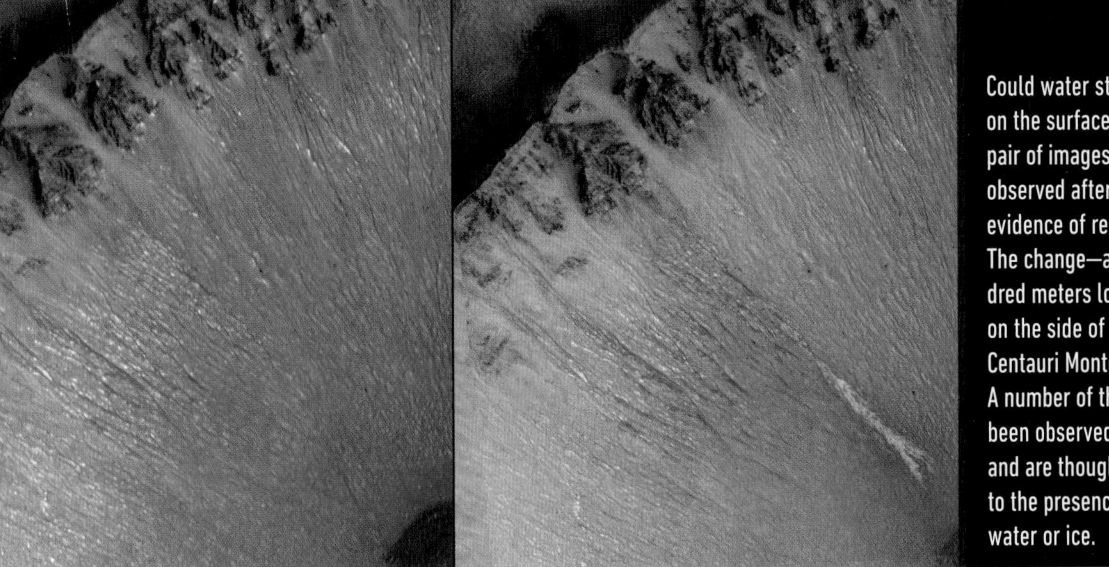

Could water still flow briefly on the surface of Mars? In this pair of images, the same region, observed after six years, shows evidence of recent water flow. The change—a gully a few hundred meters long—was formed on the side of a crater in the Centauri Montes region of Mars. A number of these gulleys have been observed on crater walls and are thought to be related to the presence of subsurface water or ice.

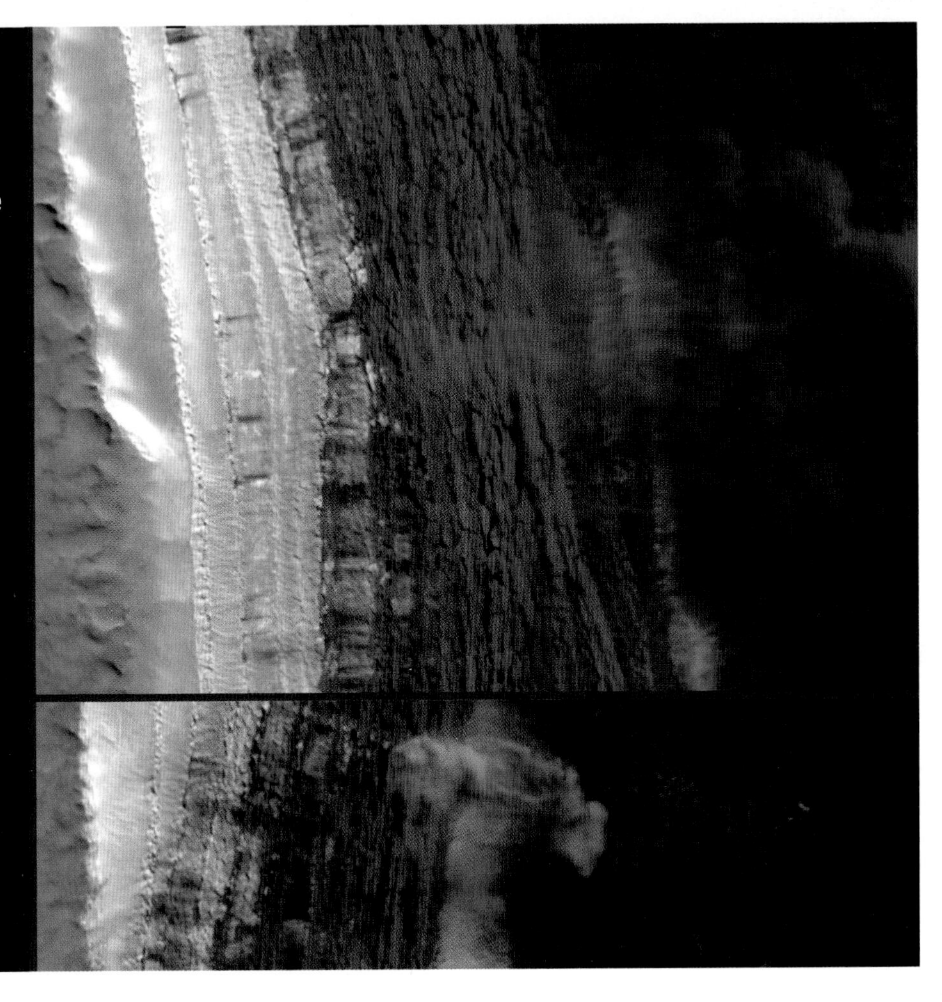

The presence of imaging satellites orbiting Mars allow for the detection of changes going on currently on the planet's surface. These images, taken with the High Resolution Imaging Science Experiment (HiRISE) camera on the Mars Reconnaissance Orbiter, show several avalanches as they occur, complete with clouds of billowing dust.

Exploration

The exploration of Mars has been done in four ways: remote observations (from Earth), orbiting satellites, stationary landers, and roving landers. Each type of exploration has revealed more detail about the planet's surface and history.

Remote Observations

Much of what's known about the global characteristics of Mars and its atmosphere were observed remotely from the surface of Earth. Spectroscopy revealed the composition of Mars's atmosphere and its low temperature and density. The fact that Mars has ice at its poles and global dust storms can also be observed from Earth. But high-resolution mapping of its surface and the discoveries that resulted had to wait for orbiting satellites.

Flyby Missions and Orbiting Satellites

Orbiters are useful for surface imaging, atmospheric studies, and as communications relays. The first successful flyby mission to Mars was the Mariner 4 mission in 1964. This flyby sent back some of the first detailed surface images, which showed the widespread presence of craters. Mariner 9, the first spacecraft to orbit another planet, made the first global maps of Mars' surface in 1971 and discovered many familiar surface features, such as the Valles Marineris. Orbiting satellites since Mariner 9 have mapped the surface of Mars in detail but have also been able to observe changes on the surface over time.

There are currently three active spacecraft orbiting Mars: Mars Odyssey (NASA), Mars Express (ESA), and Mars Reconnaissance Orbiter (NASA).

Stationary Landers

The most famous landers on the surface of Mars were the Viking 1 and Viking 2 Landers, which arrived in 1976. The Viking missions were two-part spacecraft with an orbiter and a lander. The Viking landers sent back the first images from the surface of Mars and were able to take samples from the surface and perform simple tests on them. The Viking chemical and biological tests showed the surface of Mars was sterile.

This image of water ice on the Martian landscape at Utopia Planitia looking south around the Viking 2 lander was taken in the summer of 1979. The frost is thought to form when carbon dioxide ice causes dust particles to precipitate out of the atmosphere.

Rovers

There was a long dry spell in the exploration of Mars between the Viking missions and the Mars Pathfinder missions in the late 1990s. In 1997, NASA landed *Pathfinder* on the surface, which had on board the first rover, *Sojourner*. This simple rover had cameras and an X-ray spectrometer on board to analyze the composition of surface materials. In 2004, NASA landed the more-sophisticated *Spirit* and *Opportunity* rovers. In addition to the spectrometers found on *Sojourner*, these rovers had arms that allowed them to pick up magnetic material and expose fresh surfaces on nearby rocks. The discovery of hematite spheres made with the rovers was strong evidence for water on the surface of Mars in the past.

As of 2014, there are two rovers active on the surface of the planet: the 10-year-old *Opportunity* rover and the Mars Science Laboratory rover *Curiosity*, which landed in 2012. *Curiosity* is exploring Gale Crater to determine whether the region could have ever supported microbial life.

The Mars *Curiosity* rover, shown here in a self-portrait in Gale Crater. *Curiosity* is the largest rover to make its way to the surface, and its landing required a complex ballet of parachutes, rockets, and a sky crane that lowered it to the surface in final moments. *Curiosity* has on board a variety of powerful geological and meteorological probes.

Jupiter

Statistics

MEAN RADIUS
69,911 km (10.9 Earth radii)

SIDEREAL PERIOD
11.86 years

ROTATIONAL PERIOD
9.93 hours

VOLUME
1.4×10^{15} km^3 (1,321 Earths)

SATELLITES
Four large moons and 67 known satellites

MEAN DISTANCE FROM THE SUN
5.2 AU (7.79×10^8 km)

MASS
1.9×10^{27} kg (~318 Earth masses)

ATMOSPHERIC COMPOSITION (BY VOLUME)
89.8% hydrogen and 10.2% helium, with trace amounts of methane, ammonia, and other molecules

Jupiter is the fifth planet from the Sun and the largest planet in the solar system, dwarfing Earth; in fact, it would take over 1,000 Earths to fill its volume. Jupiter's size places it squarely between the Sun and Earth, with a diameter about 10 times larger than Earth and 10 times smaller than the Sun. Along with the other gas giant planets (Saturn, Uranus, and Neptune), Jupiter is thought to have a solid core similar in size to Earth. Jupiter also has four large moons—Io, Europa, Ganymede, and Callisto—that are worlds in themselves.

Jupiter's mass had a large impact on the formation of the other planets, sweeping up much of the material in the protoplanetary disk. But despite its dominance among the planets, it has a mass that is less that 0.01 percent that of the Sun. Like the Sun, Jupiter consists of mostly hydrogen and helium.

General Observations

Jupiter's visible upper layers are dominated by rotating bands of light and dark (known as *zones* and *belts*) that exhibit complex interactions with one another. Time-lapse images also show that Jupiter's atmosphere is highly dynamic, with the only permanent feature being the Great Red Spot. Careful observations of the motion of small features in the atmosphere allowed astronomers to measure the rotation rate of the planet, which is just under once every 10 hours at the equator. Jupiter's polar regions rotate a little more slowly than at the equator, an effect known as *differential rotation.*

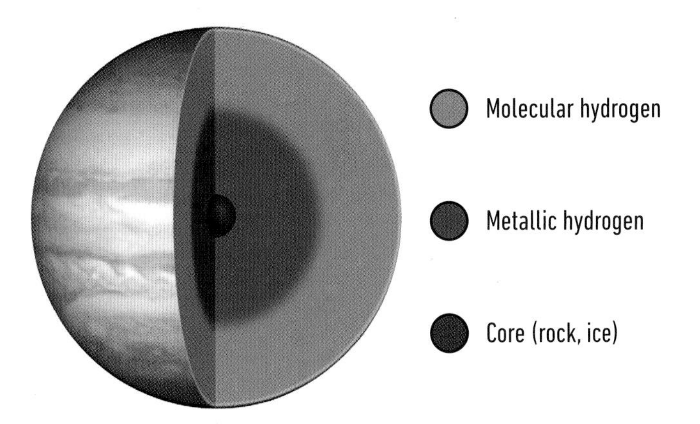

Cutaway views of the interior of Jupiter show an Earth-sized core of rock and ice, surrounded by a shell of metallic hydrogen and molecular hydrogen, and a thin upper atmosphere. The temperatures at the core of Jupiter are thought to exceed 20,000 K.

This unusual perspective, made by combining a near-infrared image and a blue image in order to simulate a true-color image, shows Jupiter's turbulent south pole as imaged by the Cassini spacecraft. You can clearly see the complex, dynamic patterns in Jupiter's alternating white and reddish bands and the Great Red Spot.

This sensitive image of the Great Red Spot was taken by the Galileo spacecraft orbiting Jupiter in 1996, almost 400 years after its discovery in 1664. The spot has been present at least since the invention of the telescope.

The Great Red Spot

This long-observed feature of Jupiter is a storm system that rotates counterclockwise with a period of about six days. Its dimensions have varied over time, but the largest dimensions have been an oval that was 40,000×14,000 km. Three planet Earths could easily fit side by side in the size of this enormous storm. The storm draws its energy from lower, warmer layers in the atmosphere, and the lack of any solid surface on the planet keep the storm from breaking up.

Atmosphere

Jupiter is mostly composed of hydrogen and helium, but its visible atmosphere is dominated by the ammonia and ammonium hydrosulfide clouds that form its upper layers. Infrared imaging of Jupiter's surface show that the light-colored zones are relatively cool (comprised of high-altitude clouds), while the darker belts are relatively warm since they are views into lower, warmer areas of the atmosphere.

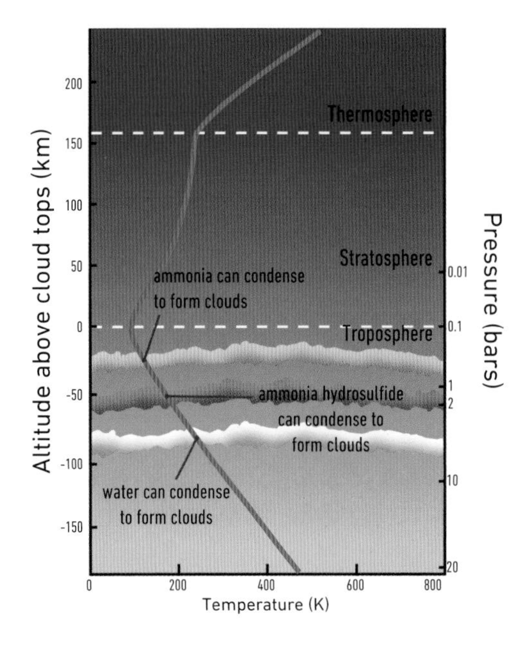

Layers in Jupiter's atmosphere. The temperature in the atmosphere increases with depth, with temperatures of about 273 K (freezing point of water) occurring at 100 km below the clouds tops.

A probe from the Galileo mission that plunged into the planet's atmosphere confirmed the layering of an upper bank of ammonia clouds above a layer of ammonia-sulfur clouds and the presence of high-velocity winds (650 km/h). The probe did not detect the lowest layer of water clouds (identified via spectroscopy), which might have been because the probe fell into a break in the water clouds.

It's believed that the fragments of comet Shoemaker-Levy 9, which collided with the planet in 1994, resulted in the high temperatures that produced hydrogen cyanide in the planet's atmosphere and exposed lower layers in Jupiter's atmosphere.

In terms of its magnetic field, Jupiter's is very powerful and is thought to be generated by the rotation of liquid metallic hydrogen in its interior. Its magnetic field lines guide material into collisions with its upper atmosphere, creating aurora. Similar to the Northern and Southern Lights (Aurora Borealis and Aurora Australis) on Earth, it leads to the formation of rings at the poles on Jupiter.

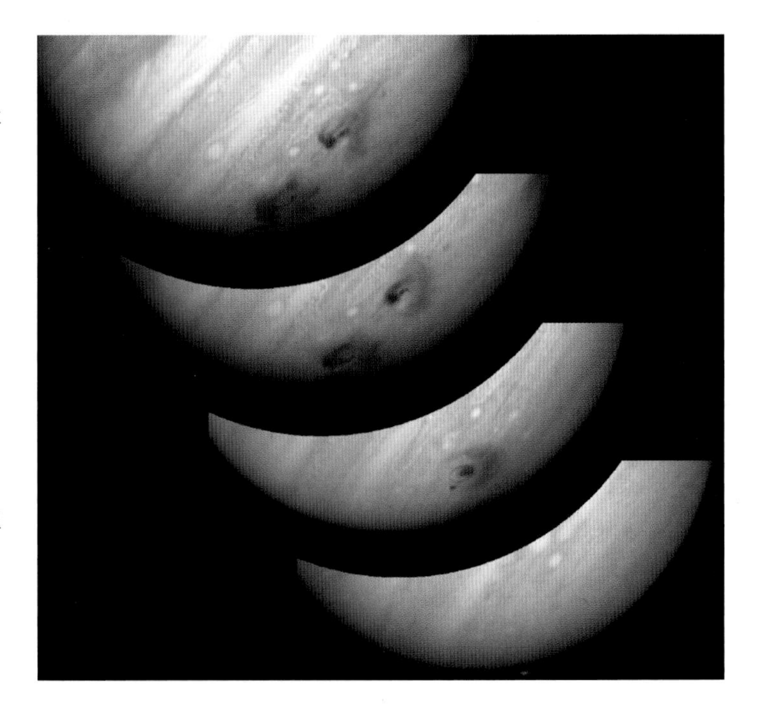

This series of images are stacked chronologically, with the lowest image being the first, showing the moment of impact of comet Shoemaker-Levy 9 with material being lifted above the edge of the planet's limb. Impacts like this are thought to have shaped Jupiter's atmosphere.

This false-color image taken by the near-infrared camera aboard the Galileo mission that orbited the planet in the 1990s shows Jupiter's Great Red Spot. Three different infrared images are overlaid in a way that highlights the varying depths of Jupiter's atmosphere. White and red areas are the highest, and blue and black areas are the lowest. The Great Red Spot appears pink, showing it is relatively high. The surrounding blue region is about 30 km lower. Smaller storms can be seen in the periphery of this image.

False-Color Images

Astronomers often image objects at wavelengths other than visible. For example, infrared observations allow astronomers to see deeper into Jupiter's atmosphere, and radio frequencies can be used to image Jupiter's powerful magnetic field. When astronomers show images of these nonvisible wavelengths, they choose a color table that makes the image easiest to interpret. Because light outside the visible portion of the spectrum does not have color, these are typically called *false-color images.*

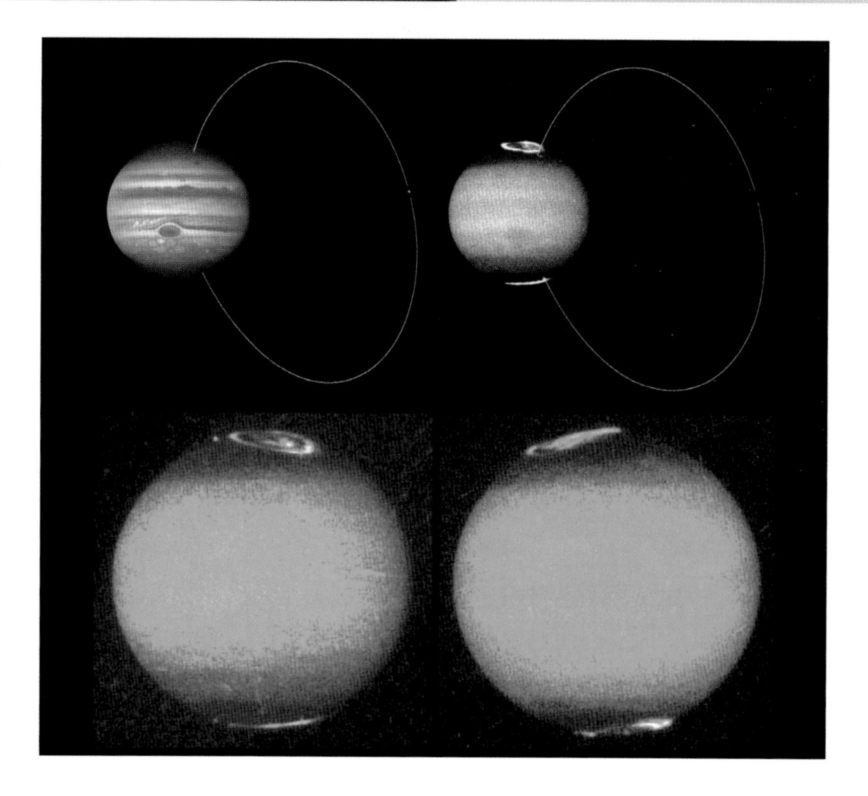

These images show Jupiter's aurorae and the endpoint of a "flux tube" that electrically connects the surface of Jupiter to its nearest moon, Io (representated by the lines shown in blue). Charged particles ejected volcanically by the moon are swept up by Jupiter's magnetic field lines and create the spots just outside the normal auroral rings. The lower images show Jupiter's aurorae in ultraviolet light as the planet rotates.

The Galilean Moons

Jupiter has 67 known natural satellites. The four largest ones were discovered in 1610, when Galileo first observed the planet through a telescope—Io, Europa, Ganymede, and Callisto, known collectively as the *Galilean moons*. Each of the Galilean moons has distinctive characteristics.

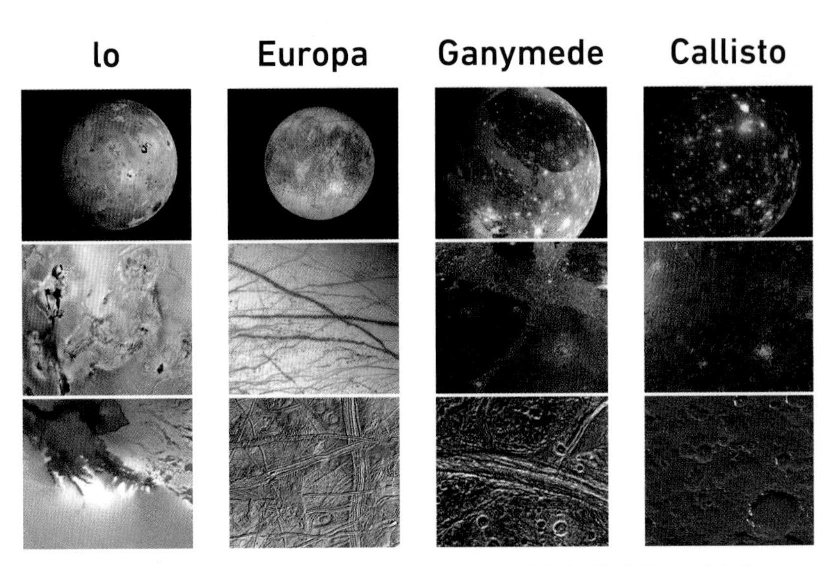

The Galilean moons of Jupiter shown both to scale and with detailed views of their surfaces. The resolution of the surface image increases from top to bottom.

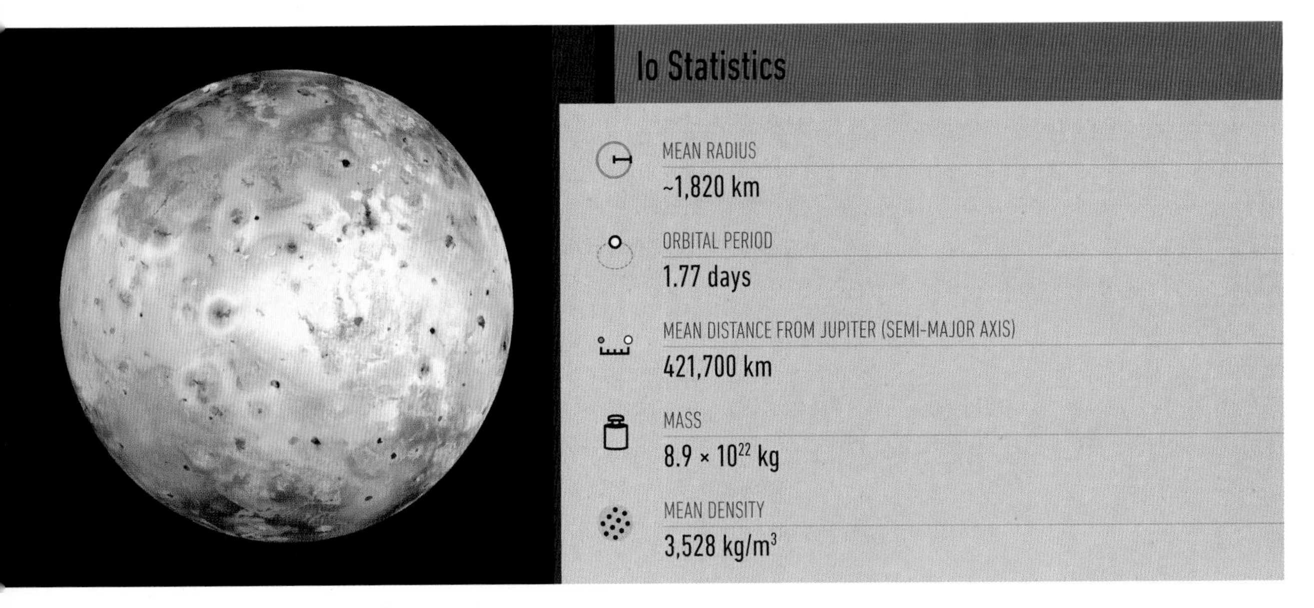

Io Statistics

MEAN RADIUS	~1,820 km
ORBITAL PERIOD	1.77 days
MEAN DISTANCE FROM JUPITER (SEMI-MAJOR AXIS)	421,700 km
MASS	8.9×10^{22} kg
MEAN DENSITY	3,528 kg/m^3

Io

Io is the most volcanically active surface in the solar system. Its geological activity is maintained by the enormous gravitational pull of Jupiter, which flexes the moon. Io's volcanoes have also been observed erupting, sending material 500 km above the surface. The variable and brightly colored surface is the result of deposits of sulfur dioxide (white areas) and sulfur (yellow and red-brown areas).

Io has a very thin sulfur dioxide atmosphere, with trace amounts of sulfur monoxide, sodium chloride, and other molecules. And as the nearest moon to Jupiter, Io orbits it the fastest, only taking 1.77 days.

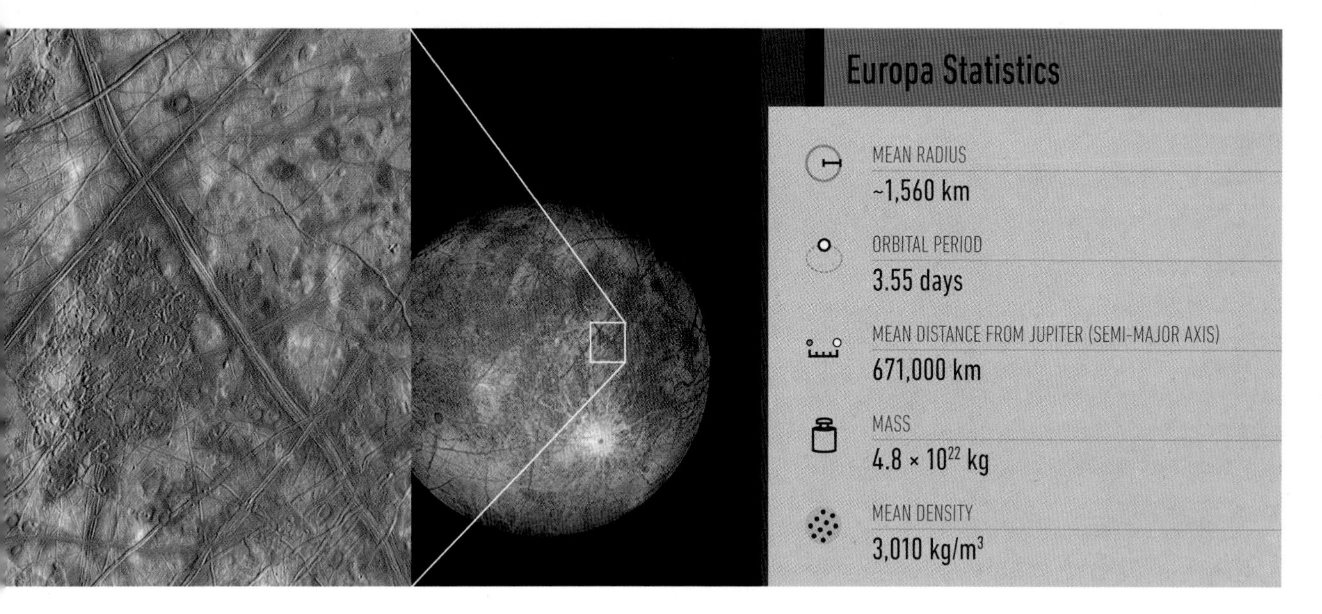

Europa Statistics
MEAN RADIUS
~1,560 km
ORBITAL PERIOD
3.55 days
MEAN DISTANCE FROM JUPITER (SEMI-MAJOR AXIS)
671,000 km
MASS
4.8 × 10²² kg
MEAN DENSITY
3,010 kg/m³

Europa

Like many other solid bodies in the solar system, Europa is composed of silicate rock and most likely has an iron core. The surface is covered in a distinctive set of cracks, known as *lineae*. These are thought to be the result of warmer material welling to the surface of Europa from deeper within, not unlike the tectonic ridges on Earth. However, craters are rare on its surface, indicating the surface is relatively young.

The minimal variation in the elevation of the surface of Europa supports the hypothesis there is a subsurface ocean. The Galileo mission confirmed the surface of Europa is covered in ice; the ocean is beneath the layer of ice, where it may be warmed by volcanic vents in the moon's solid surface. Europa's surface temperature is only about 110 K, so the water is thought to be kept in a liquid state by the tidal flexing caused by Jupiter.

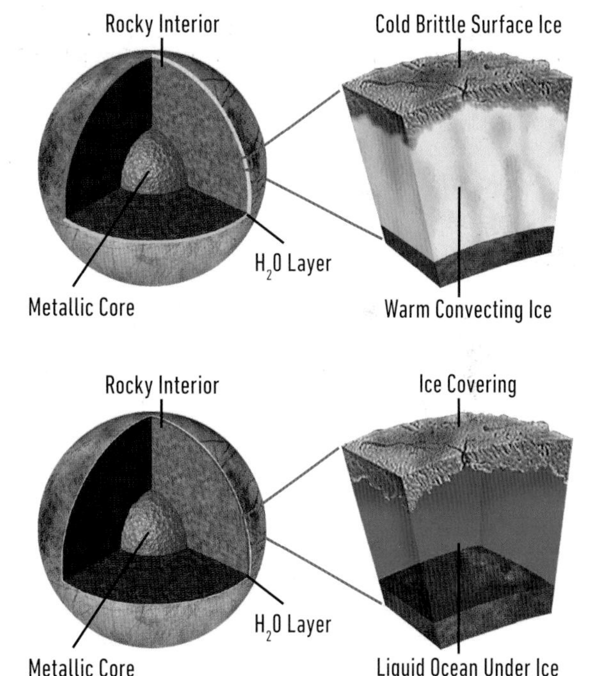

There are two models that explain the observed characteristics of Europa's surface. In the top model, the surface covers a region of warm, convective ice. In the lower model, Europa's frozen surface covers a subsurface ocean as thick as 100 km.

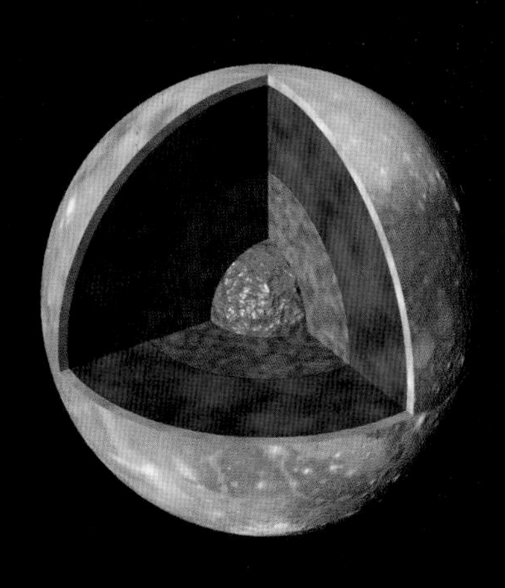

The interior of Ganymede is superimposed on Voyager images of the moon's surface. Based on surface observations alone, Ganymede's interior is consistent with either an interior mixture of rock and ice or a core of iron and rock overlaid with a layer of warm, convective ice and a hard ice crust. Galileo's measurements of Ganymede's gravitational field support the second model and the presence of a dense, metallic core.

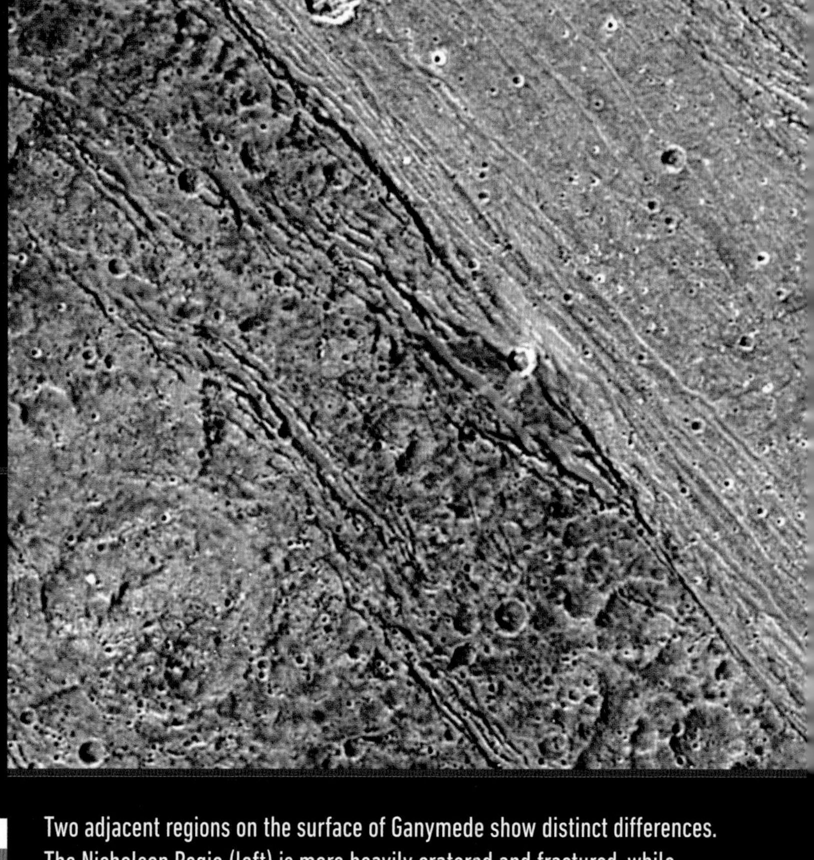

Two adjacent regions on the surface of Ganymede show distinct differences. The Nicholson Regio (left) is more heavily cratered and fractured, while the Harpagia Sulcus (right) shows fewer craters and is smoother. Tectonic processes are thought to have smoothed the region on the right. Ganymede's rigid ice surface may move about on a layer of convective ice.

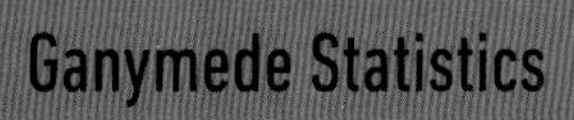

Ganymede Statistics

MEAN RADIUS
2,630 km

ORBITAL PERIOD
7.15 days

MEAN DISTANCE FROM JUPITER (SEMI-MAJOR AXIS)
1.1×10^6 km

MASS
14.8×10^{22} kg

MEAN DENSITY
1,936 kg/m^3

Ganymede

The surface of Ganymede, the solar system's largest moon, shows evidence of tectonic motion. Water ice is abundant on the surface, and the moon seems to be differentiated, with an interior of iron and rock, a silicate mantle, and a surface of water ice. The Galileo mission detected a weak magnetic field, which might be the result of motions in the liquid iron at its core. Radiation breaks apart the water (H_2O) at the moon's surface, giving it a tenuous oxygen atmosphere.

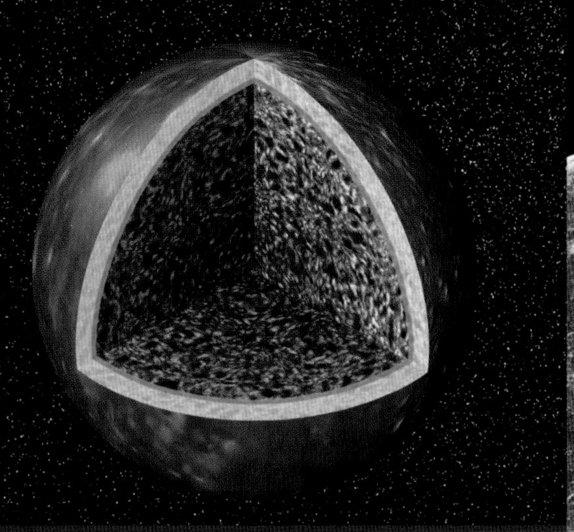

The interior of Callisto is a mixture of rock and ice. The presence of a salty ocean is inferred from variable magnetic field measurements made by the magnetometer on the Galileo spacecraft. These magnetic fields are thought to be the result of interactions between Jupiter's magnetosphere and Callisto's subsurface ocean.

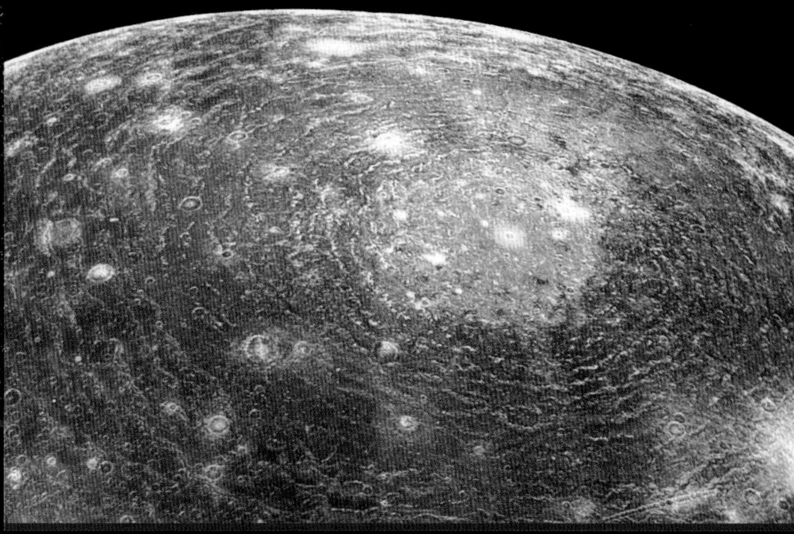

This Voyager 1 image shows the Valhalla impact structure on the surface of Callisto. The multiple rings are likely the result of the subsurface material being either ice or water, or a mixture of the two that froze as the expanding shock of the impact passed.

Callisto Statistics

MEAN RADIUS
2,410 km

ORBITAL PERIOD
16.7 days

MEAN DISTANCE FROM JUPITER (SEMI-MAJOR AXIS)
1.88×10^6 km

MASS
10.8×10^{22} kg

MEAN DENSITY
1,832 kg/m^3

Callisto

Callisto is similar in size to the planet Mercury but much less dense. The lower density results from the fact that while Mercury consists of a large iron core, Callisto is a mixture of rock and ice. At this distance from Jupiter, it is not in a resonant orbit like the other Galilean moons. The interior of Callisto doesn't appear to have differentiated as much as Ganymede.

Callisto has a very thin carbon dioxide atmosphere that is thought to be replenished by carbon dioxide ice at the moon's surface. Surface images of Callisto do not show evidence for any tectonic activity like that seen on Ganymede, as its surface hasn't been renewed or shifted.

Based on measurements made by NASA's Galileo spacecraft, Callisto may have a salty ocean beneath its crust, much like Europa. What's known is that Callisto has an icy surface that is as much as 200 km thick, on top of a layer of liquid, salty water at least 10 km thick.

Saturn

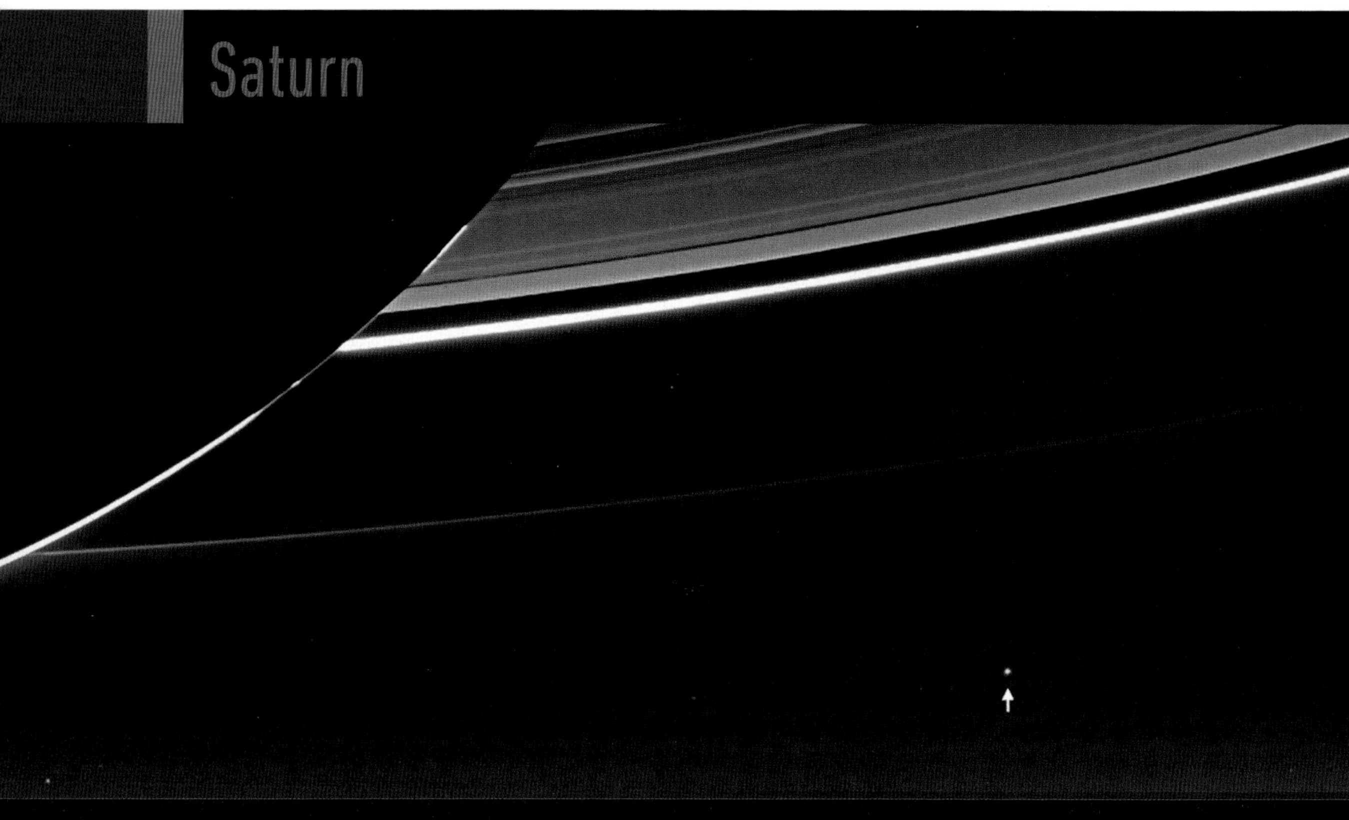

This planetary self-portrait taken by the NASA Cassini spacecraft shows Saturn and its ring system. Earth is the pale blue dot highlighted by the arrow. Sunlight can be seen shining on the limb of Saturn, which is backlit. The gaps in the bright rim are the result of shadows from Saturn's rings on its own surface.

Statistics

MEAN RADIUS
57,000 km (~9 Earth radii)

SIDEREAL PERIOD
29.45 years

ROTATIONAL PERIOD
10.6 hours days

VOLUME
8.3×10^{14} km^3 (764 Earths)

SATELLITES
Over 150 natural satellites, with Titan and Rhea being the largest, containing over 90 percent of the mass of material orbiting the planet

MEAN DISTANCE FROM THE SUN
9.6 AU (1.4×10^9 km)

MASS
5.7×10^{26} kg
(95 Earth masses)

MEAN DENSITY
687 kg/m^3

ATMOSPHERIC COMPOSITION (BY VOLUME)
96% hydrogen; 3% helium; 0.4% methane; and trace amounts of ammonia, water, and ammonium hydrosulfide ices

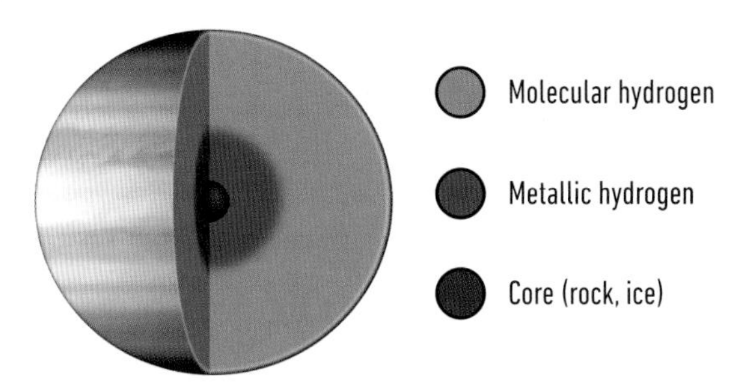

A cutaway view of Saturn shows its interior structure.

Saturn is the sixth planet from the Sun and the second-largest planet in the solar system, after Jupiter. Its interior and surface are similar to fellow gas giant Jupiter, but its rings are unparalleled in the solar system. In fact, it's those rings—which are visible through even a modest telescope—that make Saturn a favorite object for backyard astronomers.

Structure and Early Observations

Saturn, like Jupiter, is composed mostly of hydrogen and helium and has a similar interior structure. The planet's mean density is less than that of water. Along with all of the gas giant planets, Saturn is thought to have a rock and ice core, with overlying layers of metallic hydrogen, molecular hydrogen, and then its thin visible atmosphere.

Saturn and its rings were first viewed by Galileo in the winter of 1609-1610. By the middle of the seventeenth century, Christian Huygens had made improved observations of the planet and proposed that Saturn was surrounded by a thin, flat ring. Huygens also discovered the planet's large moon Titan.

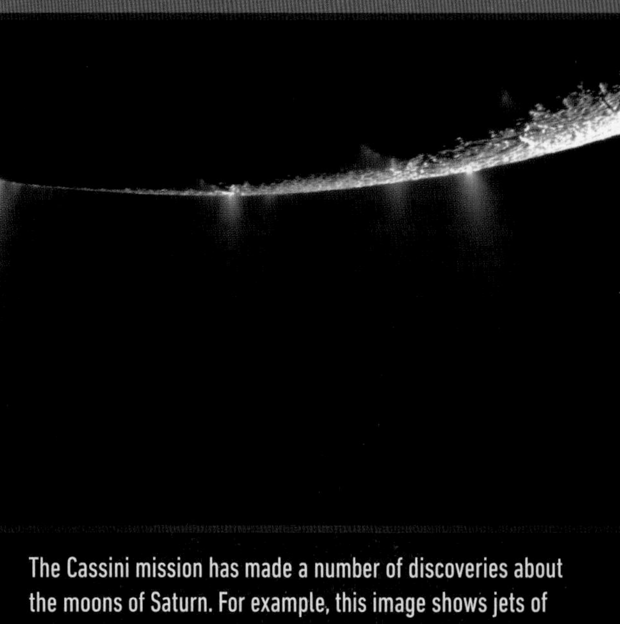

The Cassini mission has made a number of discoveries about the moons of Saturn. For example, this image shows jets of ice containing water vapor and organic compounds spraying from cracks near the south pole of the surface of the moon Enceladus. The jets indicate that Enceladus may have a region of liquid water beneath its surface.

Modern Exploration of Saturn and Its Moons

The first high-resolution images of Saturn and its moons were provided by the Voyager 1 spacecraft that flew by in 1980 on its way out of the plane of the solar system. Voyager 2 followed in 1981, capturing images before continuing on to Uranus and Neptune. There was then a 20+-year gap in images until 2004, when the Cassini-Huygens spacecraft arrived at the planet. This mission has returned stunning images of the planet and its moons and released the Huygens probe, which landed on the surface of the moon Titan in January 2005. The Cassini orbiter is expected to continue operations until 2017 or beyond.

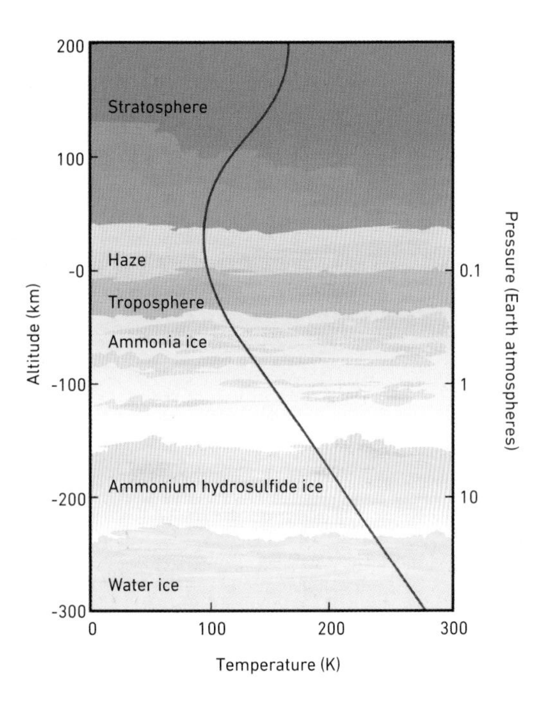

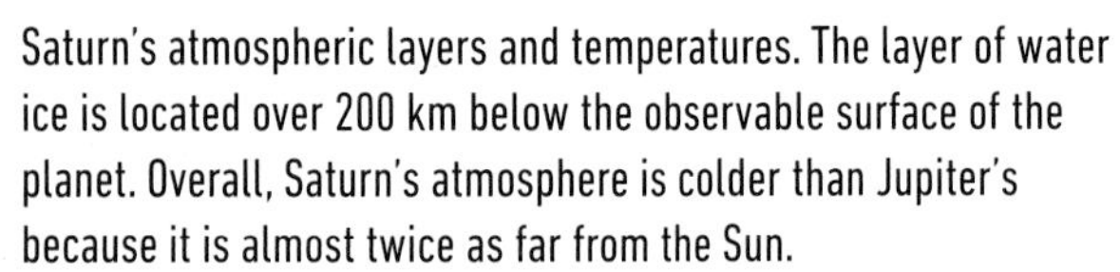

Saturn's atmospheric layers and temperatures. The layer of water ice is located over 200 km below the observable surface of the planet. Overall, Saturn's atmosphere is colder than Jupiter's because it is almost twice as far from the Sun.

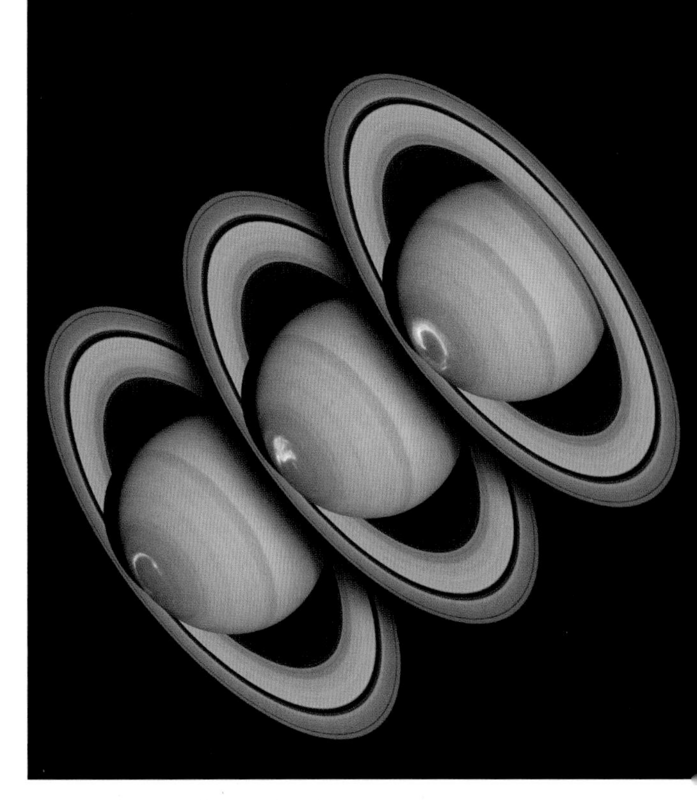

Saturn's magnetic field carries charged particles to interact with its atmosphere, creating aurorae like those seen on Earth and Jupiter. Ultraviolet images of Saturn made with the Hubble Space Telescope combined with radio frequency observations from Cassini indicate that the aurorae on Saturn respond to variations in the solar wind, as can be seen in this series of images taken over four days.

Atmosphere

To the eye, Saturn's atmosphere has many similarities to that of Jupiter, with a banded orange and yellow surface made up of light-colored zones and dark-colored belts.

Also similar to Jupiter, Saturn's upper atmosphere is made up of ammonia ice clouds, ammonium hydrosulfide ice, and water ice. However, observations of Saturn detect a much lower abundance of helium in its atmosphere than that of Jupiter. This is unusual, because the planets are thought to have formed in a very similar fashion. The most likely explanation is that Saturn, being smaller than Jupiter, cooled more rapidly. As it cooled, its helium sunk to lower layers in the planet's atmosphere beyond detection. At deeper layers, it is likely that the abundance of hydrogen and helium on Saturn are the same as on Jupiter.

While Saturn does not have a permanent surface feature like Jupiter's Great Red Spot, storms have appeared there more frequently as the northern hemisphere has entered its spring season (beginning in 2009).

This change in seasons, which each last for seven or eight years, has illuminated the northern hemisphere, allowing the Cassini mission to discover a number of new features. This includes an unusual hexagonal pattern at the north pole that rotates every 10 hours and 39 minutes, or the period of rotation of the planet's magnetic field. Centered in that hexagon is an enormous storm (called a *polar vortex*) that rotates in the same direction as the planet. Another polar vortex can be found at the south pole.

NASA's Cassini mission first detected the storm seen in this image in 2010. It was the largest storm detected by Cassini since 2004. The storm is accompanied by lightning, which was seen flashing 10 times per second. The appearance of this storm may be related to the onset of spring in the northern hemisphere.

The hexagonal pattern at the north pole of Saturn. For context, each side of the hexagon is slightly larger than Earth's diameter. The pattern rotates every 10 hours and 39 minutes, which is the period of rotation of the planet's magnetic field. The polar vortex at Saturn's north pole can be seen in the center of the hexagon (the south pole has one as well). All planets and moons that have atmospheres—including Titan—are observed to have polar vortices.

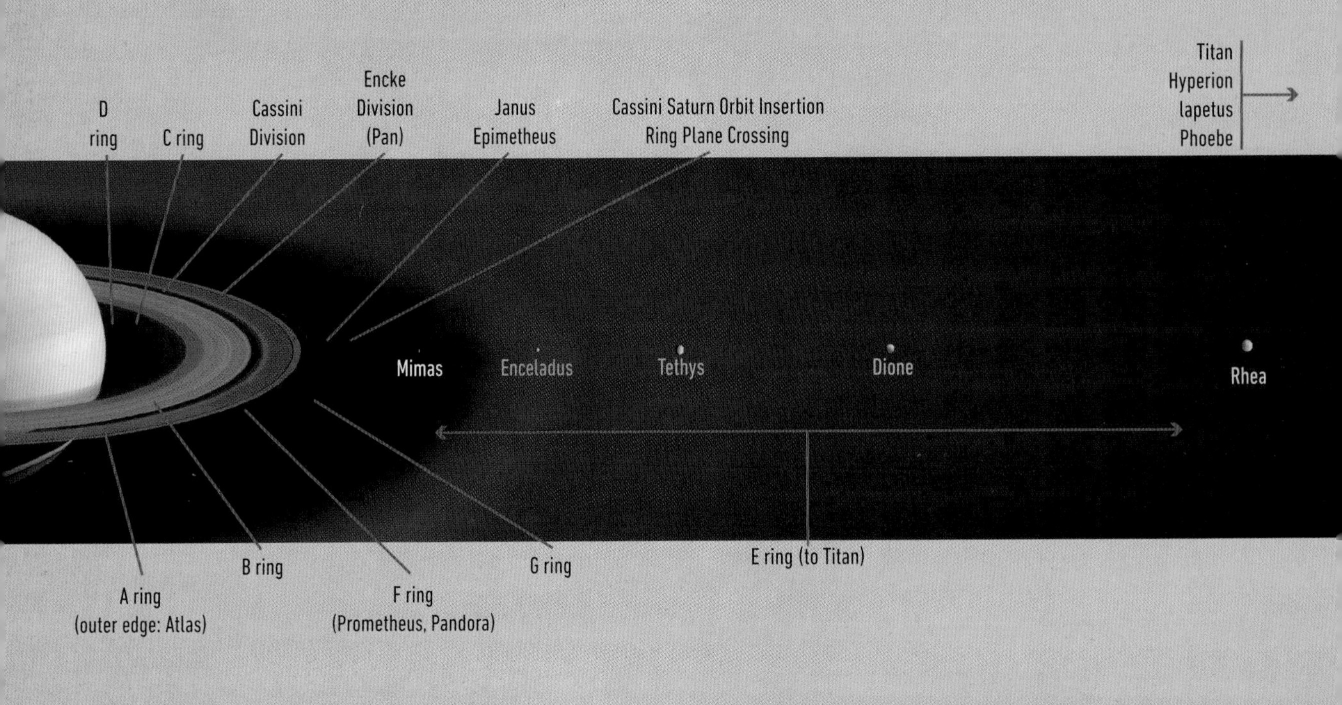

This artist's conception of Saturn's ring system shows the bright and faint rings, as well as some of the inner moons. (The orbit of Titan is beyond the scale of this image.)

Saturn's Rings

The features now known to be Saturn's rings were first observed by Galileo in 1610 but not understood until at least 1655, when Christian Huygens first suggested that the curious "ears" on either side of the planet might be a thin ring.

To early observers, the rings appeared to be a solid disk, but they are in fact made of an enormous number of small, icy objects ranging in size from less than 1 cm to over 10 m across. The objects in the ring are made of mostly water ice, with the main ring system thought to have formed from the breakup of an icy moon or an errant comet.

The rings were lettered as they were discovered, so the order of the letters from the planet outward (D, C, B, A, F, G, and E) is not intuitive. The Cassini Division is a gap between the bright A and B rings (which were discovered and named first). The smaller Encke Gap is near the outer edge of the A ring and contains the moonlet Pan. The large, faint E ring extends out almost to the orbit of Titan.

The A and B rings of Saturn (visible in a small telescope) are found between 32,000 km and 75,000 km above Saturn's surface. Despite this enormous extent, the material in the rings is very thin, only about 100 m thick at most.

Like many other planets, Saturn rotates on a tilted axis. As a result, you can view the ring system of Saturn from a variety of angles as it slowly orbits the Sun about every 30 years. This image shows a five-year portion of the full cycle of Saturn's tilt over the 30-year period of its orbit. The full cycle includes first looking down at the "top" of the rings, seeing them edge on (known as a ring plane crossing), and then seeing the rings from below.

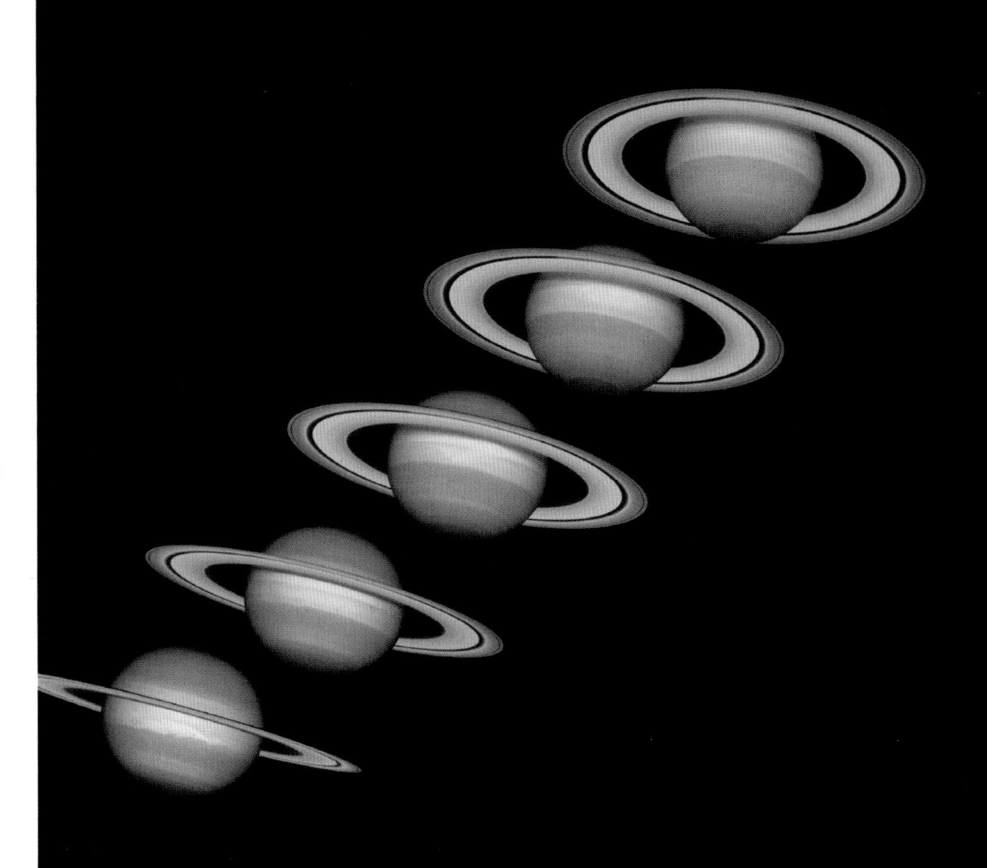

In this image, we see the Keeler Gap (the dark band) and the tiny moon Daphnis, only 8 km in diameter, casting a shadow on the rings.

Daphnis

The Cassini mission has allowed for images of Saturn's rings to be taken in unprecedented detail. One of the mission's many discoveries is the presence of a satellite, Daphnis, that orbits in the Keeler Gap at the outer edge of the A ring.

The slower-moving ring particles above the moon and the faster-moving ring particles below the moon are perturbed by the gravitational field of Daphnis. These disturbances show up as bright regions in the rings, but can also be seen by the shadows they cast on the

Early observations of the moon made in visible light gave no indication of the complex processes going on at its surface. In this Cassini color image of Titan (left), the moon appears to be a featureless, hazy orange sphere. However, the Cassini mission, equipped with near-infrared cameras that allow it to peer through the orange haze (right), show variations in brightness that are surface features on the moon.

Titan Statistics

MEAN RADIUS
~2,600 km

ORBITAL PERIOD
15.95 days

MEAN DISTANCE FROM SATURN (SEMI-MAJOR AXIS)
1.2×10^6 km

MASS
1.4×10^{23} kg

MEAN DENSITY
1,880 kg/m^3

Titan

Titan, Saturn's largest moon, is a strange world. Intermediate in size between Earth and the Moon, Titan is one of a small number of objects in the solar system with a solid surface and a substantial atmosphere. Because it orbits Saturn far out in the solar system, Titan is much colder than Earth, with the Sun about 100 times more faint than on Earth.

Composition

Titan has a surface pressure about 1.5 times that of Earth at sea level and an atmosphere that consists mostly (98.4 percent) of nitrogen. Based on the temperatures, densities, and chemical composition of the moon, Titan is thought to have lakes and oceans of liquid methane, and recent images of the surface taken by the Cassini spacecraft and the Huygens probe seem to confirm that picture.

Watching the Weather

Since 2005, Cassini has been able to watch changes in Titan. In particular, the changing seasons (as Saturn orbits the Sun) seem to have an impact on the location of weather patterns in the thick atmosphere of the moon.

For example, Saturn and its moons went through an equinox in 2009, with the Sun directly overhead at the equator. Since this equinox, Titan has been transitioning to a northern hemisphere spring. The south pole cloud cover that was seen on Titan early in the Cassini mission has disappeared and been replaced by equatorial clouds.

The muted colors in this first image from the Huygens probe on Titan (2005) show how the surface would look to the human eye. The pieces of ice or rock in the foreground are only about 10 cm across.

On a recent close approach to Titan, Cassini made this false-color radar imaging map of a 140-km-wide portion of the surface. Regions that are highly reflective are shown in shades of brown. Regions that absorb radio frequencies are shown in blue and black and are thought to contain liquid methane.

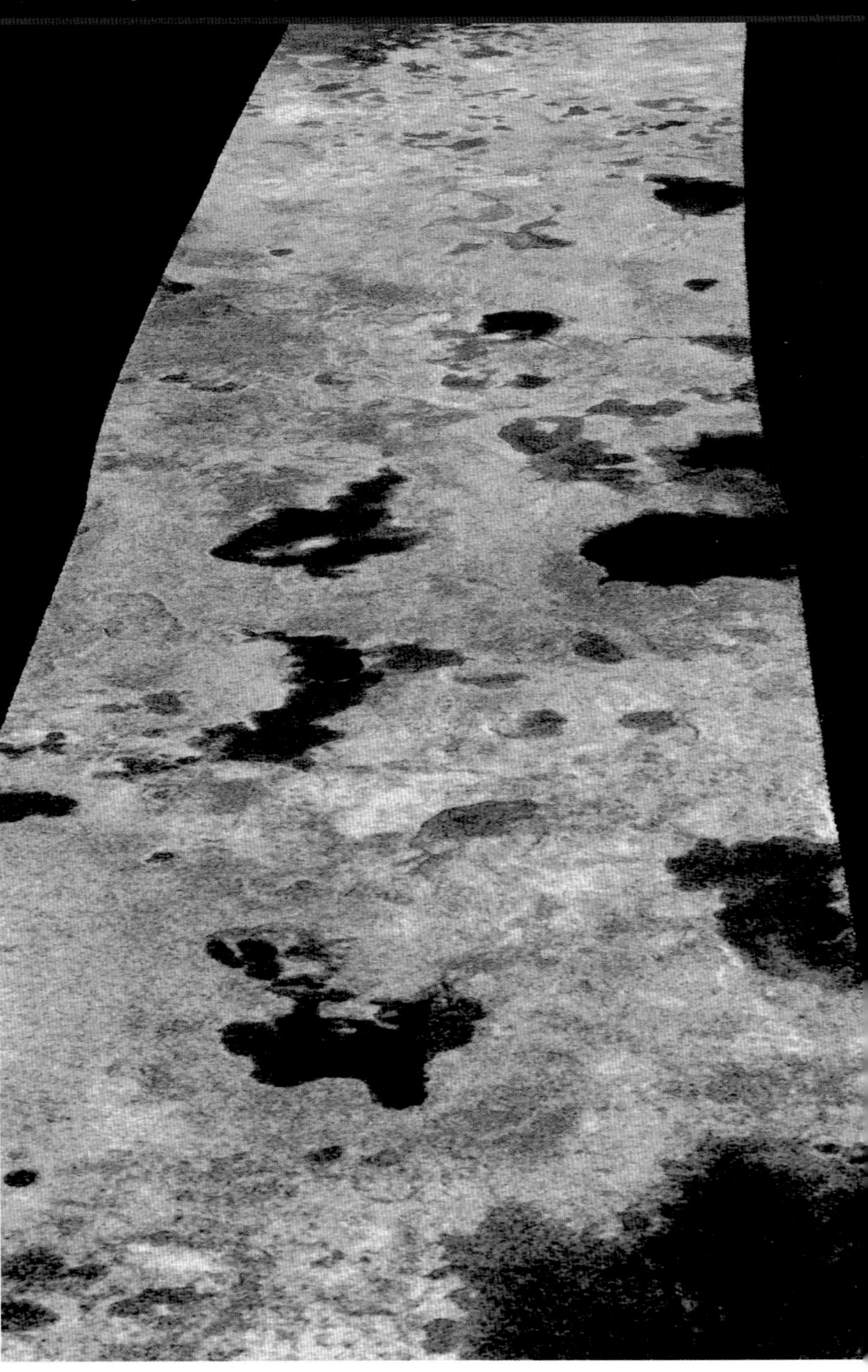

Uranus

This montage of images shows the planet Uranus and its five largest moons, as imaged by the Voyager 2 spacecraft that flew by the planet in the mid-1980s. The moons—named after characters in Shakespearean plays—are Ariel, Miranda, Titania, Oberon, and Umbriel. At visible wavelengths, the atmosphere of Uranus is an undifferentiated blue.

Statistics

RADIUS
25,362 km (4 Earth radii)

SIDEREAL PERIOD
84.3 years

ROTATIONAL PERIOD
17.3 hours

VOLUME
6.8×10^{13} km^3 (63 Earths)

SATELLITES
Five major moons and 27 known satellites

MEAN DISTANCE FROM THE SUN
19.2 AU (2.87×10^9 km)

MASS
8.7×10^{25} kg (~14.5 Earth masses)

ATMOSPHERIC COMPOSITION (BY VOLUME)
83% hydrogen; 15% helium; ~2% methane; and ammonia, water, and other ices

The planet Uranus, seventh from the Sun, is so faint that it is not visible without a telescope. It was discovered in 1781 by the English astronomer William Herschel. Uranus has a rocky core inside a shell of high-density water and ammonia. The outermost shell is liquid hydrogen and helium, with the atmosphere you observe a thin layer on top of it. The layer of liquid ammonia is thought to be the source of Uranus's magnetic field, which is about 50 times as strong as Earth's total magnetic field.

Axial Tilt

Herschel's early observations indicated its moons were orbiting in a plane nearly perpendicular to the planet's orbit around the Sun, which was the first hint that Uranus has an unusual rotational axis. Recent observations confirm the Uranus's axis is tilted by 98 degrees with respect to the plane of the solar system. (By comparison, Earth's axis is tilted at 23.5 degrees.) Because of this tilt of more than 90 degrees, the planet Uranus has retrograde rotation—that is, it rotates in a direction opposite to its orbit.

Strange Seasons

Because it is tilted onto its side, Uranus exhibits very strange seasons with extreme temperatures. As you have seen, Earth's modest tilt of 23.5 degrees is responsible for its three-month seasons, and whichever hemisphere is tilted toward the Sun experiences summer. However, because Uranus is tilted on its side, it has one side that faces the Sun for half of its orbit, or 42 years; a different side then faces the Sun for the remaining 42 years. Instead of being low in the sky in winter, the Sun actually disappears from Uranus's northern hemisphere for decades at a time.

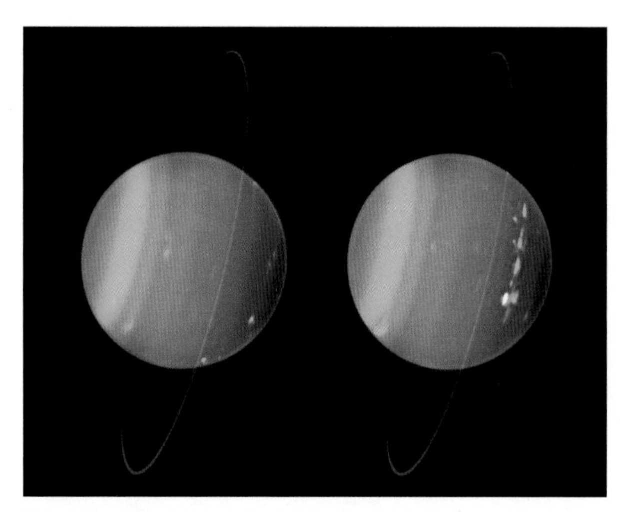

This false-color infrared Keck Adaptive Optics image of Uranus shows both atmospheric features and its thin ring system, both of which are difficult to observe optically. The north pole of the planet points to the lower right in this image.

Atmosphere and Rings

The only satellite to fly past Uranus, the Voyager 2 made close-up images of the planet and determined that its atmosphere is gaseous and consists primarily of hydrogen and helium, similar to all of the gaseous outer planets. Uranus's atmosphere has a higher percentage of methane than Jupiter and Saturn; this molecule absorbs red and yellow light preferentially, giving the observable outer layers of Uranus's atmosphere its distinctive blue color.

In 1977, Uranus was found to have a system of very faint rings. The icy material in Uranus's rings is made up of small objects that are at most a few meters in diameter.

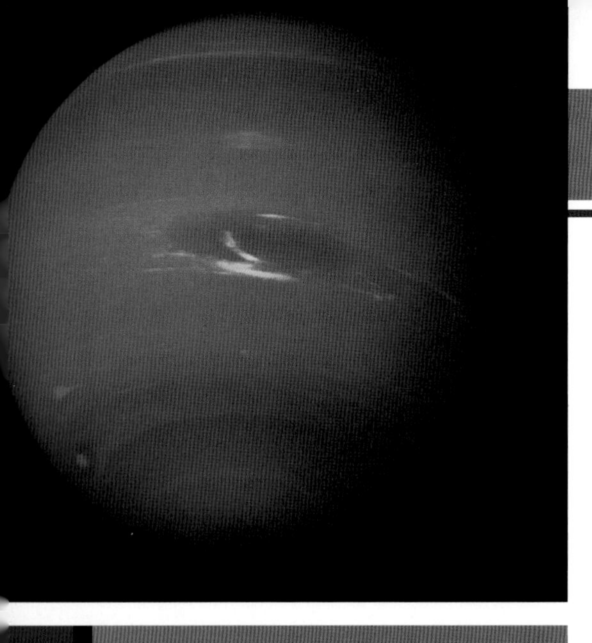

Neptune

Neptune is the eighth and most distant planet from the Sun, orbiting 30 times farther from the Sun than planet Earth. Due to this distance, Neptune only receives 1/900 as much sunlight as Earth. Like Uranus, Neptune is so faint that it can't be seen without a telescope.

Statistics

RADIUS
24,622 km (3.9 Earth radii)

SIDEREAL PERIOD
164.8 years

ROTATIONAL PERIOD
16.1 hours

VOLUME
6.3×10^{13} km^3 (58 Earths)

SATELLITES
14 known satellites

MEAN DISTANCE FROM THE SUN
30.1 AU (4.5×10^9 km)

MASS
1.0×10^{26} kg (~17 Earth masses)

ATMOSPHERIC COMPOSITION (BY VOLUME)
80% hydrogen; 19% helium; methane (~1.5%); and ammonia, water, and other ices

Observations and Discovery

Early drawings from his notebook show Galileo observed the planet in 1612 and 1613 because Jupiter (a frequent object of Galileo's interest) and Neptune were by chance in the same part of the sky. Due to its faintness and its very slow motion, Neptune could have easily been thought to be a background star.

While Galileo may have seen it first, the discovery of the planet is credited to Urbain Le Verrier (France) and John Couch Adams (England) in 1846.

This near-infrared Keck Telescope image of Neptune shows the planet and its large moon Triton. Triton's composition and retrograde orbit have led astronomers to think it might be a captured object from the Kuiper belt.

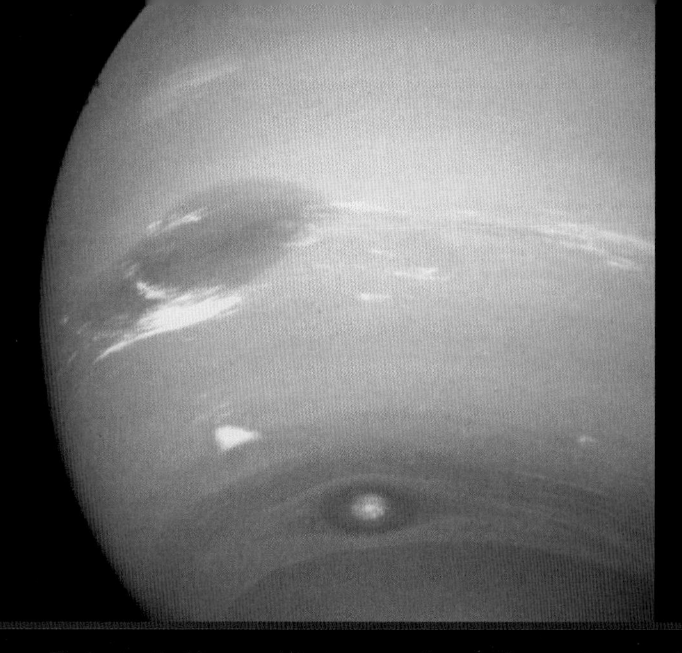

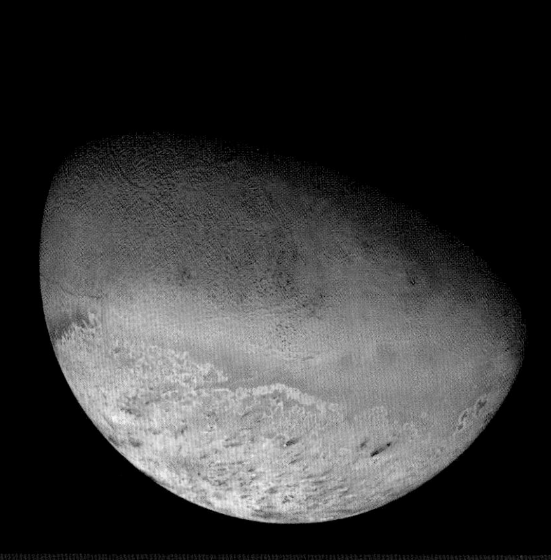

The best optical images of Neptune come from the Voyager 2 flyby that occurred in 1989. In this image, three features are obvious: the Great Dark Spot (a storm similar to the Great Red Spot on Jupiter) in the northern hemisphere, a bright cloud feature further south, and a second dark spot. The spots all move eastward (to the right) across the planetary surface.

Triton, the largest moon of the planet Neptune, is shown here in a simulated true-color image. Triton is slightly smaller than the Moon and has an unusual "cantaloupe" terrain in its northern hemisphere.

Interior

Like Uranus, Neptune is an intermediate-sized planet, having a diameter and mass that's in between the rocky terrestrial planets and the larger gas giants, Jupiter and Saturn. Neptune is thought to have a rocky core at the center of a shell of liquid water and ammonia, an outer shell of liquid hydrogen and helium, and a thin, gaseous atmosphere.

Atmosphere and Storms

Neptune's atmosphere is observed to have visible clouds made of methane ice crystals. Neptune's spots, which come and go every few years, are long-lived storms similar to Jupiter's Great Red Spot. They are dark because they result from a hole in the upper layer of methane in the atmosphere. Voyager imaged one of those dark features, called the Great Dark Spot, when it flew by in 1989.

Triton

Voyager 2 imaged Triton, the largest moon of Neptune, on its last stop before leaving the solar system. Voyager discovered that the moon has active nitrogen cryovolcanoes (or ice volcanoes) and a very thin nitrogen atmosphere. The lack of cratering on Triton indicates that the surface is young, consisting of nitrogen ice. Part of Triton's surface is covered in methane ice, probably formed by plumes detected at its south pole.

Other Moons in the Solar System

This image shows the major moons of all the planets in comparison to the size of Earth as imaged by NASA spacecraft in the past few decades. Notice that Earth has an unusually large moon for a small, terrestrial planet and that the largest moons of Jupiter and Saturn are comparable in size to the smaller terrestrial planets. Notice that Earth has an unusually large moon for a small, terrestrial planet, that the largest moons of Jupiter and Saturn are comparable in size to the smaller terrestrial planets, and that the moons of Mars are so small they can't be seen.

Moons appear to be part of the formation of planets, at least the largest ones. Of the eight planets in the solar system, six have moons. The large outer planets have many moons, some of which are as big as the terrestrial planets.

Terrestrial Moons

Of the terrestrial planets, only Earth has a large moon. Mercury and Venus have no moons, and Mars has two very small irregularly shaped moons, Phobos and Deimos, which are so small that they were only discovered in the late nineteenth century. Phobos has a mean diameter of 22.2 km, and Deimos is smaller with a mean diameter of 12.6 km. Moons have a variety of origins, with Earth's Moon likely having formed as the result of an early massive collision. The moons of Mars may be captured asteroids, or the remainder of an early ring of material. The majority of evidence, however, indicates that Phobos was formed by a large impact as well.

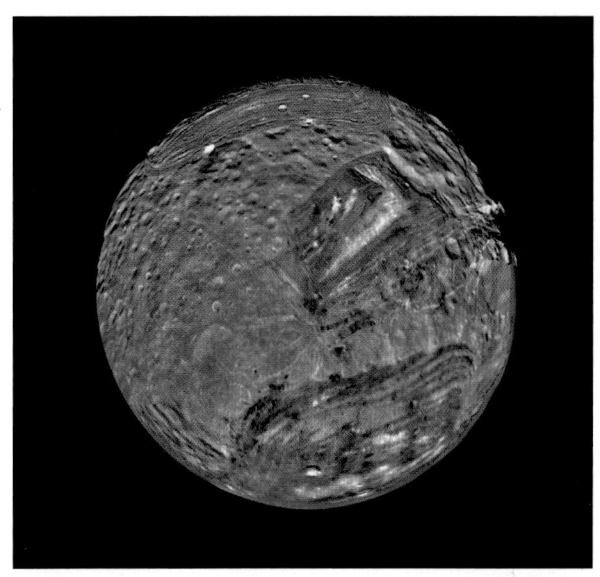

Miranda is the innermost of Uranus's major moons. Captured by Voyager 2, Miranda's surface is a jumble of old, heavily cratered, and younger terrain.

Gas Giant Moons

Knowledge of the moons of Uranus and Neptune is sparse, because only the Voyager missions have visited these planets. Hopefully future missions to these outer planets will uncover more about their icy moons.

Both Voyager 1 and Voyager 2 visited the giant planets Jupiter and Saturn. After its encounter with Saturn, Voyager 1 left the plane of the solar system and continued on its path into interstellar space. The path of Voyager 2, however, was planned to bring it into an encounter with both Uranus and Neptune, giving the only close views to date of these planets and their moons. After visiting Neptune, Voyager 2 joined Voyager 1 on a long journey to the outer reaches of the solar system and interstellar space.

A Moon in the Asteroid Belt

Even the smallest solar system objects have moons. In the asteroid belt, the asteroid Ida has a moon: Dactyl. Ida orbits in the asteroid belt and was discovered in 1884, but Dactyl is so tiny (1.4 km across) that it was only discovered when images were taken of it 110 years later.

This image of the asteroid Ida and its moon Dactyl was taken by the Galileo spacecraft that flew past through the asteroid belt in 1994 on its way to Jupiter.

Dwarf Planets

The discovery of a large and growing number of objects that are similar to Pluto in orbit and size led to the definition of dwarf planets as a distinct class. This image shows the known dwarf planets that orbit the Sun outside the orbit of Neptune.

Hubble Space Telescope images of the dwarf planet Eris and its orbiting moon Dysnomia allowed the calculation of the mass of Eris, which is both larger in radius and more massive than Pluto.

In the solar system, planetary-mass objects orbit either the Sun or another planet. If they orbit another planet, they are called a moon. But if they orbit the Sun, they are either a planet or a dwarf planet.

What Is a Dwarf Planet?

A planet was defined by the International Astronomical Union (IAU) in 2006 to refer to an object that 1) orbits the Sun, 2) is massive enough to assume an approximately spherical shape, and 3) has cleared its orbit of other objects. Dwarf planets are objects that have not satisfied one or more of these criteria.

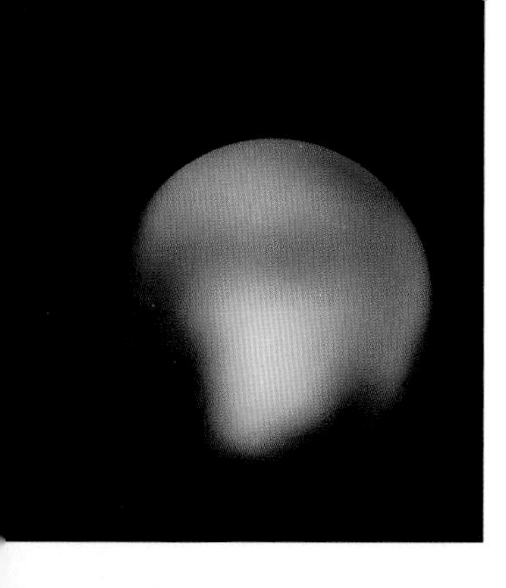

Pluto Statistics

MEAN RADIUS
1,153 km (0.329 Earth radii)

SIDEREAL PERIOD
247.68 years

ROTATIONAL PERIOD
6.39 days (retrograde)

VOLUME
6.39×10^9 km^3 (0.0059 Earths)

SATELLITES
One major moon (Charon) and five smaller satellites

MEAN DISTANCE FROM THE SUN
39.26 AUs (5.9×10^9 km)

MASS
1.3×10^{22} kg (~0.002 Earth masses)

Reclassifying Pluto

When Pluto was discovered in 1930, it was initially classified as the ninth planet in the solar system. However, this new definition changed the status of Pluto from being a planet to being a dwarf planet.

The issue arose because of the discovery of a number of objects (known as *trans-Neptunian objects*) that were in similar orbits to Pluto and larger in size. If Pluto were to be considered a planet, so would all of the other objects (and ones yet to be discovered).

After much discussion and no small amount of controversy, members of the IAU voted to adopt the new definition of a planet and add Pluto to a growing group of dwarf planets that includes Eris and its moon Dysnomia, Sedna, Quaoar, and others.

Visiting Pluto

Pluto is so far out in the solar system that even the best space-based telescopes can't take detailed images of its surface. However, the New Horizons spacecraft, launched in 2006, will fly past Pluto and its moons in July 2015. This is the first mission that will fly to Pluto and only the fifth mission to get this far from the Sun.

When the spacecraft gets to Pluto, the planet will once again be outside the orbit of Neptune. Because Pluto is in such an elliptical orbit, it occasionally dips inside Neptune's orbit. When the spacecraft arrives, it will have a close approach for one 24-hour period. Instruments aboard the spacecraft will take images and spectra for several months before and after this encounter. These images and spectra will give us our sharpest images of Pluto and its moon Charon and resolve open questions about its surface composition and temperature.

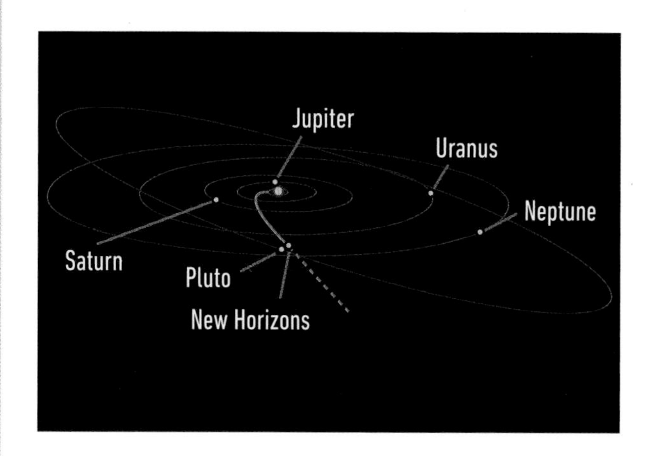

This schematic from NASA shows the trajectory and current position of the New Horizons mission as it travels to Pluto. On its way, the spacecraft made a close approach to Jupiter and its moons on a close approach, which gave it a gravitational "kick" out to Pluto.

Asteroids

Asteroids are rocky objects that orbit the Sun but are much smaller than the planets. Most asteroids in the solar system are found at the boundary between the inner, small terrestrial planets and the outer, large gaseous planets in a 1.5-AU-wide region known as the *asteroid belt*.

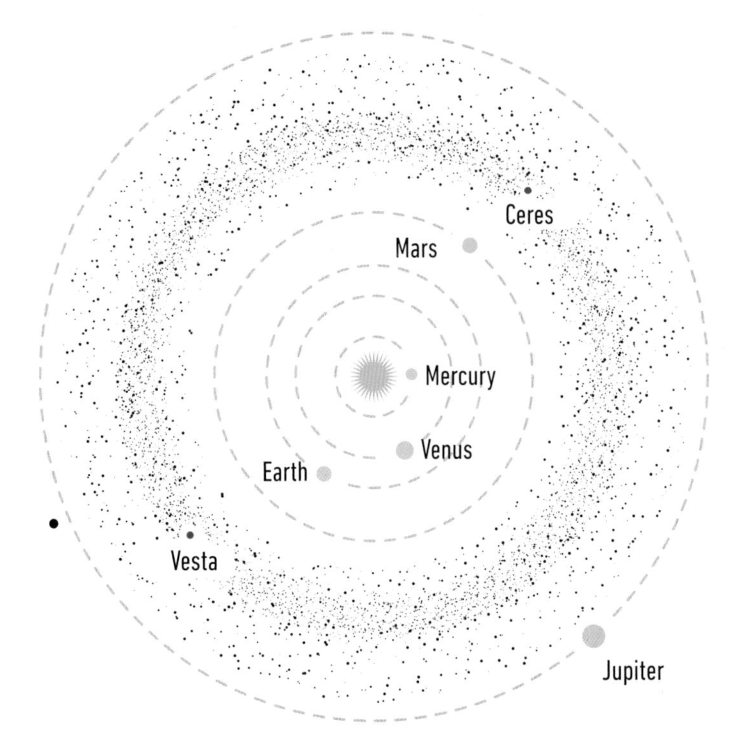

This schematic of the solar system out to Jupiter shows the location of the asteroid belt (from about 2 to about 3.5 AU) and the Trojan asteroids that are located in the orbit of Jupiter, both leading and trailing its orbit.

How the Asteroid Belt Formed

The formation of the asteroid belt instead of a planet at the same location is the result of Jupiter's strong gravity, which wouldn't allow planetesimals to combine and form planets. In fact, there may have been larger asteroids in the belt during the early years of the solar system that were ejected by interactions with the Jupiter. One such object may have been the Mars-sized planetesimal that collided with Earth and formed the Moon.

Photographic techniques used in the 1890s were a major breakthrough in accelerating the discovery of asteroids. A long exposure made by a telescope tracking a star field near the ecliptic would show asteroids as streaks across the field of stars. As of the early twenty-first century, astronomers have determined the orbits of over 100,000 asteroids.

Ceres and Vesta

Ceres, the first large object to be discovered in the region in 1801, orbits the Sun at mean distance of 2.77 AU. Ceres is now considered a dwarf planet (see "Dwarf Planets"). With about 25 percent of the mass of all the known asteroids, Ceres is clearly the largest object in the asteroid belt. Ceres may have a rocky inner core covered by a thin dust or water ice crust.

Vesta is an asteroid about the size of the state of Arizona in the United States, and has a surface of basalt, which is solidified lava.

In 2007, the Dawn mission was launched to explore both Vesta and Ceres. Dawn orbited Vesta between July 2011 and July 2012 and is expected to arrive at the dwarf planet Ceres in early 2015. Like missions to other planets, the Dawn mission has provided unprecedented views of the surface of Vesta and is likely to transform people's understanding of the surface of Ceres.

These Hubble Space Telescope images of Ceres (left) and Vesta (right) show that Ceres has characteristics that differentiate it from other asteroids. It is nearly spherical in shape and far larger than the rest of the objects in the asteroid belt.

NASA's Near Earth Asteroid Rendezvous (NEAR) mission took this image of the asteroid Eros. Like most asteroids shattered by many collisions, Eros has an irregular shape and is covered in large and small craters.

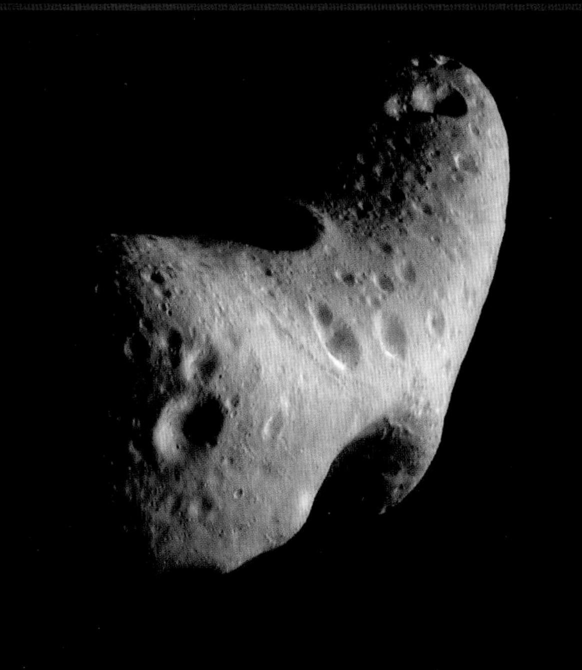

This visible light image of Vesta shows the enormous increase in detail gained by orbiting an asteroid. A mountain larger than Mount Everest is visible at the asteroid's south pole.

Comets

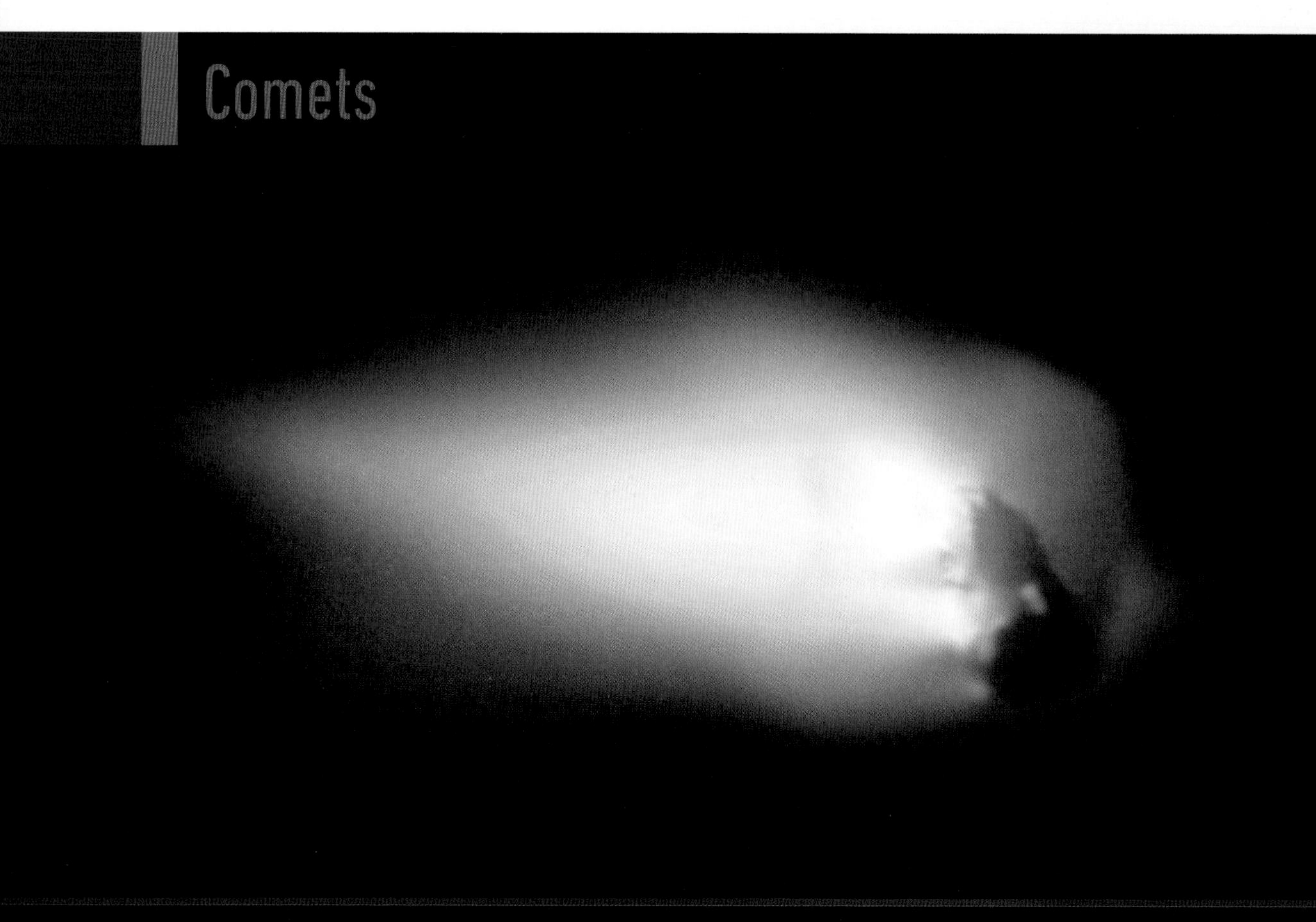

This image of Comet Halley, taken by the Giotto spacecraft as it headed away from the Sun in 1986, was the first close-up image of a comet showing both the nucleus (15 km across) and the atmosphere of expelled material, including organic molecules. As Earth crosses this path of material each October, Earth experiences the Orionid meteor shower.

Astronomers have been observing comets for as long as they have observed the Sun, Moon, planets, and stars. Before the sixteenth century, comets were thought to be nearby objects, perhaps located in the upper reaches of Earth's atmosphere. But parallax observations of a comet made by Tycho Brahe in 1577 made it clear that comets had to be very distant. A century later, Newton's laws of motion allowed Edmund Halley to calculate the size of the orbit of a comet that appeared in 1682, which allowed him to figure when it would return (1758).

Anatomy

Comets are icy objects of various sizes that orbit the Sun. Short-period comets (with orbital periods less than 200 years) are thought to originate in the Kuiper belt, while long-period comets (with orbital periods over 200 years) are thought to originate in the more far-flung Oort cloud (see "Layout of the Solar System").

Far from the Sun, comets are very similar to asteroids—dark and difficult to observe. However, as they approach the Sun, the materials on the surface and inside the comet are released, forming an atmosphere (or coma) and eventually tails.

A comet nucleus itself is typically only about 10 km in diameter. The nucleus is surrounded by a coma, or atmosphere, that is 10^6 km in diameter and a larger hydrogen envelope that can be 10 million km across. As a comet approaches the Sun, solar radiation causes the comet to form two tails—an ion tail facing radially away from the Sun (due to the solar wind) and a dust tail consisting of material lost from the comet's surface that trails behind its motion.

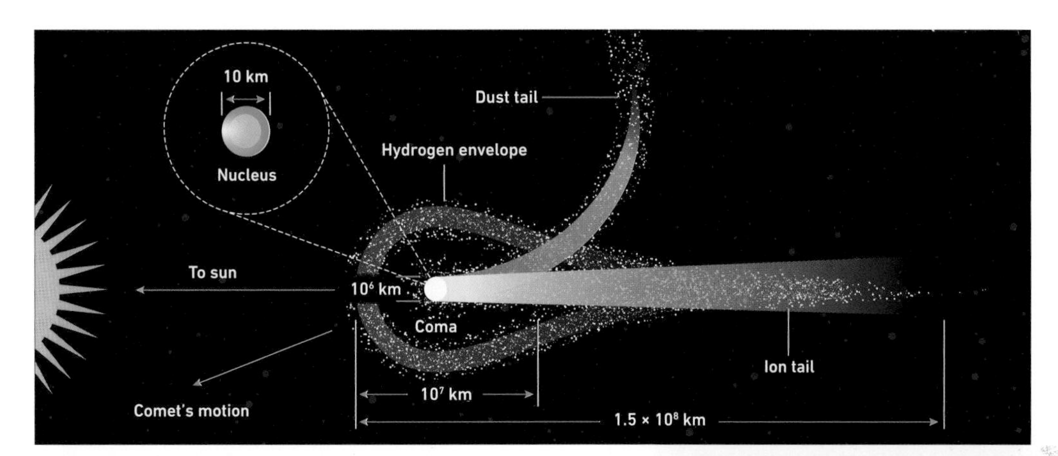

The anatomy of a comet and its tails.

Colliding with a Comet: Deep Impact

The Deep Impact mission was launched in 2005 to have an intentional collision with Comet Tempel 1 and then carefully analyze the material excavated by the impact. Analysis of the data from the impact showed that the interior of the comet contained more dust and less ice than expected and was about 75 percent empty space. After the impact experiment, the spacecraft was sent on to image another comet, Hartley 2.

Spacecraft like Deep Impact having close encounters with comets provide much more information about these occasional guests to the inner solar system.

This image of the impactor striking Comet Tempel 1 was taken about one minute after impact. Observations by other space-based telescopes confirmed that the impact site continued to outgas material for almost two weeks. Other analysis revealed that material at the comet's surface is as fine as talcum power.

Asteroid and Comet Impacts

Meteor Crater in Arizona (with a nearby highway seen to the left) is a relatively young crater, the result of an impact about 50,000 years ago. Calculations and material found at the site indicate that the responsible meteor was about 30 m in diameter and rich in iron. The crater rim has a diameter of a little over 1 km.

Craters cover the surfaces of planets and moons in the solar system, and many of these craters were formed by collisions with asteroids or comets. Depending on its level of geological activity, a planet or moon's surface shows a more-or-less detailed history of these types of impacts. For example, Venus, with its smooth, almost craterless volcanic plains, has a relatively young surface. On the other hand, the Moon—geologically inactive since the early solar system—has layer upon layer of craters evident.

This artists' conception shows the result of a 10-km asteroid or comet striking Earth near the location of the present-day Yucatan Peninsula. Impacts of this size were common in the first few billion years of the solar system.

Impacts on Earth

As disastrous as some impacts have been for life on Earth, early collisions with comets may have made the planet what it is today.

One of the most well-known impacts on Earth is the one that created the Chicxulub crater (now below the Gulf of Mexico). Near the end of the Cretaceous era—about 65 million years ago—a 10- to 15-km-diameter object struck Earth. The impact deposited large amounts of iridium (from the asteroid or comet) on the surface of Earth and created an oval-shaped impact basin near the current Yucatan Peninsula about 180 km across. This impact is widely believed to be the cause of extinction of the large dinosaurs.

Astronomers have also proposed that frequent bombardment from comets early in the history of the planet may have been responsible for filling Earth's oceans. In fact, the Herschel Space Telescope of the ESA recently analyzed the water in the coma (see "Comets") of Comet Harley 2 and found the comet contained the same ratio of heavy to light water as Earth's oceans. This observation may indicate that Earth's oceans—or a significant portion of them—came to its surface on the backs of comets.

Other Impacts in the Solar System

All of the planets have experienced and continue to experience comet and asteroid impacts. Because of their size and gaseous atmospheres, planets like Jupiter and Saturn absorb large objects without much apparent disturbance. For instance, Comet Shoemaker-Levy 9 broke into fragments and struck Jupiter in 1994, leaving a series of impact scars the size of Earth in its atmosphere (see "Jupiter").

The craters covering the southern hemisphere of Mars indicate it has suffered impacts over its 4.5-billion-year lifetime. In fact, some of the moons of the outer planets still bear the scars of massive impacts.

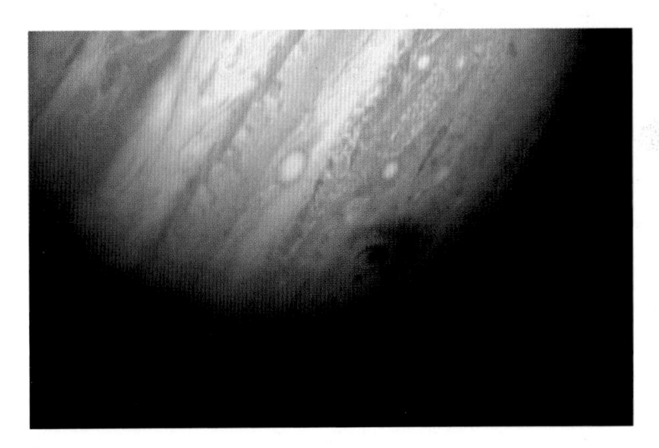

This image, taken about two hours after the impact of comet Shoemaker-Levy 9, shows the dark-colored impact site. The comet's entrance allowed astronomers to observe material from the deeper layers in Jupiter's atmosphere. The object that caused this large "bruise" is thought to have been about 1.5 km in diameter.

he Births, Lives and Deaths of Stars

The Milky Way Galaxy is filled with an amazing variety of stars, all at various stages in their evolution. Stars are born in the cores of giant clouds of molecular gas, and light up the surfaces of these same clouds as their nuclear furnaces begin making atoms. No matter their mass, they spend most of their lifetime making hydrogen into helium, and then end their lives in a variety of ways, some more spectacularly than others. Most stars in the galaxy are low-mass stars even smaller than the Sun; the more-rare high-mass stars live short, dramatic lives and have an outsized impact on their environments. In this part, you learn that stars have lifetimes, just like humans.

You examine the varied lives and deaths of stars and the probable future of the Sun (a low-mass star) as a white dwarf. Along the way, I discuss some of the more exotic remains of high-mass stars: neutron stars, pulsars, magnetars, and black holes.

Stars and Spectroscopy

At first glance, it would seem that stars would be very hard to tell apart. They are tiny points of light, immeasurably far away. But a number of technologies developed in the past few centuries have allowed people to pull apart that light and analyze it carefully.

In order to understand the nature of most stars, their light needs to be split up into its component wavelengths; this is known as *spectroscopy*. This can be done with a simple prism or diffraction grating. The function of both tools is to spread light out, showing in detail the wavelengths that comprise starlight. A prism uses the fact that different wavelengths of light bend differently in a prism, spreading out the light. A diffraction grating spreads light out by wavelength with a series of small grooves. The effect is the same: starlight comes in, and a spectrum comes out.

Spectral Types

Almost as soon as spectroscopy was invented, astronomers started to pass the light from stars through prisms and examine the spectra—first visually and then as recorded on photographic plates. What they found was that stellar spectra fell into distinct classes based on their spectral absorption lines (a "bar code" showing the presence of a particular molecule or element). For example, some stars had strong absorption lines of hydrogen, while others did not. Based on these spectral lines, astronomers began to place stars in classes.

Class	Temperature (in Kelvin)	Color	Mass (in Solar Masses)	Dominant Lines
O	> 30,000	Blue-violet	> 16	Ionized helium
B	10,000–30,000	Blue-white	2.1–16	Neutral helium, hydrogen
A	7,500–10,000	White	1.4–2.1	Strong hydrogen, ionized metals
F	6,000–7,500	Yellow-white	1.04–1.4	Hydrogen, ionized calcium and iron
G	5,200–6,000	Yellow	0.8–1.04	Neutral and ionized metals, calcium
K	3,900–5,200	Orange	0.45–0.8	Neutral metals
M	2,500–3,900	Red-orange	0.08–0.45	Titanium oxide

In the 1890s, the Harvard College Observatory produced the *Draper Catalog of Stellar Spectra,* which provided the spectroscopic classifications for thousands of stars. Williamina Fleming, working with Edward Pickering, used the letters A through N to designate classes, with the letter O used to signify stars with ionized helium absorption lines. In the early twentieth century, Annie Cannon organized the letters into the current classification system: O, B, A F, G, K, and M.

While it was not realized at the time the original classifications were made, the spectral classes divide stars by temperature, as the abundance of a certain element is related to the temperature of the star's photosphere. Thus, O stars have the highest photospheric temperatures, while M stars have the coolest temperatures.

In each class, the stars are further subdivided by numbers 0 (hotter) through 9 (cooler) to indicate smaller temperature variations. For example, the Sun's spectrum classifies it as a G2 star, meaning it's one of the hotter stars with spectral type G.

Black-Body Radiation Curve

Stars don't give off energy at just one wavelength, but at a variety of wavelengths. The curve that describes how stars give off light is called a *black-body radiation curve.* Any object with a measureable temperature will have such a curve, and the highest point on the curve (its peak wavelength) will tell you the temperature of the object. Stars are not perfect black bodies, but it is a good way to approximate their temperature.

The temperature of a star affects two of its measurable properties: its apparent color and its brightness. When a star is very hot, most of its energy comes out at shorter wavelengths and it appears bluer. When a star is cooler, most of its energy comes out at longer wavelengths and it is redder. The hottest stars are brightest in the ultraviolet part of the spectrum, and the coolest stars are brightest in the infrared part of the spectrum.

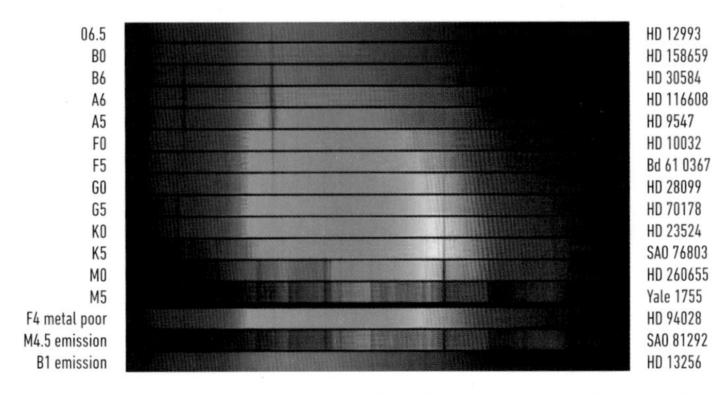

Stellar spectra from typical stars in each of the modern spectral classes, including some peculiar classes at the bottom. The names of many of the stars on the right (for example, HD 12993) are from the Henry Draper Catalog.

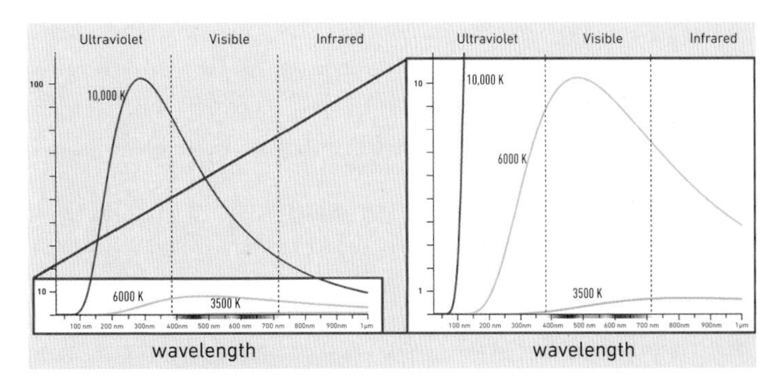

The three curves shown here represent the energy output of three stars, with surface temperatures of 3,500 K, 6,000 K, and 10,000 K. Notice that the 3,500 K star is both redder and much fainter than the 10,000 K star.

The Brightness and Color of Stars

Telescopes at the summit of Cerro Tololo Interamerican Observatory with the Milky Way seen in the background.

In your daily life, you use the words *brightness* and *color*. Astronomers use these words in very specific ways in their studies of the lives and deaths of stars.

Apparent Magnitude

Astronomers still measure the brightness of stars based on a very old system developed by the Greek astronomer Hipparchus. Stars were originally ranked by magnitude, with the brightest stars having a magnitude of 1 and the faintest stars having a magnitude of 6. The smaller the value of the magnitude, the brighter the star. The human eye can detect a brightness range of 100, so a first-magnitude star is 100 times brighter than a sixth-magnitude star, the faintest the human eye can see. That means that every step in magnitude corresponds to a brightness difference of about 2.5.

On a perfect night on Earth, the human eye can see about 6,000 stars that have magnitude brighter than 6. That number falls quickly once you take into account light pollution. As the brightness of the sky gets greater, you can only see the stars that are brighter than the sky. The brightness of a star tells you a combination of two factors: the intrinsic brightness of the star and how far away it is. The brightness of a star falls off with the square of the distance, so an object that is twice as far away is only one fourth as bright.

The modern magnitude scale has been expanded in range since the time of Hipparchos; it now includes objects brighter than first magnitude and fainter than sixth. In fact, some of the planets are so bright that they actually have negative magnitudes. The magnitudes of some familiar stars and planets are given in the following table. Notice that Neptune (not visible without a telescope) falls below the sensitivity limit of the human eye.

Absolute Magnitude and Luminosity

Often, people want to know something fundamental about a star, like how its brightness compares to other stars. If you know that two stars are at the same distance and one of them is brighter, then you know that the brighter star is hotter (see "Stars and Spectroscopy"). This type of magnitude is sometimes called an *absolute magnitude* and refers to the magnitude of a star if it were at a distance of 10 AU.

Astronomers also use the term *luminosity* to refer to the brightness of a star in absolute terms. A star's luminosity is the amount of energy it emits per unit time. The Sun, for example, has a luminosity of 3.8×10^{26} Watts.

The luminosity of a star depends on its temperature. In fact, the luminosity of a star depends on its temperature to the fourth power. So if one star is twice as hot as another, it will be 2^4 or 16 times as luminous. That is why stars like the hot Trapezium stars in the Orion Nebula are so much brighter and more energetic than all the other stars in the nebula—they are much hotter.

Color and Temperature

The color of a star also gives us information about its temperature. The redder a star is, the cooler it is, and the bluer a star is, the hotter it is. For example, when you see a field of stars like the globular cluster M15, you see observe a variety of colors and, as a result, a variety of temperatures. Because most of the stars in the field are members of the cluster, they are at the same distance (about 40,000 light-years away), meaning the only variable is the temperature of the star.

Object	Magnitude
Jupiter (max.)	-2.94
Sirius	-1.47
Saturn (max.)	-0.49
Vega	+0.03
Supernova 1987A	+3.03
Andromeda Galaxy	+3.44
M41 (open cluster)	+4.50
Uranus (max.)	+5.32
Neptune (max.)	+7.78
HST (faintest)	+31.50

The globular cluster M15 as seen by the Hubble Space Telescope (HST). The bright orange stars are red giant stars, which are both more luminous and cooler than the Sun; the bluer stars are hotter.

Stellar Lifetime Snapshot

As you learned in "Stars and Spectroscopy," astronomers use spectrometers to spread out the light from stars in order to determine the elements present in the star's photosphere. Astronomers combined this technology with photography in the late nineteenth century in order to record large numbers of the spectral lines on photographic plates.

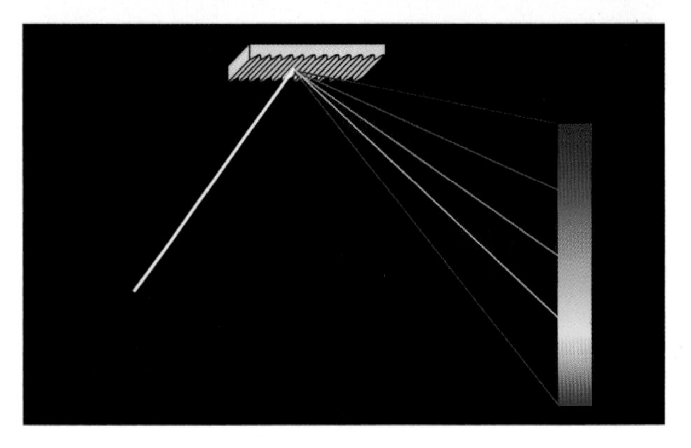

Light can be separated into its component colors in many ways. Many telescopes today use a diffraction grating to split up white light into its component colors.

Early in the twentieth century, astronomers had enough data on the brightness and temperature of stars to begin to make scatter plots of large amounts of data. Antonia Maury at the Harvard College Observatory did some of the early work of classifying stellar spectra not only by the elements present, but also by the width of the spectral lines detected.

Using data from the Harvard College Observatory and other observatories, Eljnar Hertzprung and Henry Norris Russell made the first Hertzsprung-Russell (HR) diagrams of stars in about 1910. These plots showed the absolute magnitude (or luminosity) of a collection of stars relative to their temperature (or color).

Interestingly, stars were found to populate three main areas. A broad band from the upper left to lower right (the *main sequence*) contained most of the stars observed. Then there were two other groupings—one in the upper right (now called *giants* and *supergiants*) and one in the lower left (known as *white dwarfs*). These groupings reveal much about the evolution of stars.

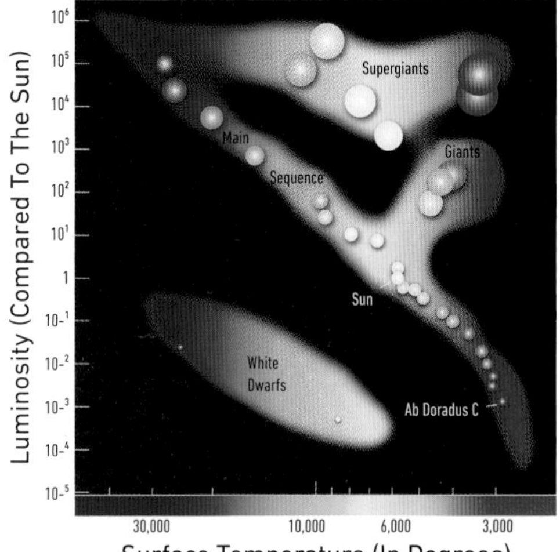

This is a scatter plot of the luminosity (or absolute magnitude) of a star relative to its temperature as compared to the Sun. Both the colors in the diagram and the scale at the bottom indicate the apparent color of the star as it relates to temperature—the cooler stars are redder, and the hotter stars are bluer.

The Very Large Telescope (VLT) of the European Southern Observatory (ESO) made this mid-infrared image of the nebula around the red supergiant star Betelgeuse in the Orion Constellation. The star Betelgeuse is late in its life and one of the brightest stars in the night sky.

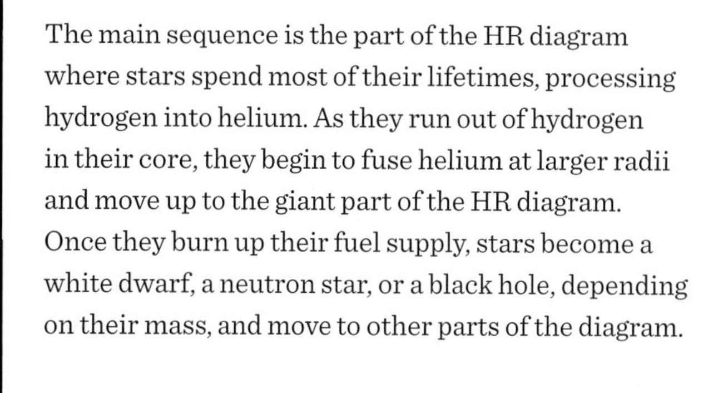

The main sequence is the part of the HR diagram where stars spend most of their lifetimes, processing hydrogen into helium. As they run out of hydrogen in their core, they begin to fuse helium at larger radii and move up to the giant part of the HR diagram. Once they burn up their fuel supply, stars become a white dwarf, a neutron star, or a black hole, depending on their mass, and move to other parts of the diagram.

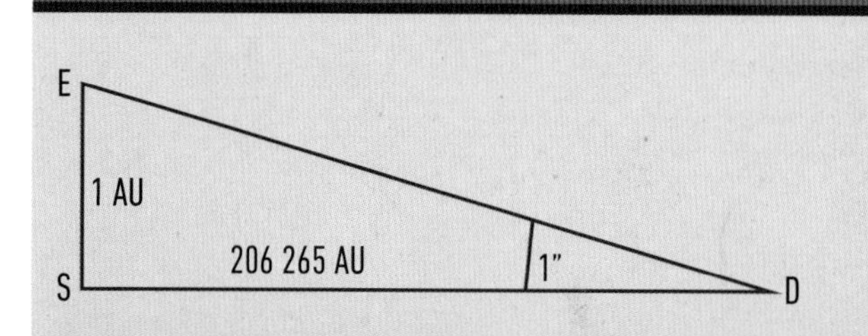

A parsec is the distance from Earth at which the Earth-Sun distance (1 AU) would appear to be 1 arcsecond. That distance is 206,265 AU.

Stellar Distances

The distances between stars are so great that astronomers need larger units than astronomical units (AU) to deal with them. For these longer distances, astronomers use light-years and parsecs.

A light-year is the distance light travels in a year, moving at 300,000 km/s (about 9.5×10^{15} m). For example, the nearest star, Proxima Centauri, is a little more than 4 light-years away.

The definition of a parsec (parallax second) is slightly more complicated. The mean Earth-Sun distance is 1 AU. If you were a parsec away from the Sun, that separation would appear to be 1 arscecond on the sky. One parsec is about 3.26 light-years.

Star Birth: The Orion Nebula

In this mosaic image of optical emission from the Orion Nebula, there are approximately 3,000 stars, most of which are low-mass stars like the Sun. The bright gas is illuminated by a handful of high-mass stars called the Trapezium stars (to the left, above the image center).

As constant as the fixed stars in the sky appear to be, they did have a beginning, some more recently than others. For example, the Sun formed about 4.5 billion years ago; if it formed like many other low-mass stars, it probably formed in a region like the Orion Nebula (also known as M42).

The Orion Nebula is found in the well-known Orion constellation, a staple of winter skies in the Northern Hemisphere. The nebula itself is important to astronomers because it is the nearest high-mass star-forming region that faces Earth. That is, optical telescopes can see into the depths of the cloud since the stars have blown out the near side. The stars have revealed an inner region that is often hidden to optical telescopes and allows a peek into the birth of stars.

Stars form as a result of gravity pulling together vast clouds of gas and dust. Star formation that has been observed in the Milky Way and other galaxies is always associated with *molecular clouds,* which are large collections of mostly hydrogen that serve as fuel tanks. In terms of the Orion Nebula, the most visible portion is the glowing gas at the surface of a molecular cloud that has been illuminated by the hot young stars that are emitting enormous amounts of radiation. This radiation lights up the clouds of material that still surround the young stars.

The formation of high-mass stars (like the Trapezium stars) is accompanied by the birth of many thousands of smaller stars. Numerical simulations show that when a cloud of cold gas collapses, it forms mostly low-mass stars and a few high-mass ones. The massive stars burn out in about 10 million years, but the small stars last for billions of years.

This colorful image shows data from the Spitzer and Hubble Space Telescopes, with different types of emission shown in different colors. The green and blue colors show emission lines from hydrogen and sulfur, which are excited by the photons streaming away from the Trapezium stars (located at the center of the nebula). The red and orange colors show infrared emission lines, which come from dust and gas that is not visible in the image.

Emission Lines

Many popular images of star formation show the sources as seen in various emission lines. Emission lines arise from atoms and molecules that have electrons in an excited state. Hydrogen is the simplest example, consisting of a proton and an electron. The electron tends to be in the ground state (the lowest state). But if a hydrogen atom is near a large energy source—like the Trapezium stars—the electron can be bumped into higher energy levels. As the electron settles back down to the ground state, it gives off emission lines at specific wavelengths. All atoms can be excited, and the emission lines seen in a region indicate what elements are there and their temperatures.

Star Death: The Crab Nebula

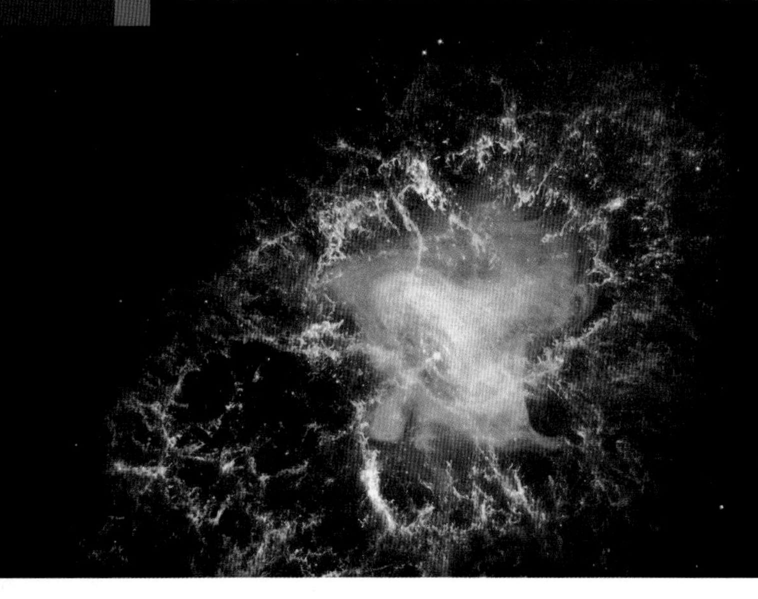

This image shows data from the Hubble Space Telescope (red and yellow) and the Chandra X-Ray Observatory (blue) overlaid. Data from the Spitzer Space Telescope is shown in purple. Being more energetic, the X-rays radiate away their energy more quickly and are seen closer to the exciting neutron star at the nebula's center. The nebula is powered by emission from the neutron star.

Stars go through their lifetimes and die in a variety of ways. Low-mass stars die rather quietly, shedding their outer layers as planetary nebulae. High-mass stars are more dramatic and end their lifetimes explosively. For example, the 10-solar-mass star that formed the Crab Nebula is believed to have exploded in 1054 C.E., when Chinese astronomers noted the appearance of a bright new star. There is evidence that the explosion was recorded in pictograms in the American Southwest as well. This nebula has allowed astronomers to study star death.

The Crab Nebula is the extended, luminous cloud that is the result of this explosion. It looks like a still image, but if you could fast-forward through the past 1,000 years, it would billow and grow like the stupendous explosion it is. The Crab Nebula is located about 6,500 light-years from Earth, about four times farther than the Orion Nebula.

At the center of the nebula is the remnant of the star that was once there: a neutron star. Neutron stars are born when a high-mass star explodes and the remaining core collapses to the density of a neutron.

When a star like the host star of the Crab Nebula explodes, the remnant enriches its local environment with elements that were formed deep inside the star. These materials move out into space and eventually get assembled by gravity into another star. That next-generation star will be enriched by the work of the stars that came before it. As a result, future generations of stars and the planets around them will have elements that were actually formed when stars like the one in the Crab Nebula exploded. So the Crab Nebula can also provide insight into how the solar system formed. The Sun and all the planets formed from the same material from earlier stars, meaning the carbon, oxygen, and nitrogen in the human body must have formed in some long-forgotten star.

The Crab Nebula, seen here in a Hubble Space Telescope image, extends over about 10 light-years.

The various colors in this detailed image of the Crab Nebula are from emission lines of different elements: hydrogen (orange), nitrogen (red), sulfur (pink), and oxygen (green).

Detail of the Nebula

In a detailed image of the Crab Nebula captured in 1995, you can see the region immediately around the remnant neutron star. The arc of bluish emission is the result of the star heating its immediate surroundings. The filaments that surround the pulsar in the nebula are in fact rushing away from it in all directions, with the yellow-green material toward the bottom of the image coming in Earth's direction. The different emission lines are seen because of temperature and density variations in the material as it expands away, a remnant of a thousand-year-old explosion.

The Birth of Low-Mass Stars

Most of the stars in the universe are low-mass stars, so an understanding of their origin is essential to understanding stars in general.

Gravitational Collapse

Stars form naturally as the result of gravity in regions of space that have high density and low internal pressure. This means that cold clouds of gas called *dark nebulae* are the primary fuel tanks of star formation. A familiar example of a dark nebula is the Horsehead Nebula. This object is part of the Orion Nebula (1,500 light-years away) and is illuminated by the massive star Sigma Orionis.

Formation starts when dark nebulae, which consist mostly of hydrogen (74 percent by mass) and helium (25 percent by mass), begin to collapse under their own self-gravity and form protostars. The gravitational collapse of a small gas cloud (sometimes called a Bok globule, with masses of about 10 solar masses) starts to heat up its gas. This conversion of gravitational energy into thermal energy is what gives a protostar its brightness early in its life.

This image of the Horsehead Nebula, taken by the Hubble Space Telescope in the near-infrared, shows an unfamiliar view of the nebula. Usually seen in absorption at optical wavelengths, the dark nebula glows in the near-infrared.

It takes stars of different masses different amounts of time to move across the HR diagram to the main sequence (see "Stellar Lifetime Snapshot"). Low-mass stars like the Sun can take millions of years before they begin their main sequence lifetime. Because most regions contain a mixture of low- and high-mass stars, this means that in star-forming regions, the low-mass stars are just getting started when the high-mass stars are almost done with their lives.

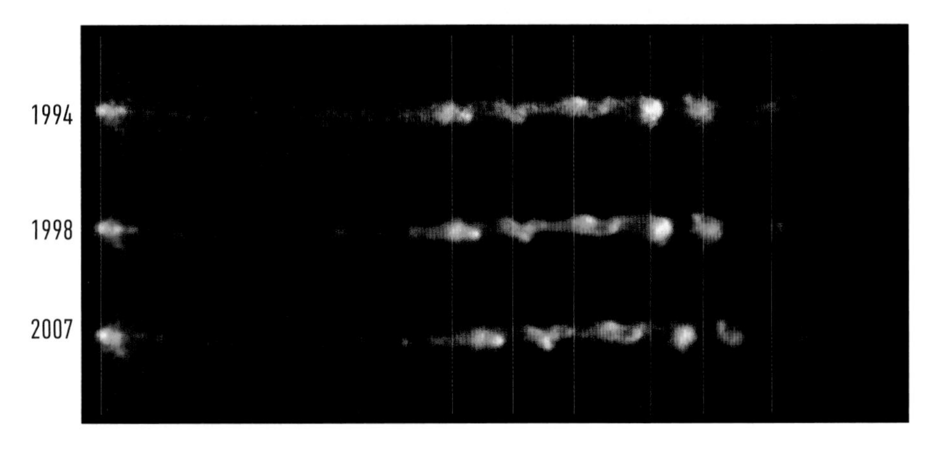

These images of a Herbig-Haro object (HH 34) about 1,350 light-years from Earth were taken between 1994 and 2007 and show its observable changes over time. The star is located on the left of each image (blue) and the knots in the jet are to the right (red). The star's magnetic field is thought to provide the force that confines the narrow jet.

Herbig-Haro Objects

It was not until about 30 years ago that astronomers began to observe that the early stages of low-mass star formation are often accompanied by the formation of long, opposite-directed jets of material called *Herbig-Haro objects.* These jets often plow into nearby gas, producing bright arcs of gas called *bow shocks.* The jets are short-lived, lasting only 10,000 to 100,000 years, but they are thought to form in conjunction with a disk of material that continues to add mass to the star (an accretion disk). Strong magnetic fields in this disk get pulled in as the disk contracts and provide a channel that guides material away from the star into a jet of material.

The jets are thought to carry away much of the angular momentum of the rotating cloud, resulting in a star that is not rotating as fast as one would predict otherwise. Angular momentum is a quantity that must be conserved, so a large object (like a Bok globule) rotating slowly will turn into a small object (like a star) rotating very fast.

After this early phase of formation, what remains of the accretion disk are protoplanetary disks (proplyds). Observations of the Orion Nebula indicate that most low-mass stars generate these types of disks as part of their formation process.

These two images show the visible (left) and infrared (right) views of a pillar of gas about 3 light-years tall in the Carina Nebula. The infrared view reveals a host of stars (bright red-orange spots) forming within the pillar, which is opaque to optical light. Two jets from young protostars (HH901 and 902) are visible.

Hard to See

Protostars that glow with the energy of their gravitational contraction are hard to see in visible light, since they are deeply embedded in a cloud of dust that's opaque to visible radiation. However, other wavelengths, like far-infrared and radio, can pass through this dust more easily.

An artist's conception of a protoplanetary disk. When a cloud fragments into smaller pieces, each one eventually will form a star surrounded by a protoplanetary disk or proplyd.

Protoplanetary Disks (Proplyds)

As you read in "Star Birth: The Orion Nebula," stars seem to form in groups, with a few high-mass stars and a large number of low-mass stars. While the highest-mass stars form, live, and die quickly, the lowest-mass stars take their time at every step of the way. As a result, after a large cloud of gas begins to collapse and fragment, the high-mass stars are getting close to the end of their lifetimes (after about 10 million years) while the low-mass stars are not yet on the main sequence.

How Star Formation Relates to Planet Formation

One of the most fascinating discoveries in the past few decades is that the formation of stars seems to inevitably lead to the formation of planets. Giant molecular clouds consist mostly of molecular hydrogen, and at a critical density, a cloud will begin to collapse because of self-gravity. As the cloud contracts and becomes denser, smaller and smaller parts of the cloud fragment and collapse independently. If you follow just one of these stellar-mass fragments, its rotation will cause it to flatten as mass collects at its center. When the density and temperature of the star increase sufficiently to begin thermonuclear fusion, it will be a main sequence star and its stellar wind will begin to disrupt the remaining disk of material, which is known as the *protoplanetary disk* or *proplyd*.

This image of the Orion Nebula includes insets of a number of proplyds forming in the same region. These protoplanetary disks will either be disrupted by the energy of the Trapezium stars or form into planets orbiting a low-mass star.

A Glimpse into the Sun's Early Life

Regions like the Orion Nebula and the high- and low-mass star formation going on there give observers a view into what Earth's neighborhood in the Milky Way might have looked like 4 and a half billion years ago. The Sun must have formed as just one of thousands of smaller stars forming in the light of nearby high-mass stars. The high-mass stars that formed with the Sun are long gone, and the family of stars born with the Sun has been long since dissipated by the rotation of the disk of the galaxy. Earth's lonely place in the Milky Way now, with the nearest stars being many light-years away, is very different from the dense nursery of stars where the Sun originated.

The Atacama Large Millimeter Array

Protoplanetary disks will be one of the main targets of the Atacama Large Millimeter Array (ALMA), now operational in Chile. This radio interferometer will be able to observe protoplanetary disks with 10 times the resolution of the Hubble Space Telescope.

The Evolution of Low-Mass Stars

This image made by the Hubble Space Telescope shows Sirius A and Sirius B. Sirius A is the brightest star in the sky and a main sequence star that's more massive than the Sun. Sirius B is the faint companion in the lower-left corner and is a white dwarf star. These two stars are some of the closest in the universe at only 8.6 light-years away.

Like all stars, low-mass stars spend most of their lifetime on the main sequence. During this portion of their lifetime, low-mass stars fuse hydrogen into helium in their hot, dense cores and maintain an equilibrium between the inward pull of gravity and the radiation pressure generated by the fusion reactions. This balance is called *hydrostatic equilibrium.* Whenever these two forces get out of balance, the star changes—sometimes in drastic ways.

When a star like the Sun gets to the end of its main sequence lifetime, it first swells into a red giant and then slowly sheds its outer layers in a series of thermal spasms.

The Smallest of the Small

For the lowest-mass stars (with masses less than 0.4 M), convection will generally mix all of the available hydrogen down to the core, allowing these very-low-mass stars to convert essentially all of their hydrogen into helium. They end their lifetimes as spheres of helium.

Becoming a Red Giant and an AGB Star

When stars like the Sun exhaust the hydrogen in their core, they leave the main sequence. The core shrinks and heats up, igniting hydrogen fusion in a shell outside the core. This fusion outside of the core causes the star to swell and cool, turning it into a red giant and moving the star into the upper-right-hand quadrant of the HR diagram. At this point, the core—enriched with helium—contracts, and core helium fusion starts. In lower-mass stars, this happens suddenly in an event called a *helium flash.* The onset of helium fusion causes the core to expand and cool. The outer layers then contract and heat up, and the star moves onto the horizontal branch on the HR diagram.

While hydrogen fusion for a typical low-mass star has lasted for many billions of years, helium fusion lasts only a few hundred million years, converting the helium in the core into carbon and oxygen. At this point, the core contracts again, releasing heat, and helium fusion begins in a shell around the core. This helium-fusing shell causes the outer layers of the star to expand and cool a second time; the type of star that results from this is called *an Asymptotic Giant Branch (AGB)* star. AGB stars have a very compact core consisting of an inert mixture of carbon and oxygen, a helium-fusing shell, and then a dormant hydrogen-fusing shell.

The Sun will enter its AGB phase in about 10 billion years. At that time, its luminosity will increase enormously, and its outer edges will reach the orbit of Earth. When this happens, Mercury and Venus will most likely be absorbed into the Sun's interior.

In an interesting reversal of brightness, when observed in X-rays with the Chandra X-Ray Observatory, Sirius B is far brighter than its companion Sirius A. Sirius B has passed through its red giant stage and has a surface temperature of about 25,000 K, which is much hotter than Sirius A.

Helium Shell Flash and Thermal Pulses

An AGB star is now primed to lose its outer layers through a series of pulsations. As the star runs through its helium, the helium shell contracts, causing the outer hydrogen shell to contract and heat up enough to begin fusing helium again. This helium sifts down into the helium shell below it, causing helium fusion reactions to begin again. This burst of energy (called a *helium shell flash*) causes the hydrogen-fusing shell to expand and cool. This process repeats again and again in what are called *thermal pulses.* Each pulse can eject the outer layers of the star into the interstellar medium.

The Death of Low-Mass Stars

When the outer layers of a low-mass star have been expelled, what remains is a hot core known as a *white dwarf*. This dense object is hot but not very bright optically. Radiation from the white dwarf illuminates the layers of gas moving away, forming a *planetary nebula.*

At the end of their lives, low-mass stars lose their outer layers as planetary nebulae. The thermal pulses that rock the center of these stars, low on fuel, eventually cause the outer layers of the star to lift off with escape velocity, exposing the remaining carbon-oxygen core. In fact, stars like the Sun lose close to half of their mass when their planetary nebulae form.

Because low-mass stars are relatively common, you might expect interstellar space to be littered with these remnants. Indeed, there are thought to be tens of thousands in the Milky Way Galaxy alone. But the abundance of planetary nebulae is limited by another factor: they do not last very long. The material in a planetary nebula is illuminated by powerful ultraviolet radiation coming from the exposed white dwarf star. These layers move away from their host star with time and remain close enough to be illuminated for less than about 50,000 years. Because of their short lifetimes, planetary nebulae are some of the youngest objects in the galaxy.

The host star of planetary nebula NGC 6302 was once about five times the mass of the Sun. A dusty doughnut of material (seen as a dark band near the image center) hides the remaining white dwarf star that illuminates the gas, while complex filaments of shocks and gas are caught up in the star's stellar wind.

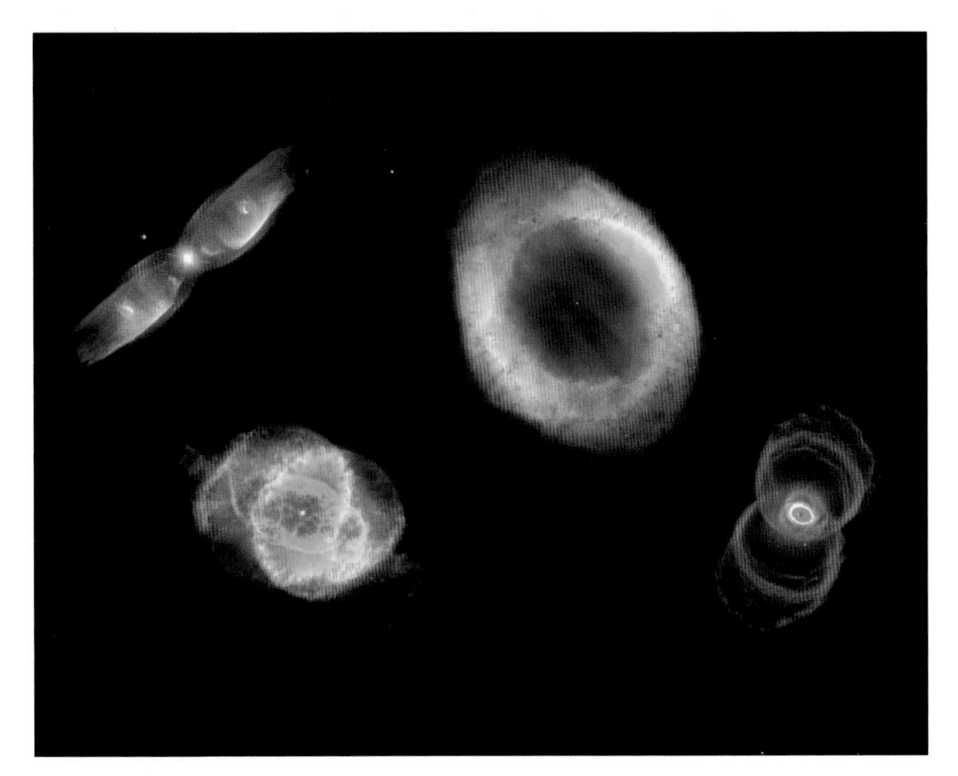

These four images of planetary nebulae from the Hubble Space Telescope show different ways in which the stellar environment can shape a nebula. The Ring Nebula (upper-right corner) has a relatively simple shape, while the Hourglass Nebula (lower-right corner) shows evidence of periodic losses of the outer layers over time.

Shapes of Planetary Nebulae

The huge variety in the shapes of planetary nebulae is the result of the environment around a given star before the nebula formed. Some nebulae form with no constriction on their expansion, making nearly spherical nebulae. In other environments, a star may have experienced low-velocity mass loss from its equatorial region. When the outer layers of that star are ejected, they encounter this slow-moving material, making the nebula take on an hourglass shape.

Properties of White Dwarfs

The stars that are left behind once a planetary nebula has been ejected are called *white dwarfs.* These stars have very high surface temperatures but are very small. White dwarfs are located on the HR diagram in the lower-left, meaning these stars are hot and not very bright. They are no longer supported by radiation pressure resulting from nuclear fusion, which has ended.

White dwarfs are strange in that they are very dense objects, with masses similar to the Sun and radii similar to Earth. This very high density is the result of gravitational forces. As a white dwarf cools, the electrons in atoms try to find lower-energy states. But if all the states are occupied, the electrons can't go to the ground state and thus still possess some energy. This phenomenon, referred to as *electron degeneracy pressure,* supports a white dwarf against gravitational collapse.

The Birth of High-Mass Stars

An artist's conception of the high-mass protostar IRAS 13481-6124, made after observations with the European Southern Observatory's Very Large Telescope Interferometer and the Spitzer Space Telescope indicated that it is surrounded by a rotating disk of gas and dust.

High-mass stars are rare and form from dark clouds much more quickly than their low-mass companions. As a result, while there are many examples of the slower process of low-mass star formation to observe, high-mass stars form so quickly that the process difficult to observe and to understand. To complicate matters, young high-mass stars are often concealed under layers of gas and dust.

Models of High-Mass Star Formation

There are two basic models of high-mass star formation: either they form like their low-mass counterparts through an accretion disk and outflow, or they coalesce from a number of lower-mass stars in a crowded region. There are a number of observations of young high-mass stars that have detected both disks and outflows.

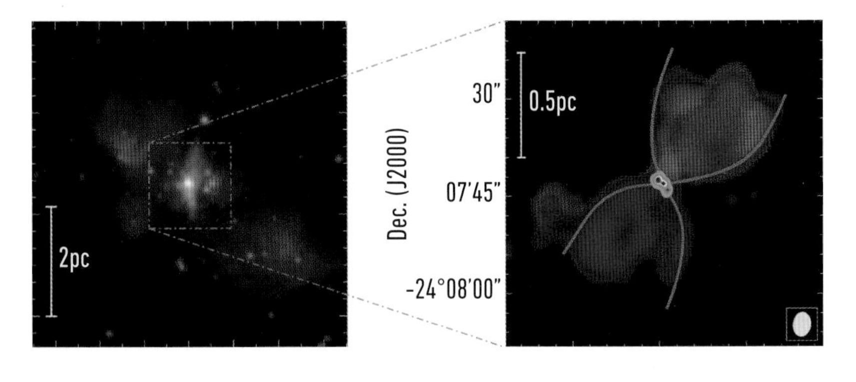

Observations of the galactic high-mass star-forming region G240.31+0.07 with the Spitzer Space Telescope (left) and the Submillimeter Array (right) indicate that in some cases, high-mass star formation looks like a scaled-up version of lower-mass star formation. In the right image, observations of the carbon monoxide (CO) molecule indicate that molecular material is flowing out in two oppositely directed lobes from this young, massive protostar.

The first model is simply a scaled-up version of low-mass star formation. The second model has the advantage of explaining the presence of very high-mass stars, which are difficult to form with the first model; very high-mass stars would blow themselves apart before they could form by the first model.

HII Regions

Because high-mass stars move on to the main sequence much more quickly than low-mass stars, the youngest star-forming regions contain a combination of high-mass stars that are fully formed and low-mass stars that are still surrounded by disks of molecular material.

High-mass stars light up the surrounding material, creating a region of ionized gas called an *HII* ("H-two") *region.* HII regions often become visible optically as the high-mass stars that have formed there push away the remaining dust and gas, revealing a family of low- and high-mass stars. But the youngest of these star-forming regions are still deeply embedded in the dark clouds from which they formed. As a result, they are not visible to optical telescopes and must be observed with longer-wavelength telescopes in the infrared and radio spectrums.

When these optically hidden regions of ionized gas are young, they are observed as very small HII regions that are particularly bright, called *ultracompact HII regions.*

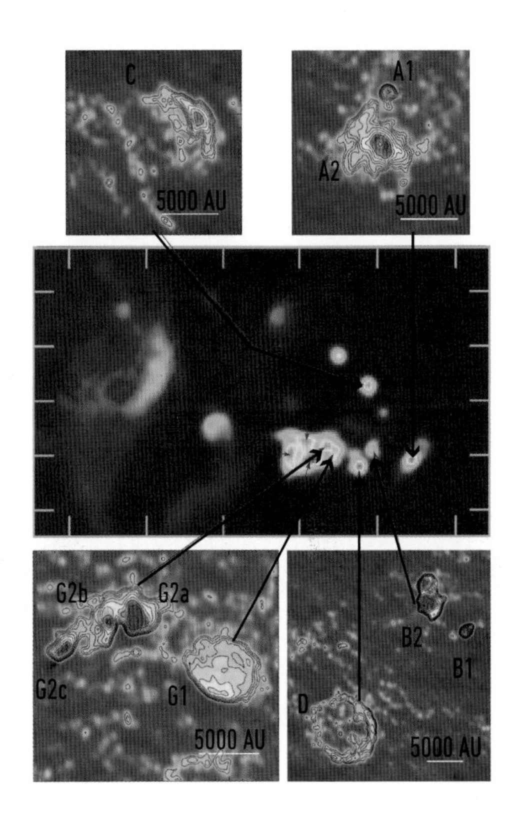

This Very Large Array image of the ultracompact HII regions in the galactic massive star-forming region W49A shows the location of ionized gas in a region that is unobservable in optical light. Each bright red-orange blob is the gas surrounding one or more young, massive stars.

How Radio Telescopes "Unveil" Star Formation

Astronomers use a variety of telescopes to explore the universe. Because the regions where stars form are often deeply embedded in clouds of gas and dust, optical telescopes are not able to see them well, unless the stars have broken out of their birth "cocoon" like the Orion Nebula (see "Star Birth: The Orion Nebula").

Submillimeter (radio) observations made with the APEX telescope at 870 micrometers (orange) highlight the cold, dark clouds in the Carina Nebula. Optical observations made with the Curtis Schmidt Telescope are shown in blue. Some of the most massive stars detected in the galaxy are in this star-forming region.

But what about other regions—clouds in which the stars have blown out in the direction away from Earth or stars that are still fully embedded in a cloud of gas and dust? Stars and protostars like these can be best observed at wavelengths longer than the optical with infrared and radio telescopes. You have seen some images of star formation as observed with infrared telescopes in the Carina Nebula. But what about radio observations? What does star formation look like to a radio telescope?

Advantage of Using Radio Telescopes

Radio telescopes are sensitive to a variety of emissions, from hot, ionized (10,000 K) gas to cold, dark clouds. The main advantage of radio observations is that they can reveal details that are hidden at other wavelengths. Electromagnetic radiation can be absorbed by material between the source and the observer. If the intervening material is gas and dust around the young star, optical photons are absorbed readily. However, infrared and radio wavelengths (which are longer) are not as strongly absorbed, and these photons make their way out of the cloud.

While some regions can be effectively observed at all wavelengths (like the Orion and Carina Nebulas), there are some regions that can only be observed at these longer wavelengths. Many of these regions lie in the plane of Milky Way Galaxy or toward the center of the galaxy, where gas and dust are most concentrated.

This view of the Galactic Center shows a 2×1-degree patch of the sky. For comparison, the Sun is about a half degree in diameter. This image shows a number of different wavelengths, including a 20-cm Very Large Array image (purple) showing hot gas and magnetic fields that aren't visible in an optical image.

Telescope Resolution

One challenge of observing at longer wavelengths is that the resolution of a telescope—its ability to see small details—is governed by the ratio of the wavelength of the light to the diameter of the telescope. Optical wavelengths are very small, so a relatively small telescope can see fine details. Radio waves are much larger, so in order to see details, the telescope diameter has to be correspondingly large.

Here is a simple example: the Hubble Space Telescope has a diameter of 2.4 meters, and typical observations with this telescope can see details as small as 0.05 arcseconds. That is very small when you consider that the full moon has a diameter of 1,800 arcseconds. In order to achieve the same resolution as the Hubble Space Telescope, the Very Large Array—a radio telescope that combines the signals from 27 separate dishes—must be much larger. In its largest configuration, it is about 35 km across.

The Very Large Array is located on the Plains of St. Augustin in New Mexico. This image shows the array in its smallest configuration. In its largest configuration, the most distant telescopes are 35 km from one another.

The Evolution of High-Mass Stars

This schematic of the interior structure of a high-mass star through various stages of its evolution, from formation (far left), through to supernova, and to formation of a black hole (foreground). Note that as a high-mass star evolves, it develops a nested structure of fusing shells.

As stars become more massive, they are able to fuse heavier elements in their core. When it comes to stars that have more than about 4 solar masses, they are able to fuse hydrogen into helium, carbon, and oxygen. However, unlike low-mass stars, high-mass stars then continue to have sufficient temperature and pressure to fuse carbon into heavier elements. This is because the strong gravitational field of a high-mass star creates conditions that allow carbon atoms (which naturally repel one another because of their positive charge) to fuse. High-mass stars fuse carbon atoms into more oxygen, neon, sodium, and isotopes of magnesium.

The highest-mass stars (with initial masses greater than about 8 solar masses) can fuse even heavier elements. At temperatures of 1.5 billion Kelvin, the cores of stars can fuse oxygen into mostly silicon, with smaller amounts of magnesium, phosphorus, and sulfur. At nearly 3 billion Kelvin, stellar cores can fuse silicon into atoms like sulfur, iron, and nickel. A barrage of neutrons is released in many of these fusion reactions, and neutrons that are captured by the nuclei of some of these atoms produce other elements.

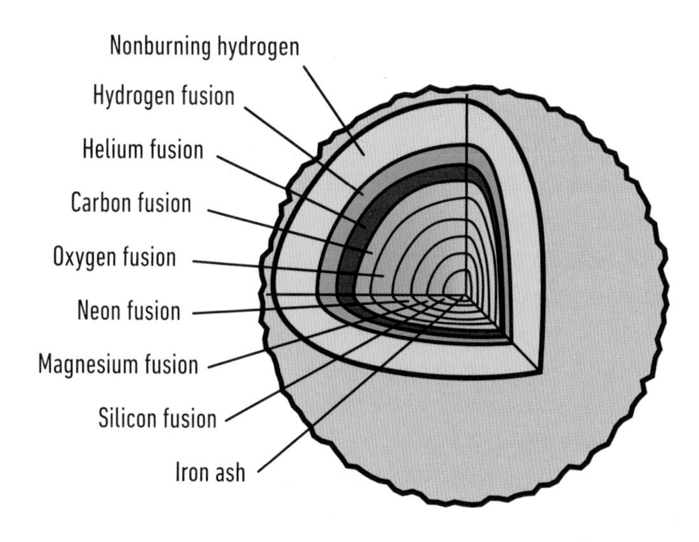

This cutaway view of the interior of a high-mass star shows how as each successive element is generated, it sinks to the center of the star. Near the end of the lifetime of a high-mass star, it has many nested layers of fusion going on simultaneously.

This Herschel image of the star Betelgeuse, a red supergiant in Orion, and its immediate vicinity show both the star and material that it has ejected into the interstellar medium around it.

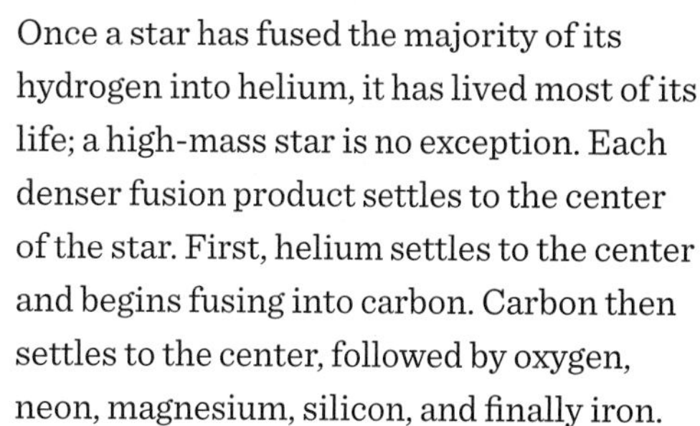

Once a star has fused the majority of its hydrogen into helium, it has lived most of its life; a high-mass star is no exception. Each denser fusion product settles to the center of the star. First, helium settles to the center and begins fusing into carbon. Carbon then settles to the center, followed by oxygen, neon, magnesium, silicon, and finally iron.

Red Supergiants

As these successive layers of nuclear fusion occur, the outer layers of a high-mass star are swelling to a tremendous size. All of the energy generated by the fusion reactions causes the outer layers to swell up and cool, creating what's called a *red supergiant*. These stars have a diameter as large as the orbit of Jupiter and burn through their remaining fuel quickly.

Running Out of Time

Each successive fusion product is made for a shorter period of time. For example, star with a mass of about 20 solar masses will fuse hydrogen to helium for about 10 million years, helium to carbon for 1 million years, carbon to oxygen for only about 1,000 years, and other heavier elements for less than a year. At the very tail end of its life, a high-mass star only fuses iron for about a day. Once the core of a high-mass star has generated iron, it is out of time. Fusing iron into heavier elements requires an input of energy, and a high-mass star has reached the end of its short lifetime.

The Death of High-Mass Stars

A star with a mass of 8 to 10 solar masses or more ends up in a suddenly unstable situation. As you learned in the previous section, the fusion of iron into heavier elements requires energy the star no longer has. At this point, the hydrostatic equilibrium of such a star—the balance between gravitational forces inward and radiation pressure outward—is suddenly unbalanced. Without nuclear fusion, there is no outward force to resist gravity, and the core contracts.

Core Collapse

The collapse of a high-mass star's core is almost instantaneous. Core collapse, during which the core density increases to that of an atomic nucleus, takes less than a second. This hot, dense material, with temperatures of billions of degrees, emits gamma rays that are energetic enough to begin to break down iron into helium and other elements, undoing much of the long work of fusion. When the core achieves a high-enough density, it is possible for electrons (e) to combine with free protons (p) to create neutrons (n) and neutrinos (ν). The particle reaction looks like this:

$$p + e^- \rightarrow n + \nu$$

The neutrinos (see "Neutrinos") generated carry away energy, and the core contracts even more until the entire mass of the star's core has contracted to a region about the size of Manhattan. The unsupported overlying layers of material rush in and come up against a suddenly rigid core unable to compress any further, having achieved the density of the nucleus of an atom. At this point, a core bounce occurs, which sends pressure waves outward, ripping the outer layers of the star to shreds and releasing the neutrinos. The interaction between the infalling layers and the outgoing pressure wave takes a few hours, but eventually the star become visible as a *core-collapse (Type II) supernova* (see "Supernova 1987A").

The gravitational energy generated by the collapse of a Type II supernova then powers the fusion of other, heavier elements on the periodic table. In fact, a supernova is one of the only possible sources of energy for the creation of the heaviest elements in the periodic table. These types of supernovae create most of the elements between oxygen and iron and about half of the elements beyond iron in the periodic table, such as silver and gold. The presence of these elements on Earth indicates the Sun formed from the debris of such an explosion.

What Is Left Behind?

Most of the material in a high-mass star created by the supernova is dispersed into the space between stars—the interstellar medium—where it is available to be pulled together by gravity to form another generation of stars. This material is known as a *supernova remnant*. The other remainder is the stellar core; in the case of some high-mass stars, that core is a neutron star (see "Neutron Stars and Supernova Remnants").

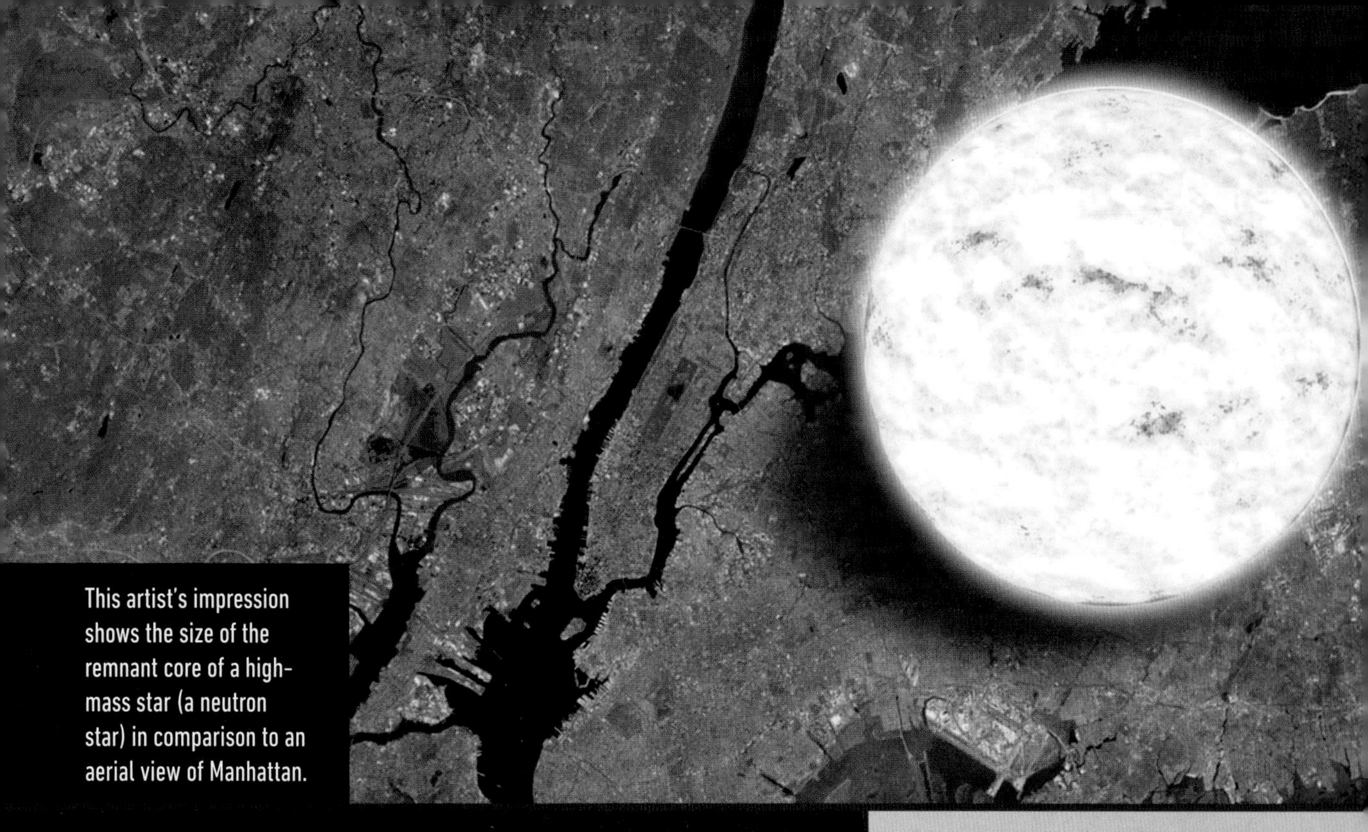

This artist's impression shows the size of the remnant core of a high-mass star (a neutron star) in comparison to an aerial view of Manhattan.

Three-dimensional computer simulations of a core-collapse (Type II) supernova give astronomers insights into how neutrinos and shock waves generate the elements in the periodic table. Results of simulations can then be compared with observations of young supernovae. The colors in this image represent different densities.

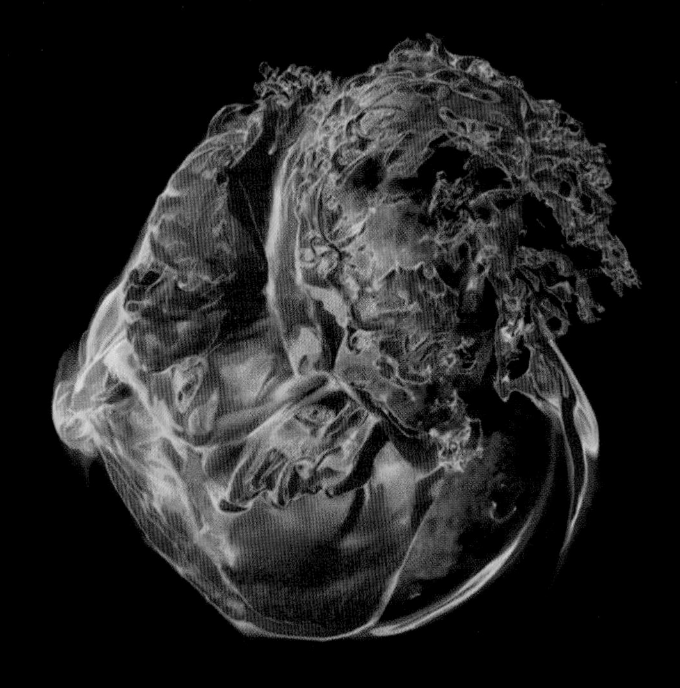

The Evolution of Type II Supernovae

Much of the understanding of the evolution of a supernova as it occurs in real time comes from powerful computer simulations that can follow the evolution of radiation, matter, and shocks in this extreme environment. These models make predictions about what elements should be observed as a supernova evolves into a supernova remnant. Astronomers also anxiously await the opportunity to record the evolution of a nearby supernova so they can compare the simulations to the real thing.

Supernova 1987A

As you learned previously, high-mass stars end their lives violently and catastrophically as supernovae. When they occur, supernovae are so energetic that they can momentarily outshine the rest of the stars in the host galaxy that they are in. The most recent optically visible supernova to occur near the Milky Way is Supernova 1987A, so named because it was the first supernova observed in 1987.

The star that blew up as Supernova 1987A was located about 168,000 light-years (51,000 parsecs) from Earth near the Tarantula Nebula in the Large Magellanic Cloud (LMC), a collection of stars, gas, and dust that orbits the Milky Way Galaxy. The supernova had an apparent magnitude of 3 at its brightest (see "The Brightness and Color of Stars"), making it easily visible without a telescope from the Southern Hemisphere. Supernova 1987A provides a nearby laboratory to trace the evolution of a Type II supernova.

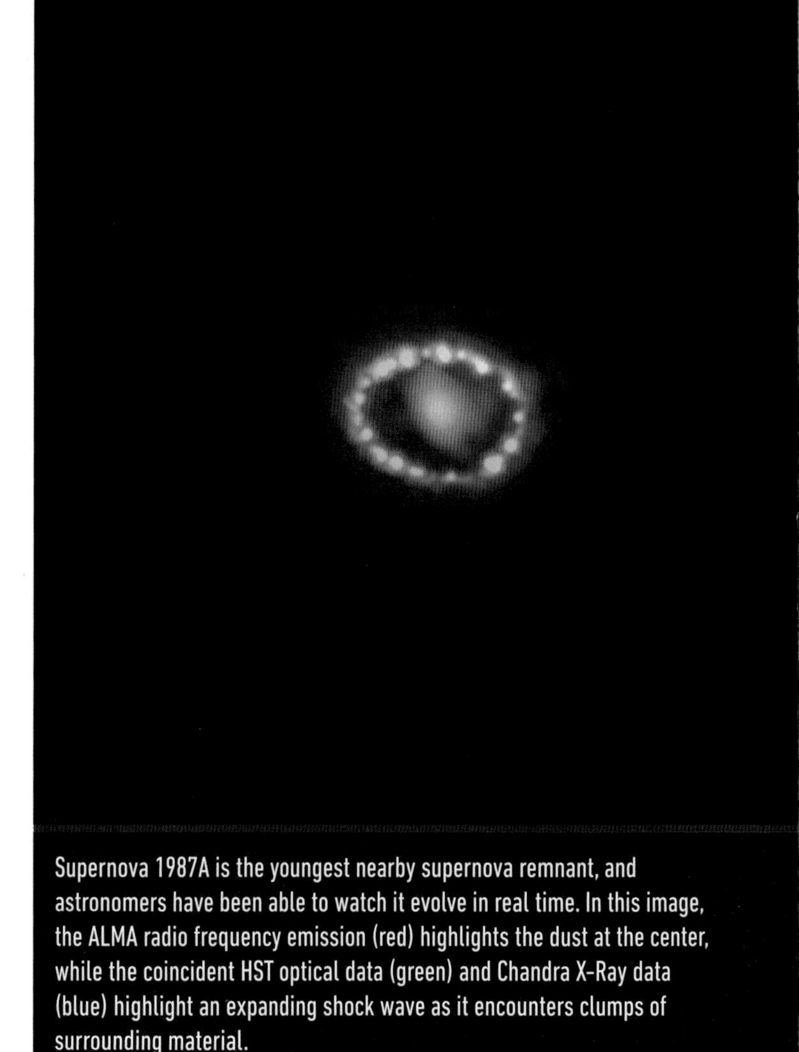

Supernova 1987A is the youngest nearby supernova remnant, and astronomers have been able to watch it evolve in real time. In this image, the ALMA radio frequency emission (red) highlights the dust at the center, while the coincident HST optical data (green) and Chandra X-Ray data (blue) highlight an expanding shock wave as it encounters clumps of surrounding material.

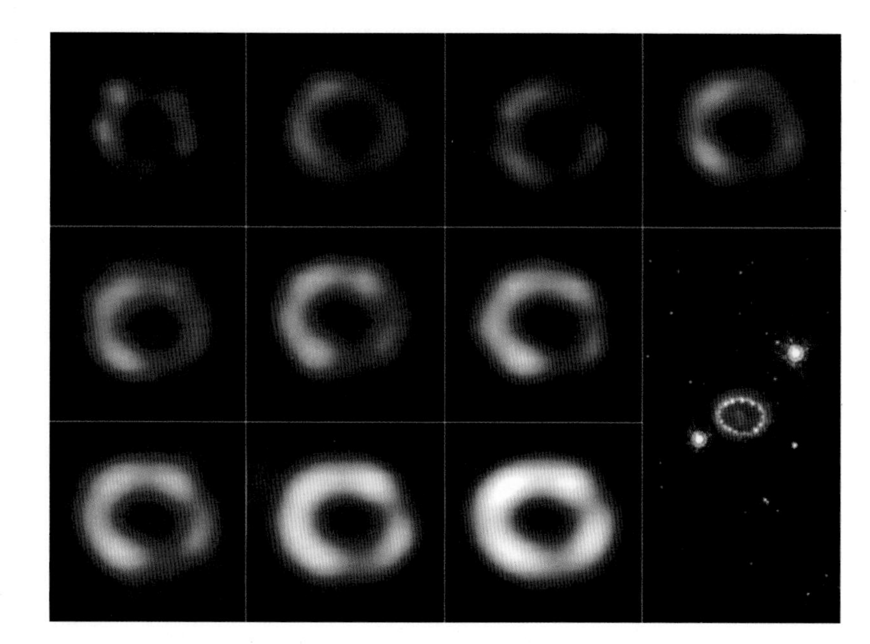

This set of images show the evolution of the X-Ray emission from Supernova 1987A as tracked by the Chandra X-Ray Observatory from 2000 to 2005. The X-Ray image is superimposed on the Hubble Space Telescope image of Supernova 1987A in the lower-right corner.

Classifications

When Supernova 1987A occurred, the world focused telescopes across the electromagnetic spectrum to record its changing brightness (called a light curve) and to classify it. Its spectra contained hydrogen absorption lines, which classified it as a Type II supernova. The evolution of its light curve, which leveled out, further classified it as a Type II-P supernova, which is thought to be the result of core collapse. The progenitor star had run out of nuclear fuel and had a core of iron. Further fusion of elements was not possible without an input of energy, so gravity won, collapsing the core of the star and causing a supernova explosion.

However, one open question in the story of Supernova 1987A is the location of the remnant of the blue supergiant star, the giant phase of the most massive stars (see "The Evolution of High-Mass Stars"), that exploded. According to supernova models, a Type II supernova should leave behind a neutron star. Whether the collapse continued to form a black hole or the remnant neutron star is still enshrouded in dust is not yet known.

The Evolution of Star SK-69

The star that exploded as Supernova 1987A had been already losing material in a stellar wind. In the current model, the star (SK-69) formed about 10 million years ago and lost its outer layers to a slow stellar wind about 1 million years ago. Before it exploded, the star blew a cavity in the gas around it. SK-69 then experienced a Type II supernova.

The inner edge of the cavity was quickly illuminated by the ultraviolet radiation from the supernova explosion and seen by the Hubble Space Telescope. The shock wave followed, and when it overtook the cavity, a series of hot spots lit up in optical and X-ray radiation. These interactions happened well after the initial explosion in 1987. For example, in 2001, supernova ejecta first encountered the inner ring and caused Supernova 1987A to increase in brightness by a factor of 3.

Neutron Stars and Supernova Remnants

Type II supernovae are some of the most energetic events in the universe and are thought to be the end of the most massive stars that are between 8 and 50 times the mass of the Sun. If the progenitor star has a mass of less than about 20 solar masses, its core collapse will leave behind a neutron star.

Neutron Stars

The collapse of the core of a massive star converts its substance from a complex mix of heavy elements into a ball of neutrons. High-energy gamma rays released in the sudden compression and heating of the core break apart the heavy nuclei into helium and other elements. This breakup continues, and even protons are transformed. In the hot, dense environment of the core, protons and electrons come together and generate neutrons, releasing a wave of neutrinos. The neutrino burst detected from Supernova 1987A confirmed that neutrinos are indeed produced in a supernova explosion.

The conclusion of this process leaves a stellar core that consists of only neutrons compressed to the density of an atomic nucleus; this is a neutron star. Nuclear fusion has ended, so the only thing supporting the neutron star from further collapse is the pressure that keeps neutrons from occupying the same state. Like the electrons in a white dwarf, the neutrons in a neutron star prop up the star. Neutron star material is unbelievably dense; an entire solar mass worth of neutron star material would only be a few kilometers across.

There are several thousand neutron stars in the Milky Way Galaxy alone, most of which were discovered as emitters of radio frequency pulses (see "Pulsars").

Supernova Remnants

Supernova remnants are the rapidly expanding remains of a stellar explosion. Because these are physical remains, they plow into other material and serve as a way to probe the environment around the host star. The shock wave can also impinge on nearby dark clouds, triggering them into collapse and further waves of star formation. In this way, the death of a massive star is intimately connected to the lives of future generations of stars, both by seeding interstellar space with a mix of heavy elements and by triggering other clouds of gas into gravitational collapse.

Supernova remnants are located at a variety of distances from the Sun, and the closest of these remnants can span huge angular sizes on the sky. For example, the Gum Nebula covers 40 degrees—almost 1/4 of the sky—but is very faint. Many supernova remnants too faint to see optically are bright at radio frequencies as the expanding shell interacts with the interstellar medium.

The Cassiopeia A supernova remnant is the expanding shell of material from a Type II supernova that occurred about 300 years ago. This false-color image shows the remnant as it appears to the Spitzer Space Telescope (red, infrared), the Hubble Space Telescope (orange, optical), and the Chandra X-Ray Observatory (blue and green). The region around what is left of the original star (either a neutron star or black hole) is visible as the small blue-green dot off center in the image.

The Cassiopeia A supernova remnant is rather faint optically. This Hubble Space Telescope image shows emission lines from several elements in the remnant, sulfur (red) and oxygen (blue).

Pulsars

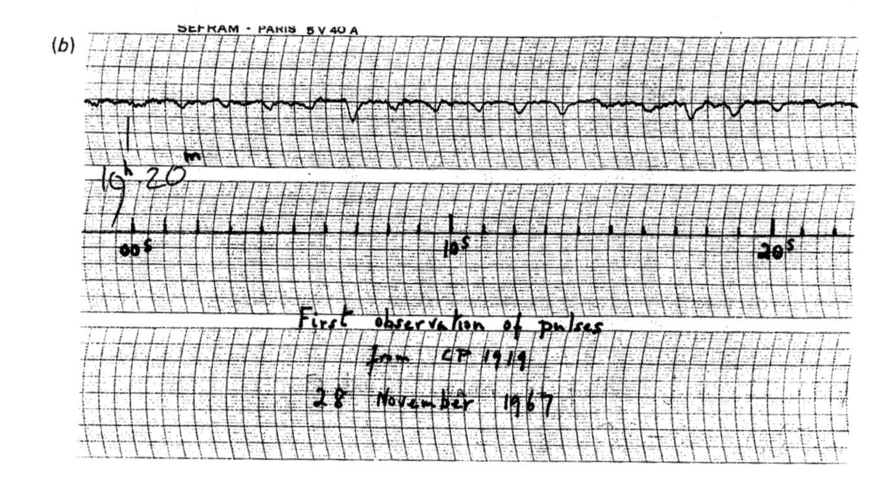

This is a copy of the chart recorder from the original radio frequency observations of CP1919 made by Jocelyn Bell and Anthony Hewish in 1967. The types of radio pulses first observed from this source were soon observed coming from other radio sources.

The story of the discovery of pulsars is one of the most intriguing in the history of astronomy because it highlights the important role of serendipity. Jocelyn Bell, a Cambridge graduate student, was making observations with her advisor Anthony Hewish using a newly commissioned radio telescope; they were looking for variations in the brightness of radio sources caused by the movements of gas between the distant source and the radio telescope.

While looking for these slight variations, Bell detected regular pulses in the radio signal from one source, CP1919. The pulses were so regular that at first she suspected the pulses were terrestrial in origin. Once it was established that the radio signals were coming from space (and not a local transmitter), some scientists even considered whether pulsars might be the result of other intelligence in the universe using some sort of beacon.

While a number of physical explanations for the pulses were considered, the discovery of the pulsar at the heart of the Crab Nebula was a big clue that pulsars were related to the death of high-mass stars. This finding also indicated that pulsars were some sort of relative of neutron stars, a possibility that had been suggested theoretically decades earlier.

Pulsars are neutron stars with a beam that sweeps across Earth as the star rotates. Pulsars have a range of rotational periods, from milliseconds to seconds.

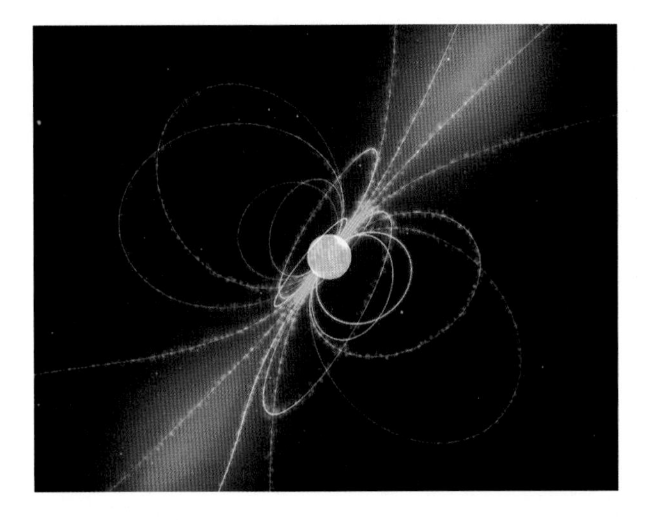

This artist's conception shows the basic structure of a pulsar. The neutron star is colored orange, and the magnetic field lines are indicated in blue. The magnetic field lines are the source of electromagnetic radiation seen in the pulses.

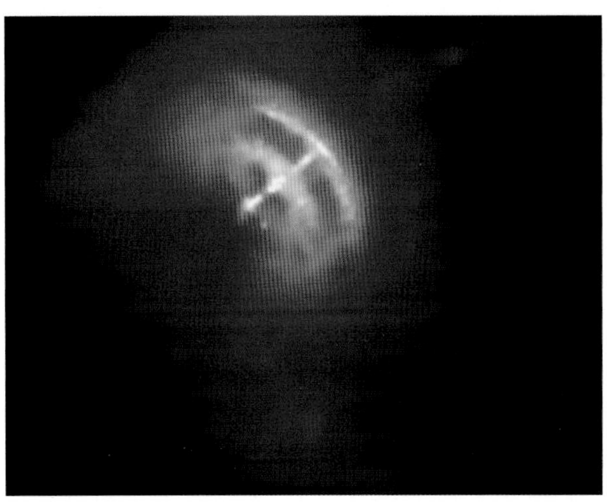

This image shows a jet of material shooting from the poles of the Vela pulsar. The pulsar is the bright spot near the center of the image, while the surrounding hot gas is shown in yellow and orange.

So what causes pulsars? As a high-mass star collapses under gravitational forces, two things happen: it gets smaller in radius and it spins faster. The first effect is the simple result of gravitational forces. The second is due to the conservation of angular momentum—as an object gets smaller, it must spin faster to conserve this quantity.

When a neutron star is formed, the host star is threaded with magnetic fields like any star. The collapsing material pulls magnetic fields in with it, amplifying the strength of the magnetic field. The result is a strong magnetic field that rotates rapidly with the newly formed neutron star, which explains both the high frequency and the strength of pulsed radio signals observed to come from pulsars. So in the case of the radio pulses that were first detected in 1967, they were the result of a beam of electromagnetic radiation from CP1919 sweeping across our field of view.

Pulsars have been compared to a lighthouse, with the pulsar "beacon" lighting up every time the beam of radiation points toward Earth. Because pulsars are thought to have two oppositely directed beams, people see a rotating neutron star as a pulsar as long as one of its beams sweeps across the field of view. Pulsars may not be the beacons of an alien civilization, but they are strange, energetic objects nevertheless.

Magnetars

Magnetars are a subclass of neutron stars with particularly strong magnetic fields. As you learned previously, when a high-mass star ends its life in a Type II supernova, its central regions compress to a tiny diameter only a few tens of a kilometers across. However, what distinguishes magnetars from other neutron stars is that some of the heat and rotational energy available during the core collapse is converted into magnetic field energy, making their magnetic fields stronger (up to 1,000 times that of a typical neutron star) and their rotation slower. So while many neutron stars visible as pulsars rotate many times per second (so-called *millisecond pulsars*), magnetars are observed to rotate as slow as once every 10 seconds.

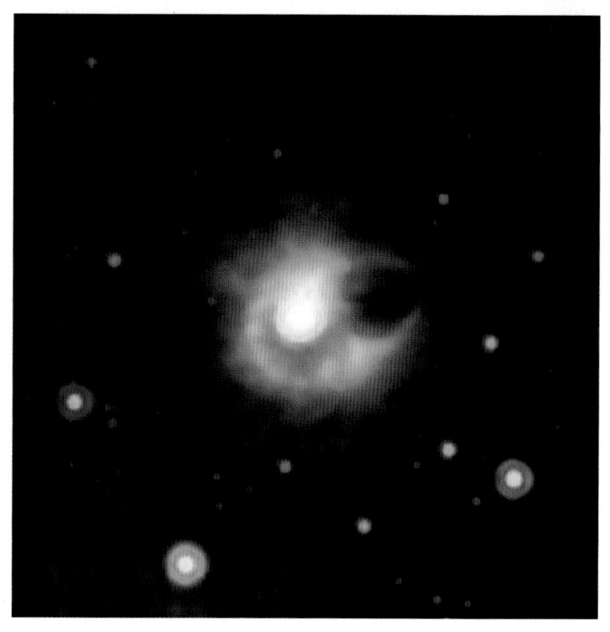

The magnetar SGR 1900+14 is located near the center of this infrared image, which highlights the expanding ring of material around it. The bright infrared source at the center of the image is not the magnetar itself, but is thought to be a cluster of young stars.

SGRs and AXPs

It is thought that the strong magnetic fields are not long-lived and decay naturally over a period of about 10,000 years. As this magnetic field decay happens, magnetars give off high-energy electromagnetic radiation, including gamma rays and X-rays. Therefore, magnetars are widely thought to be the explanation for several unusual high-energy astronomical sources, including soft gamma repeaters (SGRs) and anomalous X-ray pulsars (AXPs).

SGRs are astronomical objects that emit irregular bursts of both gamma rays and X-rays. They are called "soft" gamma repeaters, because the energy of the gamma rays is lower than other gamma ray sources. To an astronomer, hard X-rays have higher energies than soft X-rays. AXPs are astronomical objects with periodic variations in X-ray intensity.

This image shows an artist's conception of a surface outburst on a magnetar. Such outbursts are visible as variable X-ray emission.

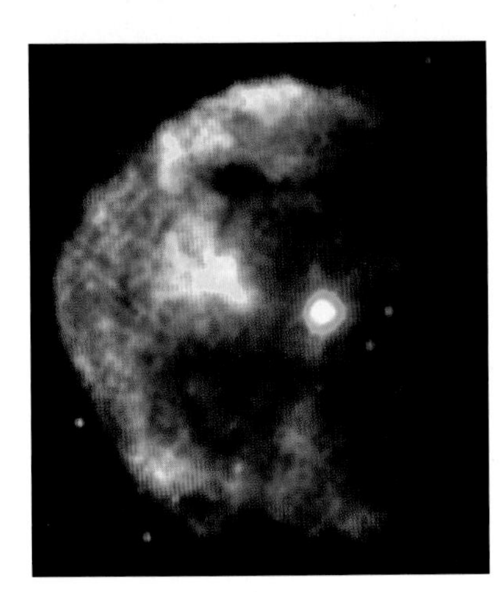

This false-color image shows magnetar 1E 2259+586 as the blue-white object in this X-ray image of the CTB 109 supernova remnant. The low-, medium-, and high-energy X-rays are in red, green, and blue, with the highest energies centered on the magnetar. Because they are a subclass of neutron stars (one possible end state of a high-mass star), some magnetars are observed in association with supernova remnants.

Studying Magnetars

The study of magnetars is new, and discoveries are frequently announced. As recently as the summer of 2013, astronomers announced the detection of a magnetar suddenly slowing in its rotation. In the middle of 2012, the source 1E 2259+586 was observed to slow abruptly. While sudden increases in the spin rate had been observed, a sudden decrease was a new and unexplained phenomenon.

By observing X-ray pulses, astronomers can track the rotation rate of magnetars carefully.

A Solid Surface?

So-called "glitches"—when the rotation rate of a neutron star suddenly increases—are thought to be the result of some of the strange properties of the surface.

Neutron stars are thought to have a solid "crust" of atomic nuclei through which electrons can move freely. Magnetic fields that are threaded through the surface drain the crust of energy, causing it to slow more than the more-fluid interior of the star. The crust then fractures, allowing astronomers to see a burst of X-rays as the surface crust begins to rotate as fast as the interior again.

Neutrinos

Cerenkov radiation, which indicates the presence of neutrinos, is visible as a blue glow in water surrounding a nuclear reactor.

Neutrinos are electrically neutral subatomic particles generated in abundance by the fusion reactions going on in the cores of main-sequence stars or by high-mass stars when they undergo core collapse during a Type II supernova event. Because they are electrically neutral and have very low mass, neutrinos are not affected by electromagnetic forces and have insignificant gravitational interactions. As a result, neutrinos pass through all material (including humans and Earth) with great ease, almost never being absorbed.

Because they interact so weakly with other matter in the universe, they are excellent probes into regions that can't be "observed" in any other way. For example, while photons of light move extremely slowly through the Sun's charged, dense interior, neutrinos stream out directly. Neutrinos come in three types: electron, tau, and muon neutrinos.

Solar Neutrinos

Neutrinos are produced in abundance in the core of the Sun as the result of fusion reactions. Solar neutrinos are created by reactions in the proton-proton chain, in which two protons (p) come together to create deuterium (^2H), a positron (e^+), an electron neutrino ($\bar{v}_e$) and a gamma ray photon. The nuclear reaction is the following:

$$p + p \rightarrow {}^2H + e^+ + \bar{v}_e + \text{gamma}$$

Deuterium is an isotope of hydrogen (^2H) that has the same number of protons as hydrogen (^1H) but one additional neutron. Because the proton and the neutron have similar mass, water molecules that have deuterium in them instead of hydrogen are called *heavy water*. In effect, the fusion reaction changes one of the protons into a neutron and generates a positron (the positively charged antiparticle of the electron) and an electron neutrino in the process.

How to Catch a Neutrino

Astronomers and physicists have devised cunning ways to catch the "greased pig" of the subatomic world. Neutrinos interact with other matter via the weak nuclear force, so two main types of detectors are used, both of which require the use of large volumes of liquid—either water or chlorine (in the form of carbon tetrachloride)—in order to detect these rare interactions.

Water detectors are surrounded by arrays of light-sensitive electronics (known as photomultiplier tubes). These detectors are sensitive to the flash of *Cerenkov radiation* that results when an inbound neutrino creates an electron or a muon—a particle with a negative charge but much more mass than an electron—in the water. This high-velocity electron causes the water molecules to radiate. The Super Kamiokande in Japan and the Sudbury Neutrino Observatory (SNO) in Canada are two water detectors that use this effect to detect neutrinos.

In *chemical detectors,* the large volume is filled with a fluidlike carbon tetrachloride, and the tank is periodically checked for the presence of argon, which is created from the chlorine in rare nuclear reactions with neutrinos. Most of the neutrinos detected on Earth in neutrino detectors come from the Sun.

Resolving the Solar Neutrino Problem

The nuclear fusion reactions taking place in the core of the Sun should produce neutrinos at a rate consistent with measurements of the Sun's luminosity. For many decades, however, the number of neutrinos detected from the Sun was about $1/3$ of the number predicted by nuclear fusion models. This was because the early measurements of neutrino flux from the Sun were made by chlorine detectors, which could only detect electron neutrinos.

In 2001, the SNO used a heavy water detector capable of detecting all three types of neutrinos (see "Fundamental Particles"). What it found was that about 35 percent of the neutrinos from the Sun are electron neutrinos, with the remaining 65 percent being tau and muon neutrinos. The current theory is that neutrinos are able to oscillate between the three types in their journey from the Sun to Earth.

Located 2 km deep in Earth in an old nickel mine, the Sudbury Neutrino Observatory (SNO) is a 12-m diameter sphere filled with heavy water. Like Super Kamiokande, the chamber is surrounded by detectors that can map Cerenkov radiation. This detector was instrumental in resolving the solar neutrino problem.

Stellar Mass Black Holes

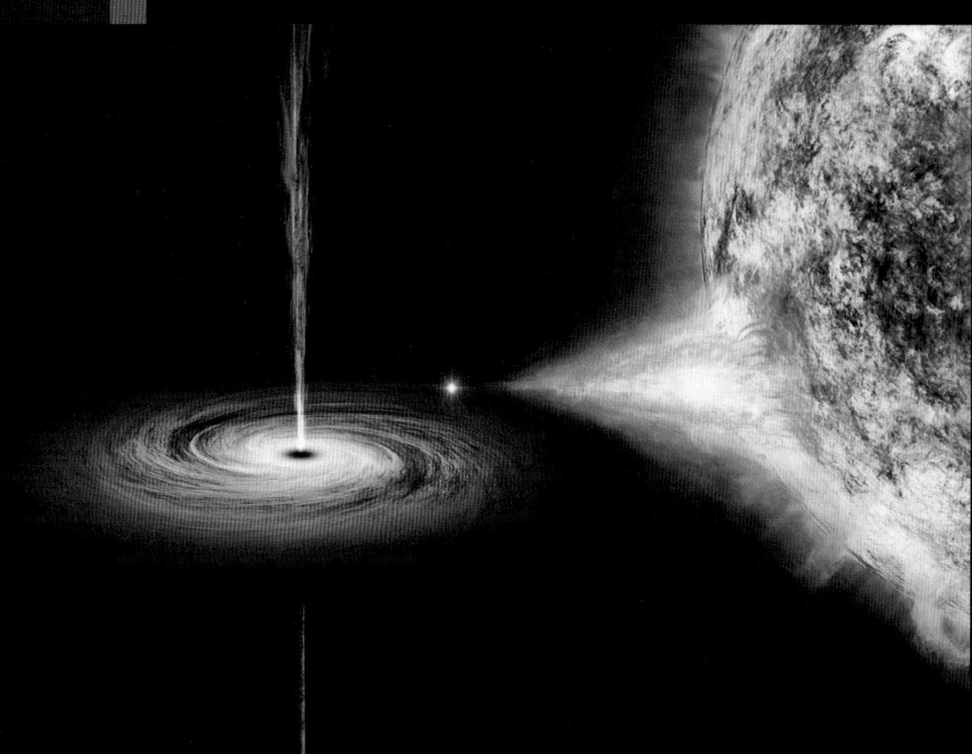

This image shows an artist's conception of the stellar mass black hole Cygnus X-1. The black hole, located in the constellation Cygnus, has a companion (the blue star to the right); the gravity of the black hole pulls material from the companion into an accretion disk (seen in red and orange).

High-mass stars have two possible end states. If the mass of the initial star is greater than about 20 solar masses, when the core collapse occurs, not even the pressure from neutrons will be able to support it. The collapse continues, and the core of the star is transformed into what's called a *stellar mass black hole.*

There are three ways that a stellar mass black hole can form:

- A neutron star in a binary system can accrete enough material to push it over the mass limit to form a black hole.
- Two stars in a binary system can merge to form a black hole.
- The core collapse of a star with a mass above 20 solar masses can create a black hole.

The strange thing about black holes is that they can't be observed directly—they can only be observed secondhand through their gravitational effects and their effect on nearby matter. This is because unlike a neutron star, a black hole itself emits no radiation. It is "black" in the sense that no electromagnetic radiation of any frequency can get away from it and its mass is concentrated in such a small volume that nothing—not even light—can escape.

Because of their gravitational field, black holes exert forces on nearby material and are surrounded by an accretion disk. Astronomers are able to detect X-rays from the hot, inner, white shaded region of the accretion disk.

Black Hole Misconceptions

One common misconception about black holes of all sizes is that they are like giant vacuum cleaners, sucking in everything around them. This is true only near a black hole's event horizon. However, if you are far enough away from that, the only effect you will feel is its gravity, and the gravitational effects are no different than the gravity of any other type of matter.

For example, if the Sun were replaced with a black hole of the same mass, all of the planets in the solar system would orbit in exactly the same way. The solar system would be a much colder place without the energy output of the Sun, but the planets would continue in their present orbits.

However, despite being unobservable directly, black holes still have mass. Because of this, their gravitational effect on nearby objects can be detected, meaning their mass can be calculated from the orbits of nearby stars.

Black holes of a given mass are associated with a size, though it is not the size of a physical object. How is that size calculated? Every black hole has an event horizon. Imagine a sphere centered on the black hole—this defines a region around a black hole from which light can't escape. The radius of the sphere is called the *Schwarzschild radius*, first calculated almost 100 years ago by Karl Schwarzschild. A good rule of thumb is that the Schwarschild radius of a star is equal to 3 km times the mass of the star in solar masses. Thus, a 5 solar mass black hole will have a Schwarzschild radius of 5 × 3 km or 15 km.

Binary Stars

Because low-mass stars are more numerous than high-mass stars, the most common type of star in the Milky Way is a solitary red dwarf star. However, while $^2/_3$ of the galaxy's stars are solitary, the remaining $^1/_3$ are found in binary systems. In a binary star system, two stars orbit a common center of mass. The more massive of the two stars is the *primary,* and the less massive is the *secondary*.

Because so many stars are found in binary pairs, these stars provide useful ways to probe a number of stellar properties, including mass and composition.

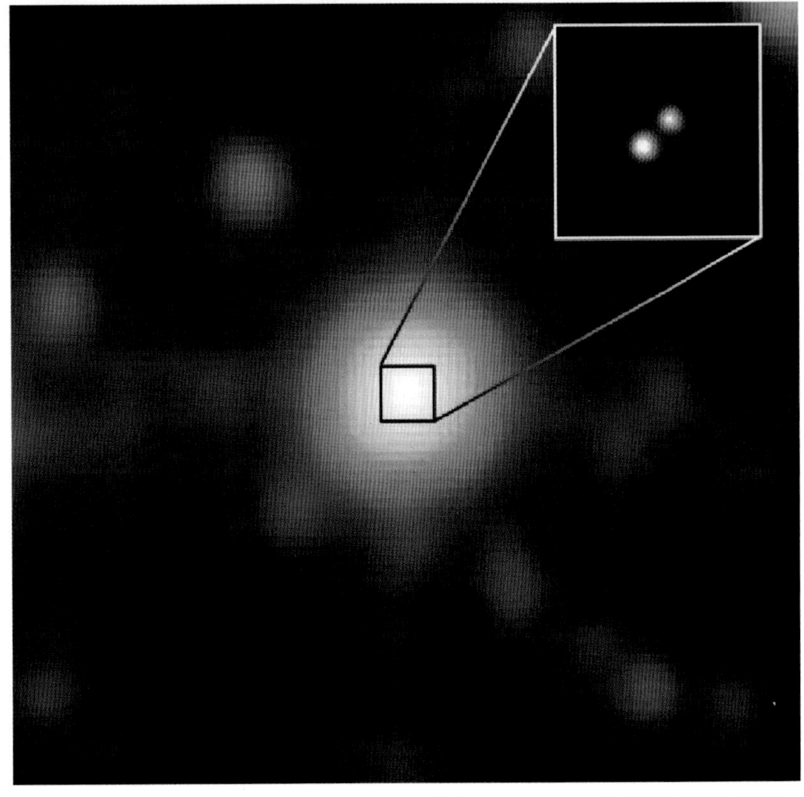

The third-closest star system to the Sun—at 6.5 light-years away—is WISE J1049-5319, recently revealed by the Gemini South telescope to be a visual binary star system of two brown dwarf stars.

Types of Binary Stars

Visual binaries are binary star systems where the separation between the two stars is great enough that you can view them in a telescope. These are the most difficult to spot, because the distances to stars are so great that the distance between a binary pair becomes (as an angular separation) very small. Also, the stars in a visual binary may have very different brightness, making it difficult to observe both (like Sirius A and B).

Binary systems are more commonly detected by their spectra; these are known as *spectroscopic binaries.* As discussed in "Stars and Spectroscopy," every star has a distinct spectrum determined by its mass and therefore its temperature. When the spectra of some stars are observed, they are seen to have a combination of two spectra that move relative to one another, which is due to the stars orbiting a common center of mass. This motion causes a Doppler shift in the spectral lines of the two stars, revealing a binary star system.

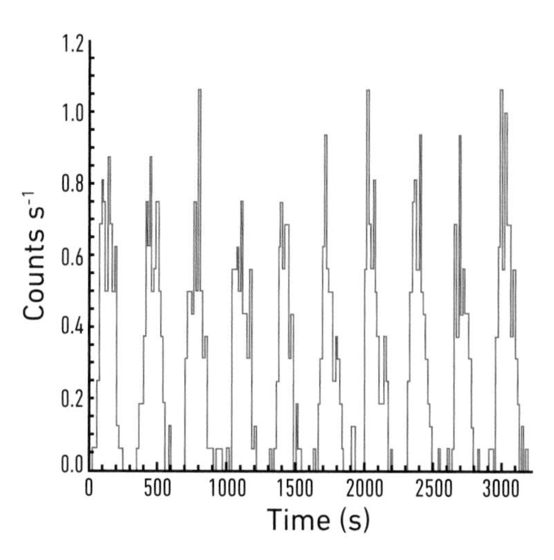

Observations of the system RX J0806.3+1527 indicate that its X-ray emission varies periodically, brightening every 321.5 seconds. Astronomers have proposed that this emission comes from a binary pair of white dwarf stars orbiting one another with this period. According to this model, the two stars orbit closer to one another than Earth and the Moon.

Spectroscopic binaries provide a valuable probe of the masses and composition of the two stars.

Eclipsing binaries may also appear to be a single star in a telescope. However, careful monitoring of the brightness of the star will reveal the star's brightness varies in a regular way, as first the primary and then the secondary star passes in the front of the other. This type of light curve can be used to determine some characteristics of the binary stars, including period and radius. A similar technique is used to detect planets orbiting other stars.

Finally, there are *astrometric binaries*. In these systems, a star seems to orbit empty space when, in fact, the star is orbiting another star. The other star may not be visible for a number of reasons. For example, it may be very faint, or it may not emit electromagnetic radiation at the same wavelengths as the first star.

Determining the Mass of Binary Stars

The distance between binary stars and the time it takes for them to orbit one another (the period) can help you determine the total mass of the binary pair using Kepler's Third Law:

$$(m_1 + m_2)P^2 = R^3$$

In this simple equation, m1 and m2 are the masses of the two stars (in solar masses), P is the period of orbit (in years), and R is the separation between the binary pair (in AU). If you have the spectral type of one of the stars, you can then determine the mass of the second star from the combined mass.

PART 4

The Milky Way Galaxy

The Milky Way Galaxy is Earth's home, containing several hundred billion stars, of which the Sun is but one, orbiting about halfway out in the galaxy's thin disk. In this part, I teach you about this galaxy's place in an even-more-vast universe.

First, I review the basic anatomy of the Milky Way, from its central regions that harbor a supermassive black hole, to its spiral arms, to the dark matter that seems to dominate its mass. You also learn about the Milky Way's two companions—the Large and Small Magellanic Clouds—and how they both actively form stars and interact with the Milky Way. At the end of this part, I delve into the search for life in the Milky Way—in particular, the very productive searches for planets around other stars (exoplanets) that have been made in the past two decades.

The Milky Way in the Night Sky

This image shows the 8-m-diameter Gemini South telescope in operation at Cerro Tololo, Chile. In the background, the band of the Milky Way rises. The center of the Milky Way Galaxy is best viewed from the Southern Hemisphere of the planet. The bright beam of light is part of the Gemini South Adaptive Optics (AO) system that significantly improves the resolution of ground-based telescopes.

The Milky Way, which looks to the naked eye like a band of haze across the sky, has been observed since humans first looked up. Because the Sun is one of hundreds of billions of stars in the galaxy, the Milky Way surrounds Earth in a great circle. Depending on the time of year and the darkness of the skies, you can see different parts of the Milky Way rise and set with the stars, showing it clearly has something in common with the "fixed" stars.

Ancient Ideas About the Milky Way

People now know the Milky Way is a view of the galaxy that's home to the solar system, along with billions of other stars and planets. In ancient times, however, many cultures described it as a river due to its hazy, meandering nature.

Before the invention of the telescope, Persian astronomers proposed the haziness of the Milky Way was the result of a large number of stars very close together. Some ancient astronomers also suggested the Milky Way might consist of distant stars.

The true nature of the Milky Way's structure did not become clear until after Galileo's first use of the telescope in 1609. His observations made it clear that when observed with better resolution, the haze of the Milky Way was indeed filled with individual stars, too close together for the human eye to discern.

This panorama of the Milky Way Galaxy from the ESO's Paranal Observatory shows its complex bright and dark bands and the accompanying Large and Small Magellanic Clouds to the left of the image.

Current Views of the Milky Way

Today, for many people living in the world's great cities, the Milky Way is invisible. Light from human activities turns the skies above cities from black to blue-gray, erasing the hazy band of the Milky Way. When that light competes with natural light from the Moon, planets, and stars, it is called light pollution.

Many amateur astronomers use the Bortle scale to compare sky brightness in different locations. The scale runs from 1 (excellent dark-sky site) to 9 (inner-city sky). The Milky Way disappears at a value of 6, typical of bright suburban skies.

Though not considered as often, the sky can be polluted with radio frequency transmissions as well. The explosion in radio frequency transmissions in the past few decades (cell towers, wireless networks, satellite radio, and so on) have polluted the radio spectrum and made it harder for radio astronomers to work.

However, there are some spots that allow astronomers to view and study the Milky Way and other objects with little radio interference. For example, the National Radio Astronomy Observatory (NRAO) facility in Green Bank, West Virginia, is the largest radio quiet zone in the United States.

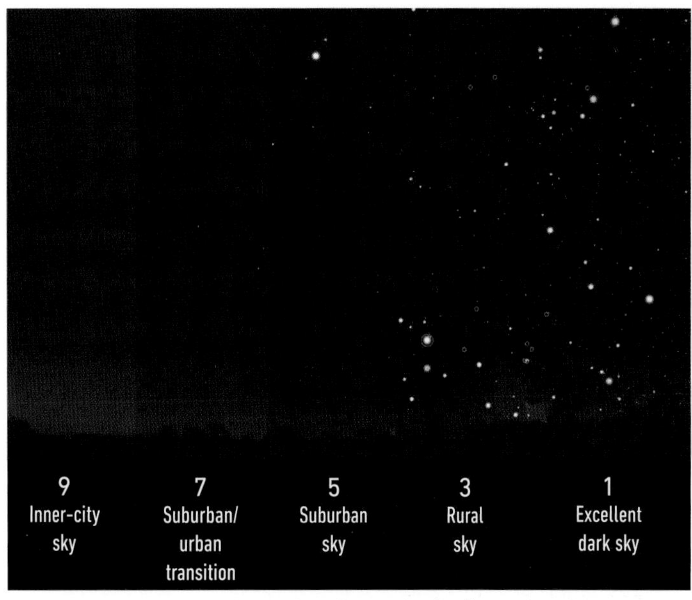

This simulation of the night sky from inner-city (left) to an excellent dark-sky site (right) shows how city lights erase the view of stars, planets, and the Milky Way.

The Structure of the Milky Way

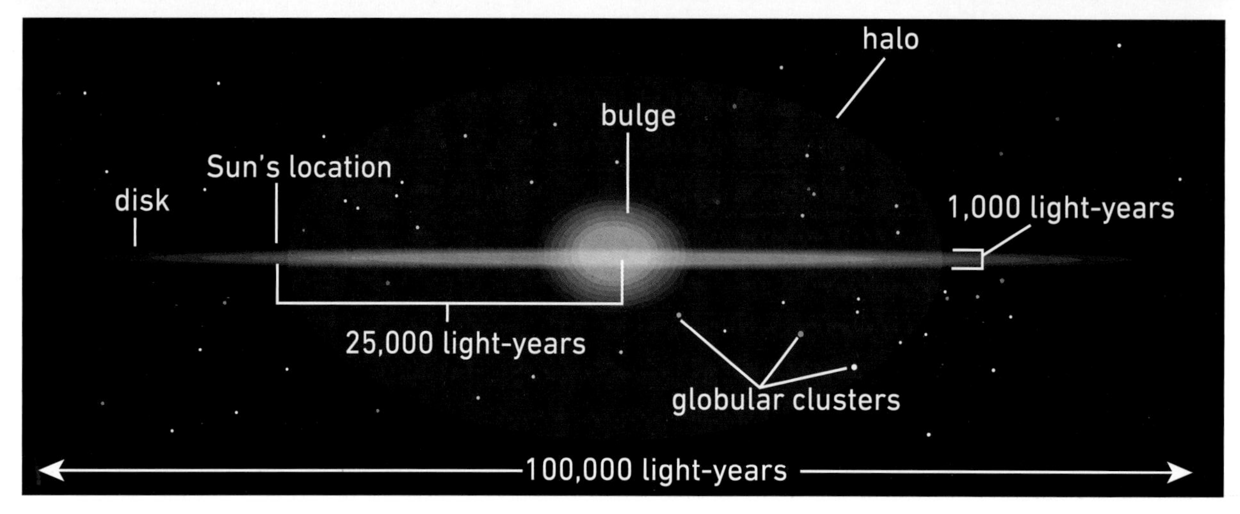

This "side view" of the Milky Way shows its major features, including the location of the Sun, which is about halfway between the Galactic Center and the outer edge of the optical disk. As you can see, the disk of the galaxy is very thin.

The Milky Way is the home galaxy of the solar system. However, understanding the shape and layout of the galaxy has been challenging, since Earth is embedded in the disk itself.

Early telescope observations made it clear that the Milky Way was a concentration of stars (see "The Milky Way in the Night Sky"), but the shape and structure of the galaxy were not well understood until radio observations of the hydrogen atom (the HI or "H-one") line were made in the second half of the twentieth century.

The diameter of the Milky Way (about 100,000 light-years) is much larger than its thickness (less than 1,000 light-years). The disk of the galaxy is comprised of spiral arms containing a mixture of dust, gas, and high-mass and low-mass stars.

Above and below the disk of the galaxy—known as the *halo*—are globular clusters, which are collections of 100,000 or more old, low-mass stars (the high-mass stars have long since formed, lived, and died in these structures). The globular clusters in the halo are thought to have formed long ago, before the galaxy formed into a flattened disk. As a result, the globular clusters retain the orbits of those early stars. Most of the mass of the galaxy is located in the halo, but its exact composition is still a mystery (see "What Is Dark Matter?").

The center of the Milky Way contains an elongated structure of orbiting stars called the *bar*. The main spiral arms begin at the ends of this structure and contain both young and old stars. The *galactic bulge,* also located at the center of the galaxy, contains a high concentration of high- and low-mass stars. At its center is a supermassive black hole.

In terms of the solar system's place in the Milky Way Galaxy, the Sun is thought to be located in a spur of stars, gas, and dust located between two spiral arms, Sagittarius and Perseus, about 25,000 light-years from the center of the galaxy. The Sun and all its planets orbit the center of the Milky Way very slowly, about once every 250 million years.

Gamma-ray emissions
X-ray emissions
Milky Way
50,000 light-years
Sun

The Milky Way Galaxy contains a large bubble-like structure. The two bubbles—imaged with NASA's Fermi Gamma Ray Telescope—extend 25,000 light-years above and below the plane of the disk. The structure is thought to be millions of years old and related to either an outburst of energy from the black hole at the Galactic Center or a past wave of massive star formation.

This image, based on data from the Spitzer Space Telescope, shows a bird's-eye view of the current model of the Milky Way, dominated by two major spiral arms (Scutum-Centaurus and Perseus).

The Galactic Center

To find the center of the Milky Way Galaxy, you need only to look for the constellation Sagittarius. To the northwest of the famous "teapot" of the constellation is the Galactic Center.

The center of the Milky Way Galaxy, about 25,000 light-years away from Earth, is obscured by large amounts of gas and dust. While the central bulge is visible in optical images of the Galactic Center, the heart of the galaxy can only be viewed in wavelengths that are longer—in the infrared and radio part of the electromagnetic spectrum (see "The Milky Way at Different Wavelengths").

The densest gas and dust in the Galactic Center was opaque even to infrared radiation (seen in the dark filaments observed in the 2MASS image), so the clearest views of the Galactic Center region are provided by radio telescopes. Radio waves have wavelengths that are so long they are not affected by the small particles of gas and dust that scatter shorter-wavelength visible and infrared photons.

This 2MASS (2 Micron All Sky Survey) image covers a large area, about 10×8 degrees. The bright region of the Galactic Center (upper left) is home to a supermassive black hole. The scattering of light by dust in the galactic plane, still visible as filamentary dark areas in this image, reddens the color of stars.

The very center of the Milky Way is the bright white region to the right of the image center (about the size of the full moon). The X-ray emission (in blue and pink) near the Galactic Center is thought to arise from a combination of the accretion disk near the black hole itself and energetic outflows from other nearby stars.

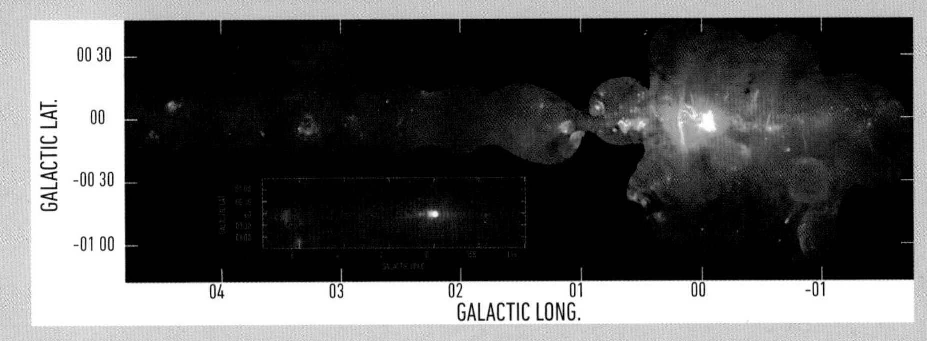

Data from the Very Large Array (VLA) and the Green Bank Telescope (GBT) created this radio frequency image of the Galactic Center region, shown in green false color. The inset in red shows the GBT data alone. The increase in resolution provided by the VLA is clear in this pairing of images. The supermassive black hole at the Galactic Center is located on the right hand side of the image.

Sources of Radio Emission

Two important sources of radio frequency emission are hot gas and strong magnetic fields. In the Galactic Center, long filaments connecting to the bright region are thought to be the result of electrons spiraling around magnetic field lines. Several of the bright blobs to the left (East) of the Galactic Center region are regions forming massive stars. These hot, young stars ionize the material around them.

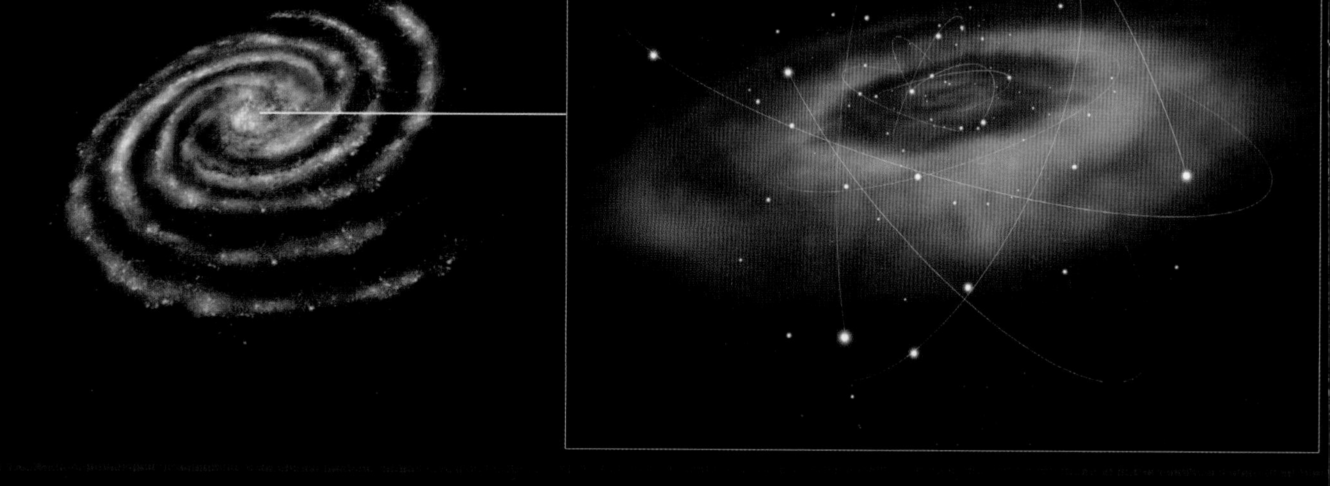

An artist's impression of the Galactic Center region shows how the clouds of gas near the black hole might be distributed in a torus (ring-shape volume) about 15 light-years across.

Supermassive Black Hole

By definition, a black hole is a region of space that has such an enormous concentration of mass that nothing—not even light—can escape its gravity. If no photons can escape, then the region is black. Therefore, the supermassive black hole located at the Galactic Center is not directly visible.

However, astronomers can study the black hole by observing the regions around it. These regions are visible at a variety of wavelengths, from radio, to infrared, to X-rays. These observations typically highlight the material that is falling toward the black hole, achieving very high temperatures. Plus, careful monitoring of the Galactic Center region has allowed astronomers to witness periodic flares in brightness at many wavelengths as material approaches the black hole and heats up.

The exact location of the Galactic Center black hole was first pinpointed by radio frequency observations that identified Sgr A*, or the brightest radio source in the constellation Sagittarius (Sgr). Stellar orbits have confirmed the location of the black hole and helped to determine that the black hole must have the mass of 4 million Suns.

Recent observations by the Herschel Observatory of the ESO have also confirmed that gas located near the black hole is very hot, at almost 1,300 K. Shock waves in the gas are thought to cause the exceedingly high temperatures.

And according to recent observations with the Chandra X-Ray Observatory and the Very Large Array, there may be a jet associated with the Galactic Center black hole. The shock fronts in connection with the jet might be formed when the jet interacts with clouds of gas and dust located near the Galactic Center.

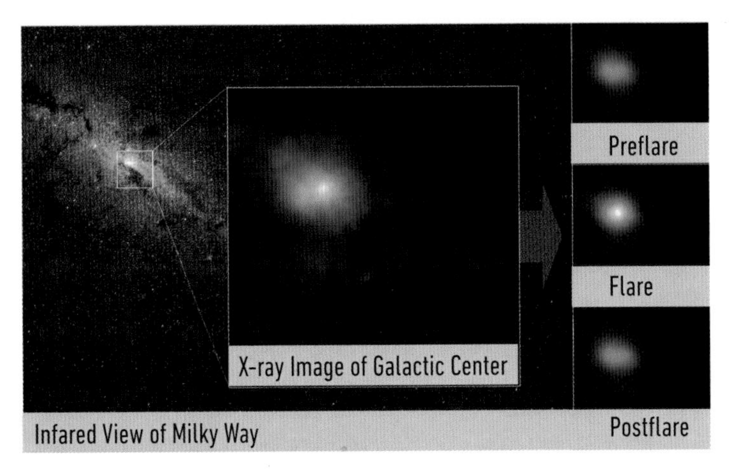

In a series of images, NASA's NuSTAR (Nuclear Spectroscopic Telescope Array) instrument detected a high-energy X-ray flare that occurred over a period of two days. The bright emission in blue to the far right shows the X-ray emission and comes from gas heated to 100 million Kelvin, hotter than the core of the Sun. The central image shows X-ray emission at four different energies.

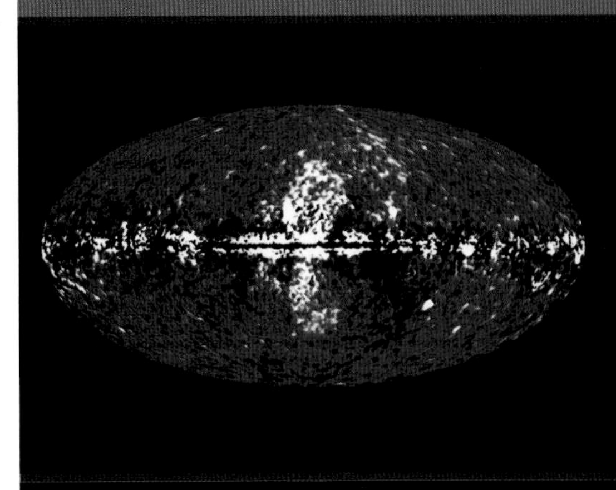

This image by NASA's Fermi Gamma-Ray Space Telescope shows enormous lobes of high-energy emission located above and below the plane of Milky Way. While the origin of the Fermi bubbles is still being debated, they may be the result of activity from the supermassive black hole at the galaxy's core, perhaps in the form of a jet of particles.

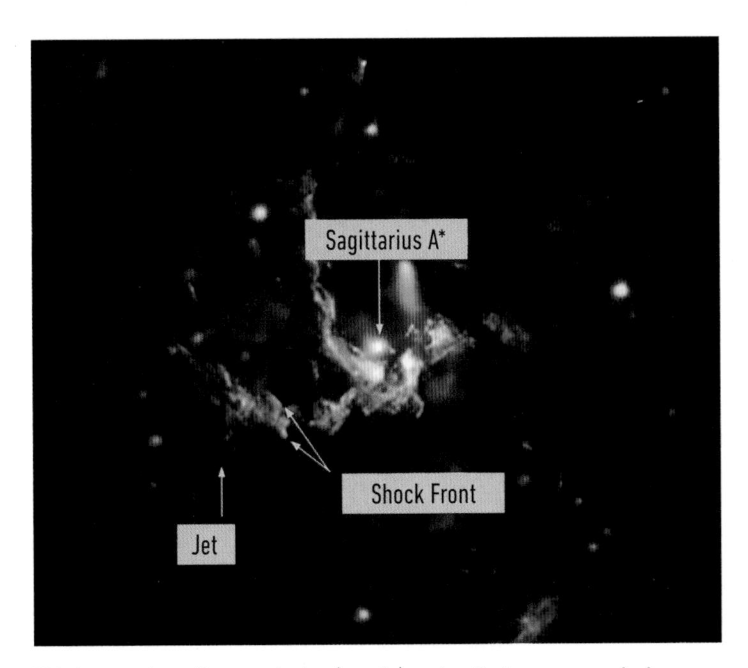

This image shows X-ray emission (purple) and radio frequency emission (blue) from the Galactic Center. The image is labeled with the position of Sgr A*, the potential jet, and the shock front associated with the jet.

Periodically Active?

The black hole in the center of the Milky Way, like those observed in many galaxies, is thought to be active periodically. In fact, the 50,000-light-year-long emission structure observed by the Fermi Gamma-Ray Space Telescope might be related to a jet emitted by the Galactic Center black hole. An alternative to this is that it's a burst of star formation that occurred in the crowded central region of the galaxy (also known as a ***starburst***).

Stellar Orbits

In the past few years, astronomers have spoken with increasing confidence about the presence of a supermassive black hole located at the center of the Milky Way Galaxy. What observations give them the ability to study an object that gives off no electromagnetic radiation directly? Their ability to determine the black hole's size and location comes mostly from careful observations of the surrounding region and the orbits of stars located there.

But observations of the Galactic Center region face many challenges. The region is highly obscured by gas and dust, and its great distance means that very high-resolution observations are required to successfully measure the motions of the stars there. Ground-based infrared telescopes are limited in their resolution by Earth's atmosphere; however, resolution can be remarkably improved using a relatively new technology called *adaptive optics (AO).* This technology reduces the blurring effect of Earth's atmosphere, giving astronomers a clearer view of stellar positions.

With accurate stellar positions enabled by AO, astronomers have been able to track the trajectories of stars in the central arcsecond of the galaxy. These orbits can be used to calculate the mass of the object the stars are orbiting, which has been found to be 4 million times the mass of the Sun.

In 2013, a large cloud of gas had an encounter with the supermassive black hole at the Galactic Center. As it neared the black hole late in 2013, the gas cloud's velocity was observed to increase as it was stretched by the intense gravitational field near the black hole.

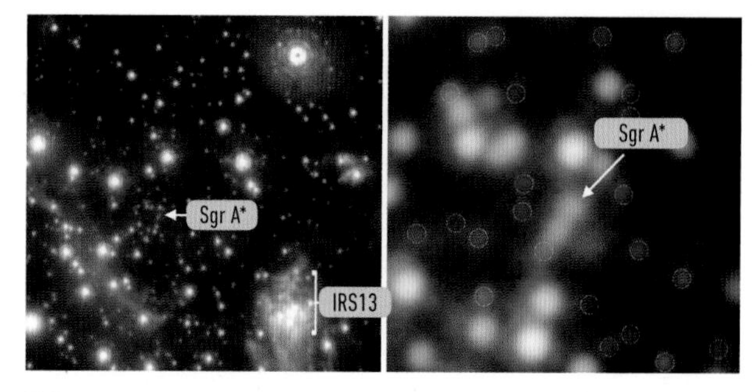

This infrared image comes from combined observations at 1.8 microns, 2.2 microns, and 3.8 microns. For nearly 20 years, the stellar positions have been observed in the Galactic Center. The stars are orbiting the radio source Sgr A* at the position marked by the white arrow.

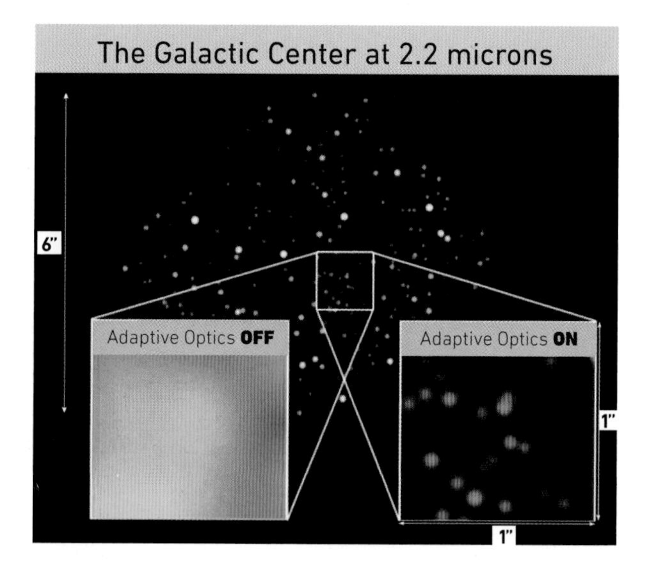

These near-infrared images from the Keck Observatory show the impact of AO technology on observations of the Galactic Center. The lower-left image shows the Galactic Center region without AO at 2.2 microns, while the lower-right image shows the same field with AO enabled. Clearly, AO technology makes tracing the orbits of individual stars possible.

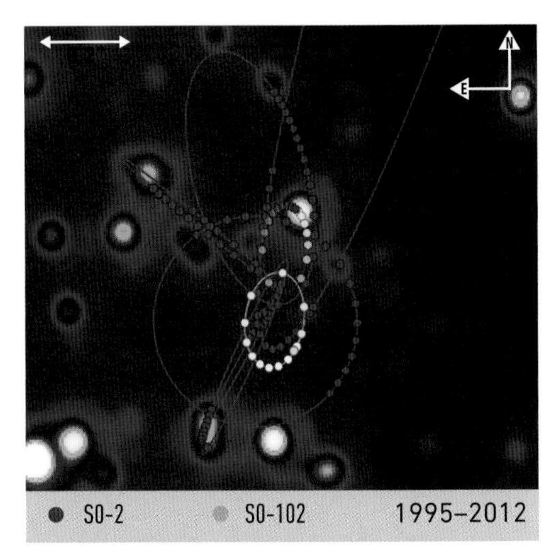

The false-color background shows an infrared image of the stars in this region taken in 2012, overlaid with trajectories of a number of stars, all of which appear to be in Keplerian orbits around the center of the image. The two orbits highlighted are the stars (S0-2 and S0-102) that are observed to be in the smallest and thus fastest orbits.

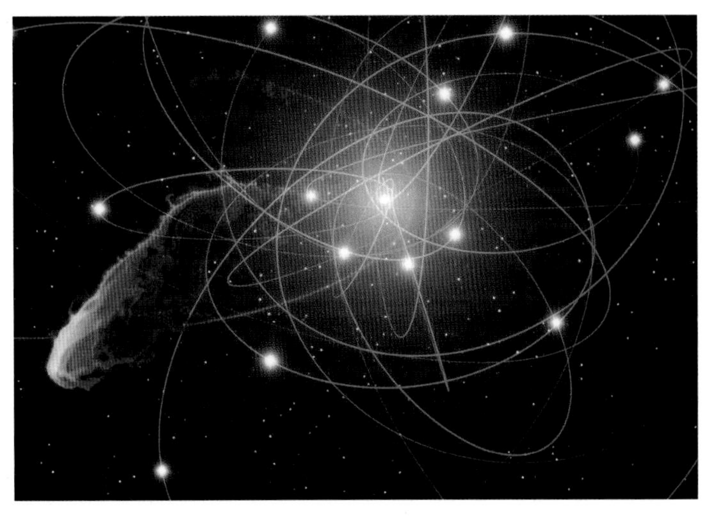

This artist's impression shows the trajectory of the gas cloud (red) and the orbits of the stars near the Galactic Center (blue) in the year 2021

Accuracy and Precision

These terms are often used interchangeably, but they refer to different characteristics of a measurement. The *precision* of a measurement relates to repeatability—that is, if a number of measurements all have close to the same result, they are precise. The *accuracy* of a measurement tells you how close the measurements are to the true value. In the case of the Galactic Center, measuring the orbits of many stars allows astronomers to make both a more accurate and a more precise estimate of the mass of the black hole.

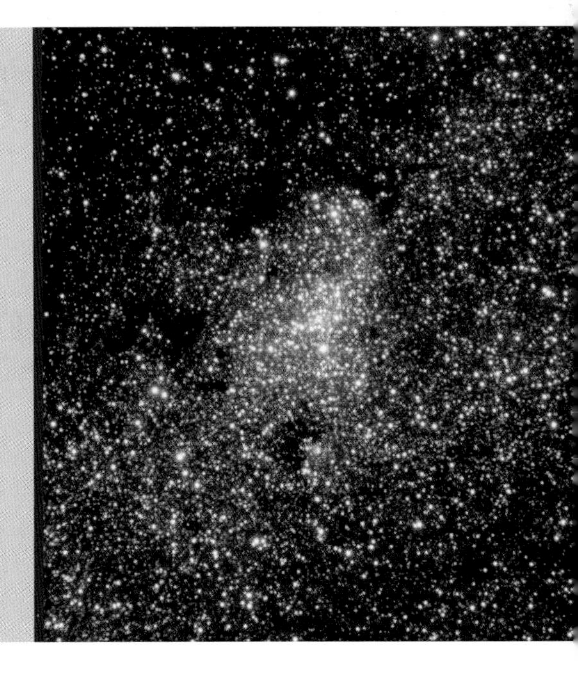

The Milky Way at Different Wavelengths

The following images are made at nine different wavelengths used to observe the Milky Way, highlighting different properties:

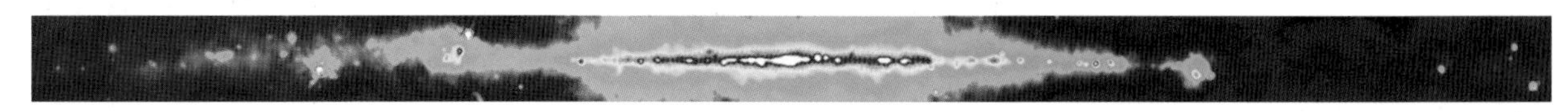

Radio Continuum

408 MHz Bonn, Jodrell Banks, & Parkes

This emission comes from electrons associated with hot gas and magnetic fields, such as those seen in star-forming regions and supernova remnants.

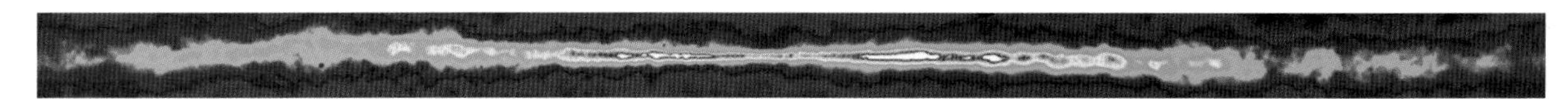

Atomic Hydrogen

21 cm Leiden-Dwingeloo, Maryland-Parkes

This emission from neutral atomic hydrogen (HI) shows the location of the warm interstellar medium, a fuel tank for future star formation.

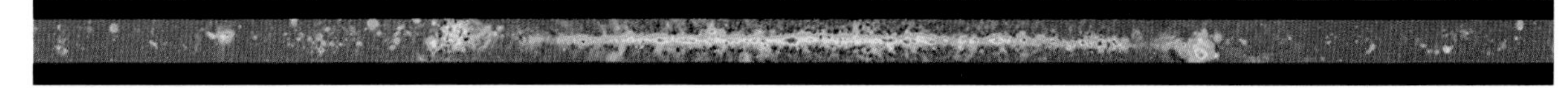

Radio Continuum

2.4–2.7 GHz Bonn & Parkes

This emission essentially shows a higher-resolution picture of the Milky Way at radio wavelengths. The Galactic Center, star-forming regions, and supernova remnants are some of the brightest sources at this wavelength.

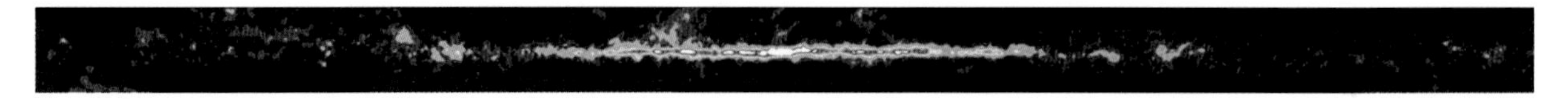

Molecular Hydrogen

115 GHz Columbia-GISS

The hydrogen molecule (H^2) is difficult to observe, but this image of the carbon monoxide molecule made at millimeter wavelengths shows the location of molecular clouds. The new Atacama Large Millimeter Array (ALMA) operates at similar wavelengths to this but will see the galaxy in much sharper detail.

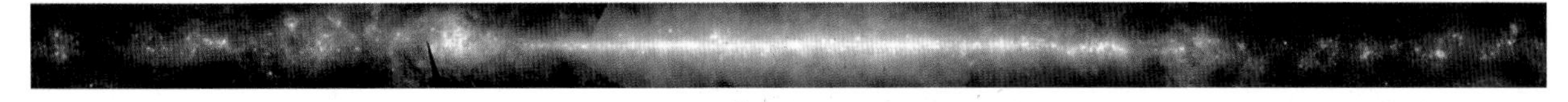

Infrared

12, 60, 100 μm IRAS

The mid- and far-infrared (12, 60, and 100 microns) shows the location of interstellar dust and star-forming regions—note the similarity to the high-resolution radio continuum image.

Near Infrared

1.25, 2.2, 3.5 μm COBE/DIRBE

This composite (1.25, 2.2, and 3.5 microns) highlights the positions of stars smaller and cooler than the Sun.

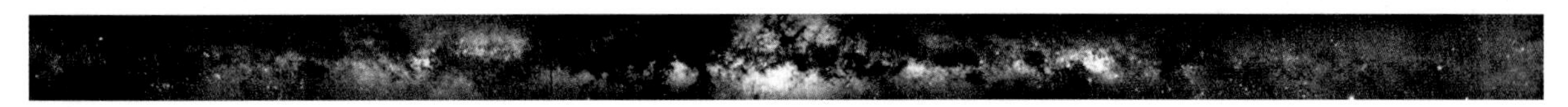

Optical

Laustsen et al. Photomosaic

Because material in the galaxy absorbs optical photons very efficiently, most of the stars shown at this wavelength are within a few thousand light-years of the Sun. The Galactic Center region is 25,000 light-years away and not visible optically. The dark blobs and filaments are the result of the gas and dust clouds better seen in the infrared and molecular hydrogen wavelengths.

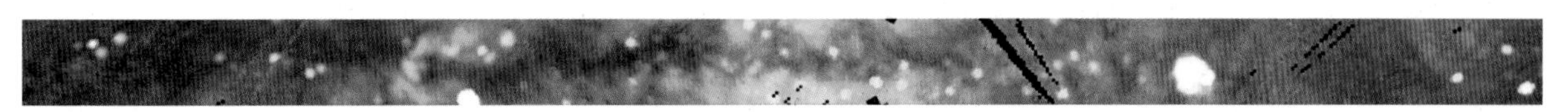

X-Ray

0.25, 0.75, 1.5 keV ROSAT/PSPC

The colors shown relate to the energy of the detected X-rays, with the lowest energies in red and the highest energies in blue. Many of the point sources are supernova remnants.

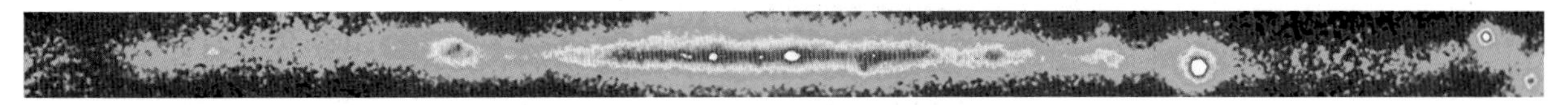

Gamma Ray

>100 MeV CGRO/EGRET

When cosmic rays collide with atomic nuclei in interstellar clouds, they generate gamma rays. This interaction explains the diffuse gamma ray emission. Many of the visible point sources are associated with pulsars or distant background galaxies.

How Infrared Telescopes Unveil the Milky Way

Telescopes sensitive in the infrared part of the spectrum can't operate productively from the surface of Earth. In order to make sensitive, high-resolution images, infrared telescopes have to be located at high elevations, aboard high-flying aircraft, or in orbit around Earth, as Earth's atmosphere blurs images and absorbs infrared radiation. The following are a few of the infrared telescopes used to observe the Milky Way and how they do it.

This pair of images shows the M17 star-forming region as seen in the infrared (upper image) and the visible (lower image). Notice the broad, dark bank in the lower right of the visible image is brightly illuminated in the infrared. This image supports the idea that as molecular material passes through a spiral arm, it begins to form stars.

The Keck Telescope

The *Keck Telescope* has a number of instruments that operate in the near-infrared part of the spectrum (1 to 5 microns), which make it through Earth's atmosphere. The adaptive optics system at Keck (see "The Galactic Center") can correct for turbulence in Earth's atmosphere to create high-resolution images and spectra in the near-infrared.

SOFIA

In order to observe longer infrared wavelengths (far-infrared), telescopes must be located above most of Earth's atmosphere. *SOFIA,* an abbreviation for the *Stratospheric Observatory for Infrared Astronomy,* is a 2.5-meter-diameter telescope that flies in a modified Boeing 747 at altitudes between 12 and 14 km. Above Earth's water vapor, instruments on the telescope are sensitive to wavelengths from the visible part of the electromagnetic spectrum out to 650 microns. These longer wavelengths, which reveal structures deeper in molecular clouds, do not make it to the surface of Earth.

The SOFIA telescope is visible here in its modified Boeing 747SP, with the bay door open. The high-altitude flights of this aircraft allow the telescope to make observations not possible on Earth's surface without the expense of an orbiting spacecraft.

The Spitzer Space Telescope

The Spitzer Space Telescope, launched in 2003, is the fourth of NASA's Great Observatories, which include the Hubble Space Telescope (optical), the Chandra X-ray Observatory (X-ray), and the Compton Gamma-Ray Observatory (gamma rays). The Spitzer spacecraft is in an Earth-trailing orbit, which keeps it away from Earth's heat and allows it to view a broader area of the sky. In order to function, the longest-wavelength instruments aboard Spitzer needed to be cooled to very low temperatures. Long wavelength observations can reveal the cores of star-forming regions and are largely unaffected by intervening material.

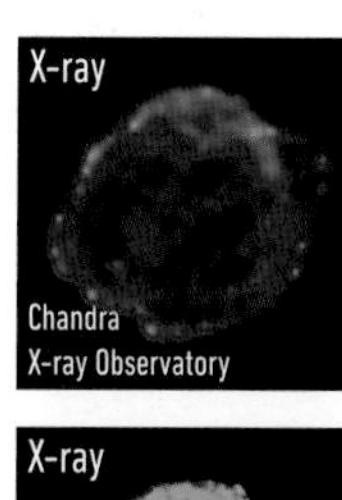

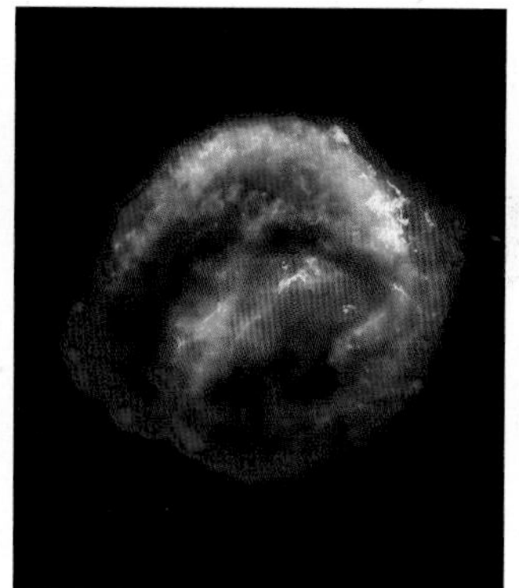

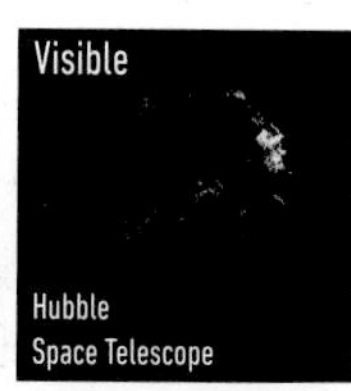

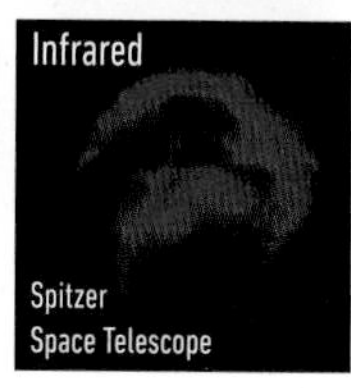

These false-color images show a supernova remnant (about 14 light-years across) as observed by three of NASA's Great Observatories in 2004. The optical emission from the remnant is faint, but the X-ray and infrared emission show an expanding bubble of material. The Spitzer Space Telescope, which captured the infrared (red) image, is sensitive to dust particles heated by the passing shock wave.

The Interstellar Medium

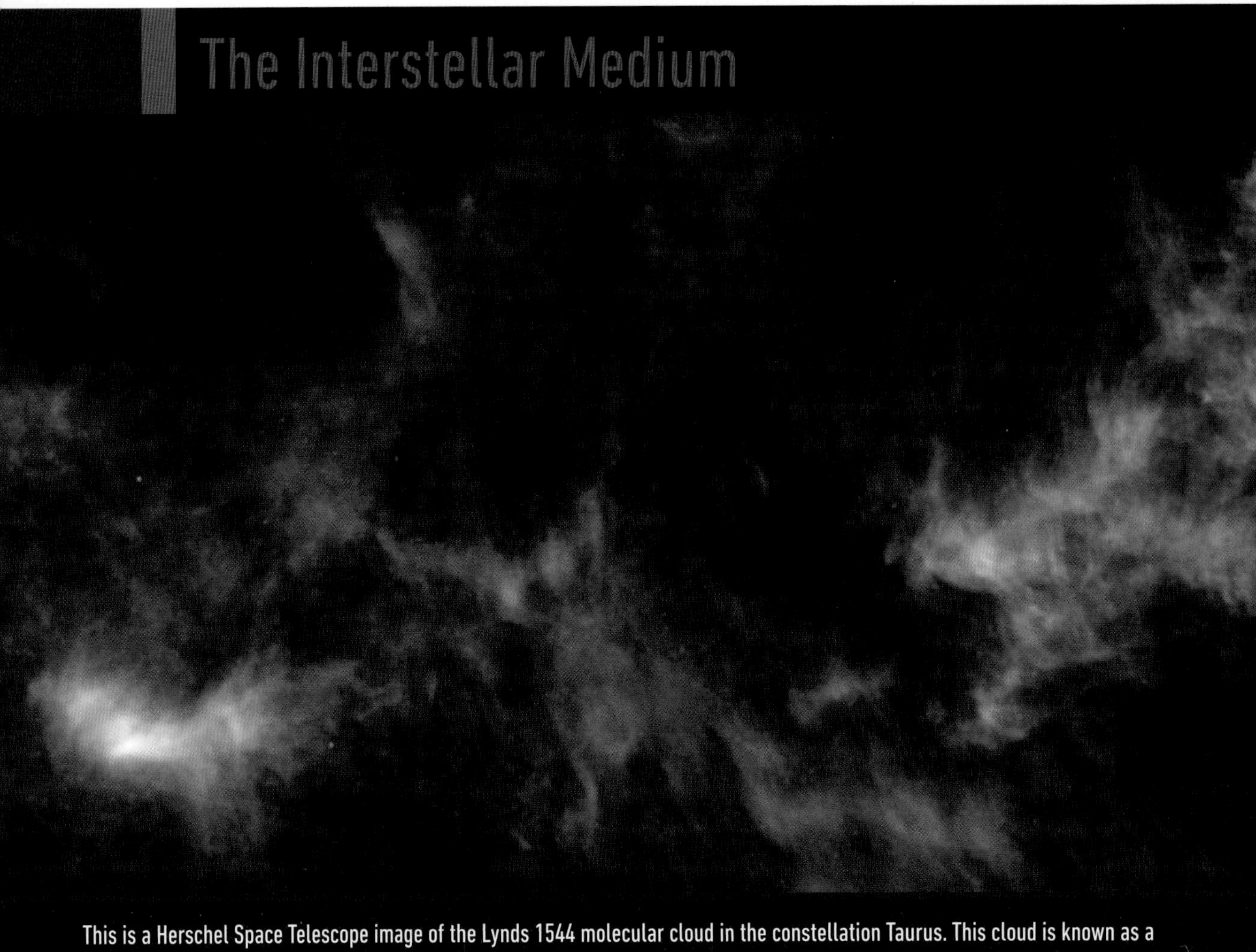

This is a Herschel Space Telescope image of the Lynds 1544 molecular cloud in the constellation Taurus. This cloud is known as a prestellar core and is an example of a cold molecular cloud. The internal motions of gas in the cloud suggest that it is just beginning to collapse.

The interstellar medium (ISM) consists of the matter located between the stars in a galaxy. By mass, the ISM of the Milky Way is mostly gas (99 percent) in the form of hydrogen (89 percent), helium (9 percent), and a small percentage of other elements. Elements other than hydrogen and helium are the result of stellar cores fusing heavier elements during their lifetimes and then ejecting that material into the ISM in the form of planetary nebulae and supernova remnants.

Temperatures in the ISM are the main way astronomers distinguish between its various components:

Cold molecular clouds have temperatures of only ~10 K and densities up to 1 billion parts per cubic meter. This cold, dense material is often observed at radio and infrared wavelengths.

The *cold neutral medium* consists of clouds of neutral hydrogen, or HI. This material has temperatures of ~100 K and densities of 100 million parts per cubic meter. The HI line is visible in the radio part of the spectrum at 21 cm.

The *warm neutral medium* has temperatures of up to 7,000 K and densities of a few hundred thousand parts per cubic meter. It's not quite hot enough to be ionized.

The *warm ionized medium* has temperatures of 10,000 K and densities of a million parts per cubic meter. Gas at this temperature is ionized and visible as HII regions.

Finally, the *hot ionized medium* has very low density (only about 10,000 parts per cubic meter) and very high temperature (~1 million K). The material in this state is referred to as *coronal gas,* because of its similarity to the hot, low-density material in the Sun's corona. The coronal gas makes up most of the volume of the ISM.

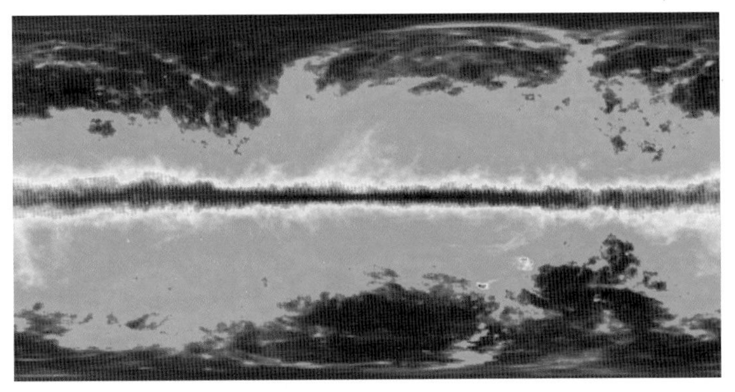

This false-color image of the cold neutral medium in the Milky Way Galaxy was taken from a survey of the sky at a wavelength of 21 cm. As you can see, HI is mostly concentrated in the plane of the galaxy but is located at great distances above and below the disk as well.

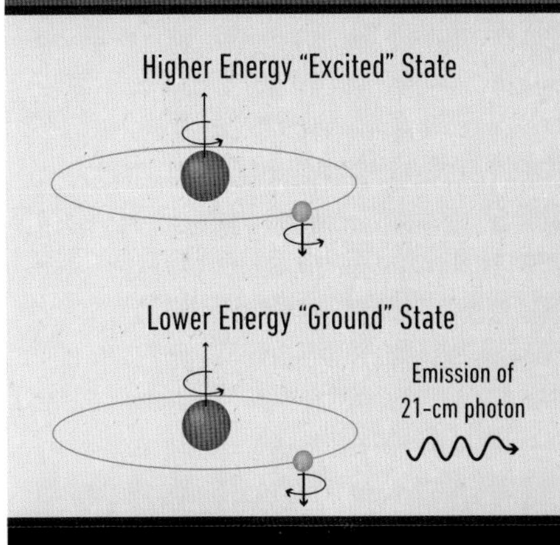

The hydrogen atom is shown here in its two possible spin states: one with higher energy (top) and one with lower energy (bottom). When the electron's spin flips, it emits a 21-cm photon, which can be detected with radio telescopes.

The 21-cm Line

A neutral hydrogen atom consists of a proton and an electron. Both of these particles have a property called *spin* that can be in one of two directions: up or down. If the spins are aligned (both up), the atom has slightly more energy than when they are not aligned (one up and one down). Very rarely, the electron in a neutral hydrogen atom will flip its spin spontaneously. When this happens, the atom has slightly less energy, and this energy comes out in the form of a long-wavelength (21-cm) radio wave. Because the Milky Way is filled with hydrogen atoms, detectors sensitive at this wavelength have been able to make images of the cold neutral medium, which is found mostly in the disk of the galaxy.

Galaxy Detail: Molecular Clouds

The Vela-C molecular cloud, located in the plane of the Milky Way, is seen here at visible (upper) and far-infrared (lower) wavelengths. Bright spots in the filamentary structure are regions that are just beginning to form new stars. The HII regions are also visible as the blue structures, associated with the shortest wavelengths in the infrared image.

Molecular clouds are the fuel tank from which current and future generations of stars are generated. These clouds are found in the disk of the Milky Way and consist mostly of molecular hydrogen (H_2). Molecular clouds are generally confined to a very thin disk only a few hundred light-years in "height."

Molecular clouds are composed of cold molecular hydrogen and helium, both of which are difficult to observe because of their symmetry. As a result, these clouds are often detected in carbon monoxide (CO), which is a molecule contained in these clouds that is easier to detect at radio frequencies. Known ratios between the CO and H_2 molecules are then used to determine the amount of molecular hydrogen present in a cloud.

Giant Molecular Clouds (GMCs) contain masses of 10^5 to 10^7 solar masses and diameters of 50 to 300 light-years. They are associated with ongoing star formation in the disk of the Milky Way. Bok globules are smaller clouds with masses of less than a few hundred solar masses. The view into these clouds is severely limited at visible and even near-infrared wavelengths.

Some molecular clouds are filled with activity, including outflows from young stars (see "The Birth of Low-Mass Stars"). Dark clouds are thought to be dissociated by the outflows from stars, as well as by the ionizing radiation if the cloud contains more massive stars (see "Galaxy Detail: HII Regions"). The detailed structure of molecular clouds is best revealed in far-infrared and submillimeter (radio) images.

The Bok globule known as Barnard 68 is shown in two views: on the left in visible and near-infrared, and on the right in visible (blue), near-infrared (green), and infrared (red). Barnard 68 is only about 400 light-years away and is similar in size to the Oort cloud that surrounds the solar system (~10,000 AU).

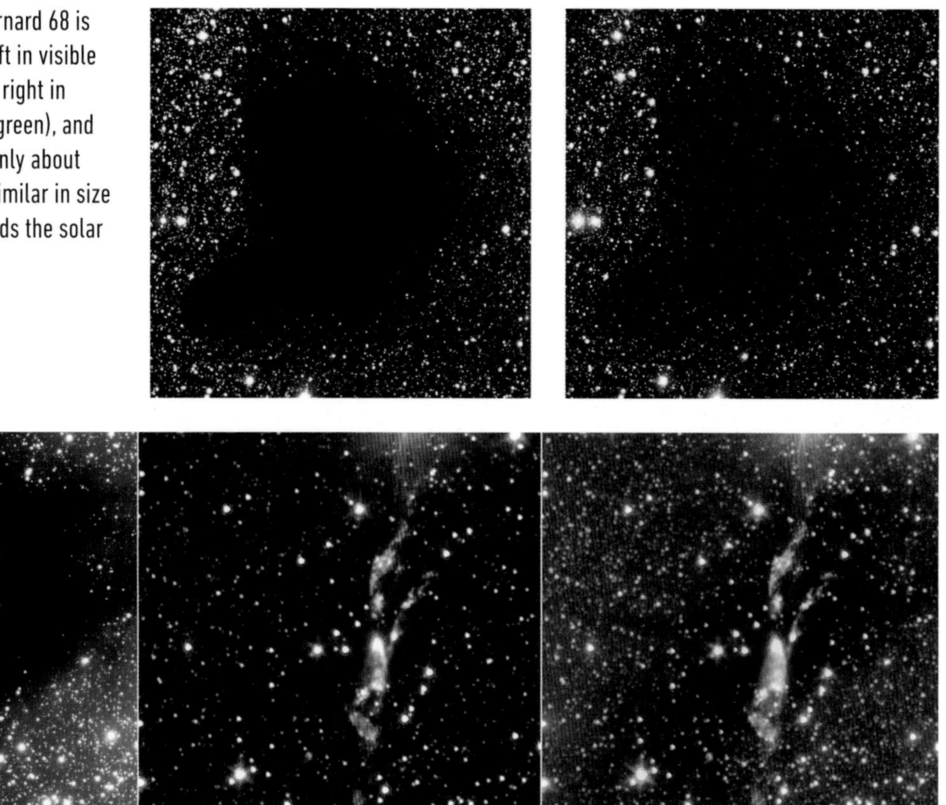

BHR 71 is a molecular cloud containing two young stars that are blowing away the surrounding material that formed them. In the visible image (left), we see the obscuration of background stars caused by the dense gas in the cloud. An infrared image (center) reveals two young stars and their associated outflows.

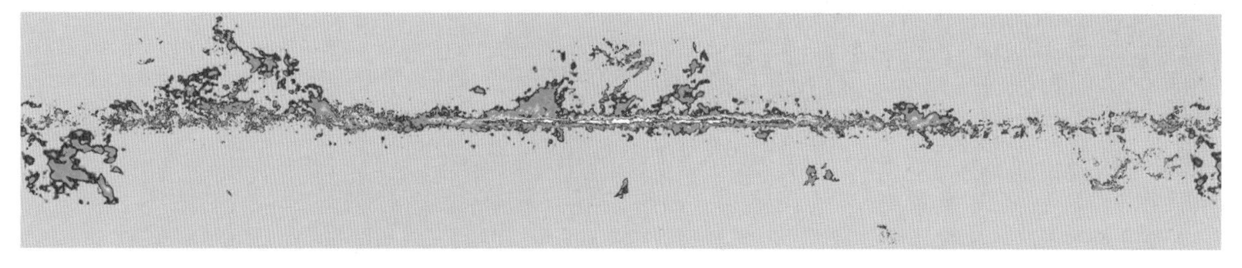

This false-color image shows the plane of the Milky Way as seen in CO, with blue emission being the faintest and white the brightest. The plane of the Milky Way runs from left to right across this image, centered on the Galactic Center.

Galaxy Detail: Supernova Remnants

As you learned in Part 3, supernova remnants are the remains of the outer layers of a high-mass star. High-mass stars go through their lifetimes very quickly, in about 10 million years, and their cataclysmic ends send their outer envelope hurtling into the interstellar medium.

Remnants in the Milky Way can be observed at a variety of wavelengths, depending on their location. Because supernovae are associated with high-mass stars, and these stars are found in the plane of the galaxy, only the nearest supernova remnants can be observed optically.

The following are a few supernova remnants located relatively nearby in the Milky Way:

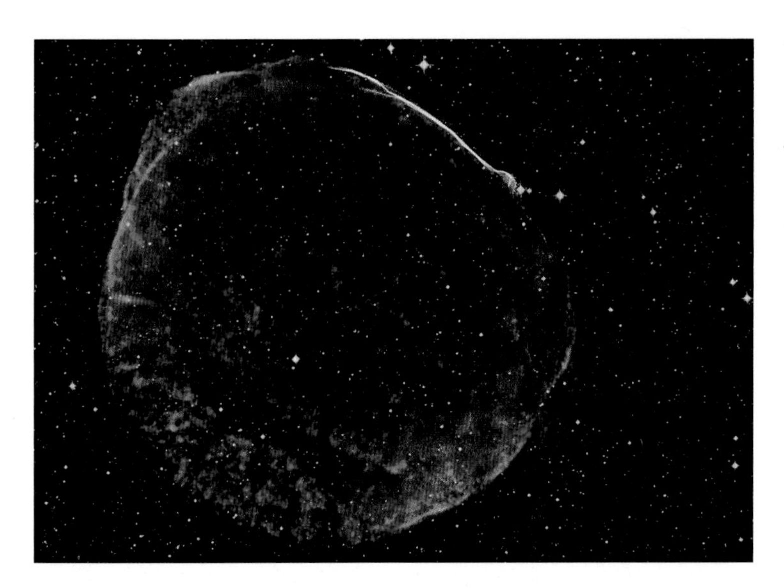

This image of SN 1006 is a combination of data from the Chandra X-ray Observatory (blue), the 0.9-m Curtis Schmidt telescope at CTIO (yellow), the Digitized Sky Survey (orange and light blue), and the Very Large Array (red). In the upper right of the image, the shock wave of the supernova is interacting with nearby material.

Supernova remnant SN 1006 is over 1,000 years old; the supernova that created its expanding shell of material was observed by astronomers in China, Japan, Europe, and the Middle East. Studies of this remnant classify the supernova as a type Ia supernova, which is the result of a white dwarf that captured enough mass from a companion star to explode.

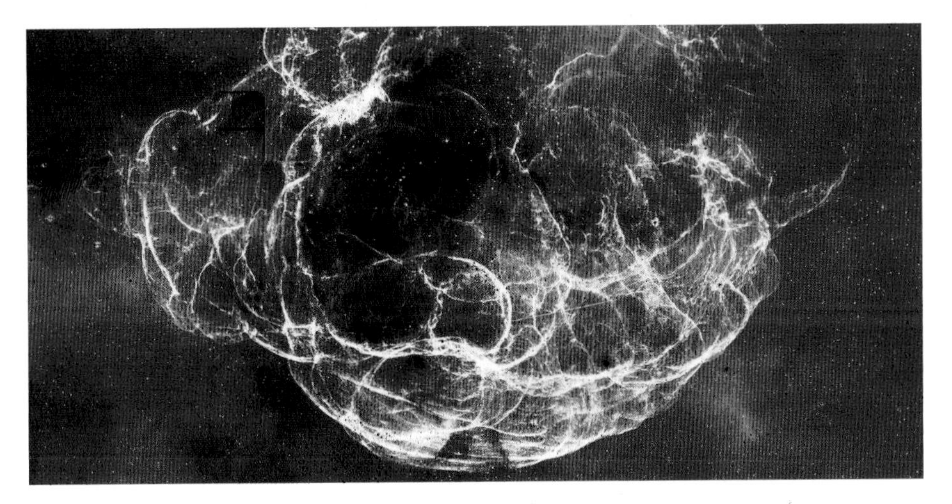

This image of the supernova remnant S147 was taken in H-α emission with the Isaac Newton Telescope (INT). The H-α spectral line is strong in regions containing hot gas, typically at about 10,000 K. Because the supernova remnant is very large (the image is approximately 4×5 degrees), this image is built from many separate images.

Supernova remnant S147 looks like a freeze frame of an enormous explosion. H-α is a spectral line of hydrogen in the visible part of the spectrum located at 656.3 nanometers. This line is strong in regions containing hot gas, typically at about 10,000 K. This supernova remnant is very large on the sky (the image is approximately 4×5 degrees), so this image is a mosaic, built up from many separate images in this part of the sky.

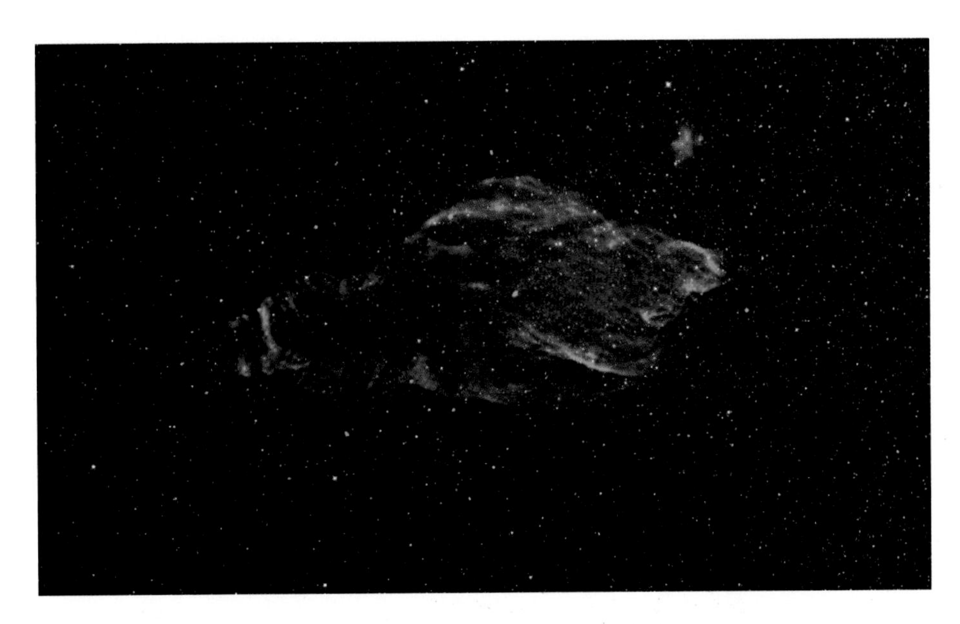

This Very Large Array (VLA) image of the supernova remnant W50 (Westerhout Catalog 50). The elongation of the nebula to the left and right are the result of the jet pushing remnant material aside. Like many visible nebulae, this radio supernova remnant has been given a poetic name: the Manatee Nebula.

Supernova remnant W50 is 700 light-years along its long axis and spans about 2 degrees on the sky, the equivalent of four full moons. The remnant is thought to be about 20,000 years old. At its center is likely a black hole, pulling gas from a companion star. The accretion disk around the black hole is the source of a powerful jet, named SS433. The complex patterns in the remnant are thought to be related to interactions between the remnant and the jet.

Galaxy Detail: Globular Clusters

Globular clusters, or collections of stars that orbit the Galactic Center as satellites, are some of the oldest objects observed in the Milky Way. In fact, they are so old that the ages of stars in these objects are used to determine the age of the universe itself. They are thought to have formed when the galaxy was young, so they still retain their orbits out of the plane of the galaxy due to their origins. Most globular clusters contain about 100,000 stars.

Most globular clusters were catalogued from optical telescope data, so the infrared survey images of the Spitzer Space Telescope revealed a previously unknown cluster (9,000 light-years away). Stars are shown in blue, while dust is in red. The Spitzer Space Telescope image is shown with the visible image of the same field inset.

The Decrease of Globular Clusters in the Galaxy

There were probably a few thousand globular clusters when the Milky Way was first forming, but only about 150 to 200 of them remain now. Many globular clusters have been dissipated by repeated crossings of the plane of the galaxy.

The globular cluster M72, shown here as imaged by the Hubble Space Telescope, is about 50 light-years across. M72 is located in the constellation Aquarius and is 50,000 light-years away, or twice as far away as the Galactic Center.

Blue Stragglers

Because high-mass stars begin to leave the main sequence first, the highest-mass stars still on the main sequence in the HR diagram (see "Stellar Lifetime Snapshot") of a globular cluster can be used to determine the age of a cluster. But sometimes there are stars that "lag behind" and are bluer and more massive than the other main-sequence stars in the cluster; these stars are called *blue stragglers*.

One explanation for blue stragglers is they are the result of stellar collisions and captures that happen in the very dense centers of globular clusters. Supporting this idea, blue stragglers are most commonly observed in globular cluster cores. Recent studies show that many blue stragglers have a white dwarf companion.

This globular cluster, imaged by the Hubble Space Telescope and named NGC 1846, contains an intriguing object near the bottom of the image. Believed to be a planetary nebula, this green shell is material cast off by a dying low-mass star. The globular cluster orbits the Large Magellanic Cloud (LMC) that in turn orbits the Milky Way.

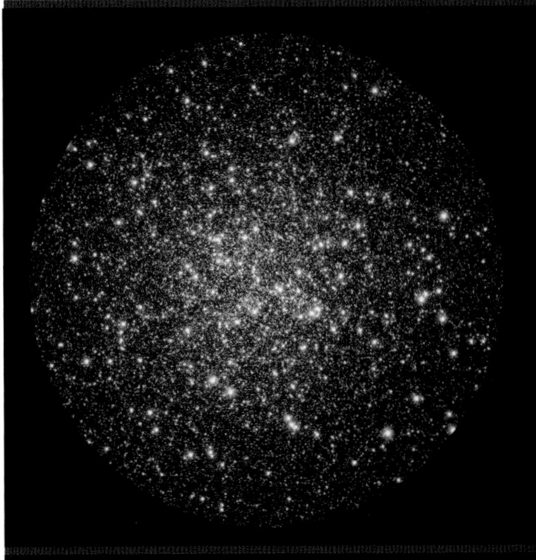

M13 is a bright globular cluster that contains several hundred thousand stars. Located at about the same distance as the center of the galaxy (25,000 light-years), M13 is about 150 light-years across.

What Is the M For?

Many objects that amateur astronomers observe in the night sky start with the letter M. This is an abbreviation for *Messier*, after the eighteenth-century French astronomer Charles Messier. The Messier catalog of objects contains sources that were considered "fuzzy" and not starlike yet could not be classified as comets—instead, they appeared to be fixed among the stars. These 110 Messier objects have turned out to be a variety of very different physical objects, from galaxies, to planetary nebula, to globular clusters.

Galaxy Detail: Spiral Arms

The galaxy UGC 12158, a barred spiral, is thought to have a structure very similar to that of the Milky Way. In barred spiral galaxies, spiral arms start at the end of a bright "bar" of stars orbiting the galactic center and curl outward to the edges of the visible disk.

It's easy to observe the spiral arm structures of other galaxies as outside observers. Determining the structure of the Milky Way is difficult, however, since observers are embedded in the middle of it, about halfway between the Galactic Center and the outer edge of the galaxy's disk.

The proposed structure of the Milky Way has varied over the years based on different observations. In the 1950s, observations of clouds of neutral hydrogen gas (HI) gave people the first evidence that the solar system was located in a spiral galaxy. The location and velocities of these clouds of hydrogen were built into the first maps of the galactic disk, showing the outlines of four spiral arms.

Observations of low-mass stars with the Spitzer Space Telescope seemed to indicate that stars were located in only two spiral arms instead of four. However, the most recent study using radio telescopes to map the positions and velocities of high-mass star-forming regions (which are only located in spiral arms) supports the original model of four spiral arms. One reason for the discrepancy between the models could be that the recent radio frequency work used observations of high-mass stars, which live much shorter lifetimes than the low-mass stars catalogued in the Spitzer Space Telescope data and have less time to spread out.

Because galaxies like the Milky Way rotate, spiral arms can't be static structures. Galaxies rotate at different speeds, depending on the radius. Material closest in to the center of the galaxy rotates the fastest, with material farther out rotating more slowly. As a result, any concentration of stars in the disk (like spiral arms) would be quickly smoothed out by rotation. The current understanding of spiral arms is that they are density waves, or compressions, moving around the disk of the galaxy.

In this artist's rendition of the Milky Way, the galaxy appears to have just two major spiral arms, which start at the ends of its central bar.

Density Waves and Star Formation

The density waves that propagate around the disk of the Milky Way are thought to be the cause of star formation. First, material in the disk overtakes the slow-moving spiral density waves. As material enters the spiral arm region, it is compressed, in some cases to the point where it begins to collapse and form molecular clouds into families of high- and low-mass stars. The spiral arms may also cause some clouds to collide, thus triggering star formation indirectly. The massive stars first form and then light up the spiral arms with their ionizing radiation.

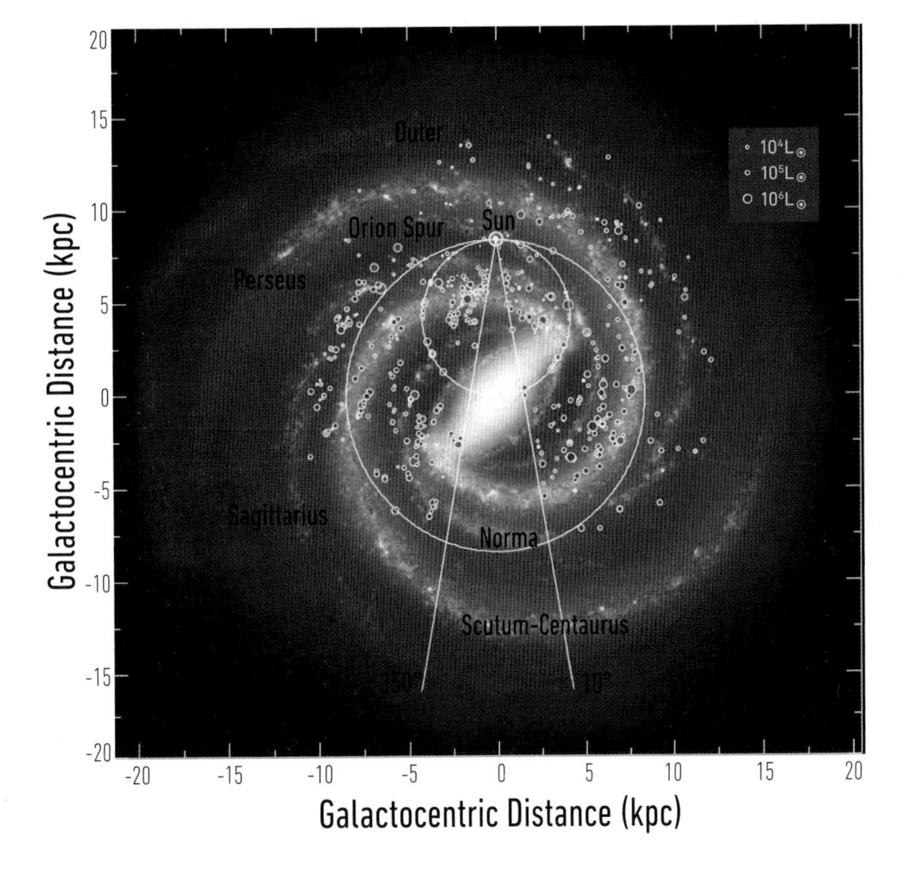

The colored dots in this image show the locations of compact and ultra-compact HII regions in the disk of the galaxy. The larger circle in the image shows the location of the Sun's orbit around the Galactic Center, a journey that takes roughly 250 million years. Radio frequency observations like these indicate the presence of four spiral arms.

Galaxy Detail: HII Regions

HII (pronounced *H-two*) regions are clouds of hot, ionized gas found in and near the plane of the Milky Way Galaxy where star formation has recently taken place. The Orion Nebula is one of the most familiar HII regions (see "Star Birth: The Orion Nebula"), but most HII regions are not visible optically, as most of their radiation is absorbed and scattered by the gas and dust in the Milky Way. The source of energy to ionize an HII region comes from the ultraviolet photons streaming away from young, hot, high-mass stars.

While it is visible optically, even the Orion Nebula has much to reveal at other wavelengths. Near-infrared images of the Orion Nebula show a host of lower-mass stars that are embedded in the molecular material that has not yet been dissipated by the high-mass Trapezium stars. In HII regions, the high-mass stars clear out the area around them and illuminate the surface of the molecular cloud near them. The formation of a few high-mass stars accompanied by many hundreds or thousands of lower-mass stars is probably typical of star formation.

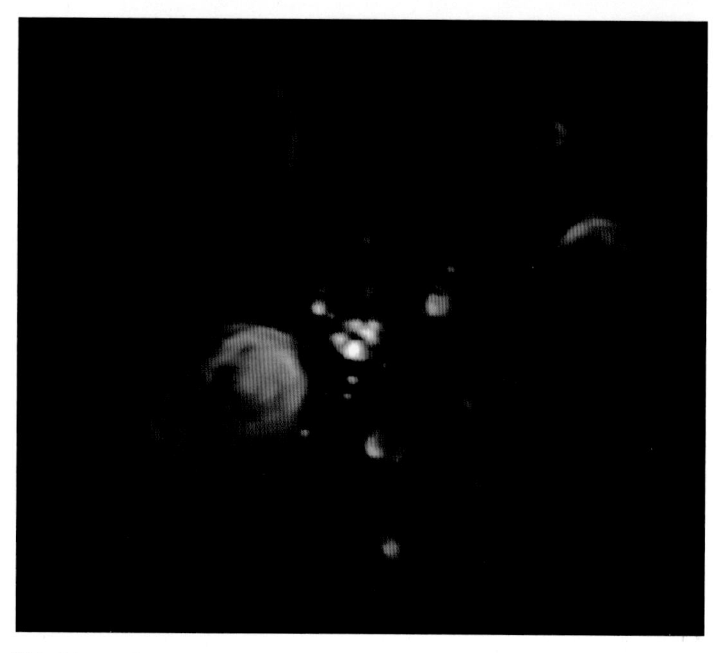

This false-color image shows ionized gas around a large collection of high-mass stars in the Sagittarius constellation (Sgr B2). Each bright blob is made up of hot (10,000 K) gas surrounding one or more high-mass stars. This region is about 25,000 light-years away in the plane of the Milky Way, less than a degree from the Galactic Center.

What's in a Name: HI, HII, and H_2

Hydrogen, with the chemical symbol H, is the most abundant element in the solar system, the Milky Way, and the universe. Neutral hydrogen, made up of one proton and one electron, is designated as HI. If the electron is stripped away or ionized, the hydrogen is designated as HII. If two hydrogen atoms bond together in a molecule—something that can only happen in cold, dense molecular clouds—it can form H_2, a hydrogen molecule.

In some cases, star formation can become self-propagating, with the formation of HII regions triggering molecular clouds into further collapse. In many HII regions, like RCW 120, the expansion of a region of ionized gas around a high-mass star can cause surrounding molecular material to collapse and form new stars. Infrared observations can highlight the position of newly formed massive stars.

The area around the Trapezium stars (the source of the ionizing radiation that lights up the Orion Nebula) is shown in visible light (left) and near-infrared (right), showing hundreds of lower-mass stars.

RCW 120 is an HII region about 10 light-years in diameter. In this false-color image, visible light is shown in red and blue, highlighting the background stars. The HII region is the central diffuse red emission, from the SuperCosmos H-α survey. The emission from the surrounding molecular material is shown in light blue.

Magnetic Fields in the Milky Way

Magnetic fields are responsible for many effects you see in the solar system, from Earth's auroras to the powerful solar flares that leap above the surface of the Sun.

The Milky Way Galaxy contains its own magnetic fields; however, like other characteristics, the magnetic field of the galaxy is hard to measure because Earth is located in the middle of the galaxy's disk. For this reason, in order to observe the galaxy's magnetic field, astronomers need to observe wavelengths that can pass through the galaxy easily: radio waves.

Electrons generate two types of radio emission: thermal and synchrotron radiation.

Thermal radiation occurs when an electron is slowed down by its interaction with another charged particle in a relatively dense environment. Thermal radiation is associated with ionized gas like that found near hot, young stars.

Synchrotron radiation results from electrons spiraling around magnetic field lines and giving off radiation because of their changing direction. Synchrotron radiation is observed from strong magnetic fields. It has been seen coming from the Sun; jets of young, low-mass stars; and even very large scales above and below the plane of the galaxy. The presence of synchrotron radiation, plus measurements of the polarization of the radio waves, can tell astronomers about the strength and orientation of the Milky Way's magnetic fields.

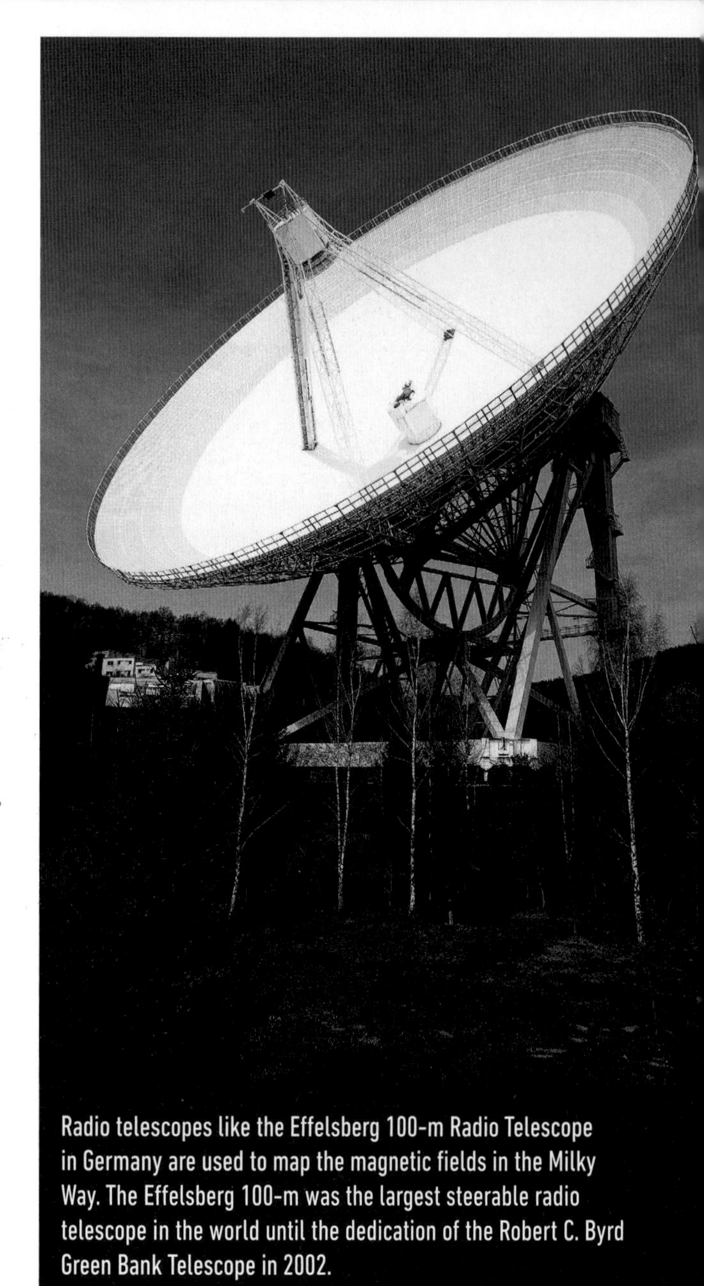

Radio telescopes like the Effelsberg 100-m Radio Telescope in Germany are used to map the magnetic fields in the Milky Way. The Effelsberg 100-m was the largest steerable radio telescope in the world until the dedication of the Robert C. Byrd Green Bank Telescope in 2002.

As polarized radio emission from distant galaxies passes through the disk of the Milky Way Galaxy, the magnetic fields cause the polarization of the light to rotate, an effect called *Faraday rotation.* The amount the polarization is rotated depends on wavelength; as a result, observations made at several radio frequencies of polarized light can measure the strength of magnetic fields between you and the source of polarized light. In order to measure the strength and orientation of magnetic fields throughout the galaxy, astronomers can combine observations of background sources throughout the Milky Way.

Based on these types of observations, the magnetic field lines of the Milky Way appear to be consistent with a spiral, parallel to the plane of the galaxy's disk. The origin of these galactic-scale magnetic fields are still debated. However, recent polarized radio frequency observations made with the CSIRO 64-meter Parkes Radio Telescope lend credence the idea that the gamma-ray emission "bubbles"—which are associated with the large-scale magnetic fields—were generated by a vast wave of star formation near the Galactic Center.

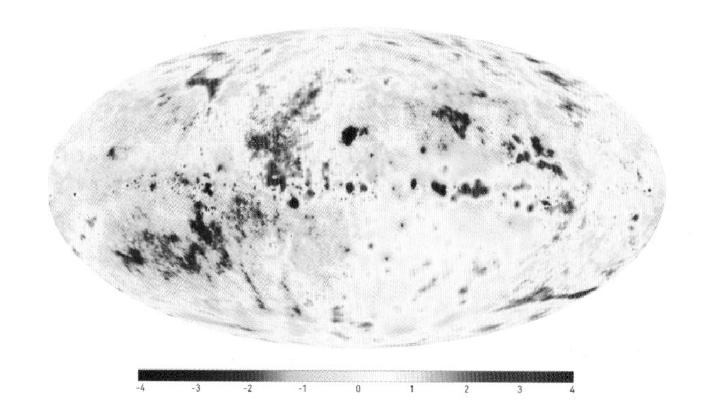

Data from over 40,000 observations were combined to make this image of magnetic fields in the Milky Way. Magnetic field lines have directions; red indicates areas where the galaxy's magnetic field points toward Earth, and blue indicates where it points away from it.

Recent observations of the Galactic Center region show that gamma-ray emission stretches far above and below the plane of the Milky Way Galaxy in two enormous bubbles. The gamma-ray emission appears to be related to large radio structures in the same part of the Milky Way.

The Large and Small Magellanic Clouds

This image shows a panorama of the Milky Way and the LMC and SMC (left side of image) as seen from Very Large Telescope (VLT) site in Chile. The housing of VLT 1 is seen to the right of the image.

The Large Magellanic Cloud (LMC) and Small Magellanic Cloud (SMC) are satellite dwarf galaxies to the Milky Way that are easily visible from the Southern Hemisphere. These companions to the galaxy were mentioned by Persian astronomers and are named after the Portuguese explorer Ferdinand Magellan, who first circumnavigated the globe. Both are classified as dwarf irregular galaxies, or galaxies without a distinct regular shape that have a smaller number of stars than galaxies like the Milky Way.

The LMC and the SMC are much smaller than the Milky Way and are located hundreds of thousands of light-years away. That might seem far away until you consider the nearest spiral galaxy (Andromeda) is about 2.5 million light-years away. The LMC and SMC contain about 10 billion solar masses of stars and gas each, though the LMC—with a diameter of 14,000 light-years—is about twice the size of the SMC. The LMC is about 10 degrees across in the night sky, while the SMC is about half that size.

The LMC and SMC are connected to the Milky Way by a vast arc of material called the *Magellanic Stream.* This stream of gas is thought to have formed because of gravitational interactions between the Milky Way and its two companions.

This black-and-white optical image shows the plane of the Milky Way Galaxy (top to bottom on the left of the image) along with the Large and Small Magellanic Clouds (to the right). The LMC and SMC are about 160,000 and 200,000 light-years away respectively from the center of the Milky Way.

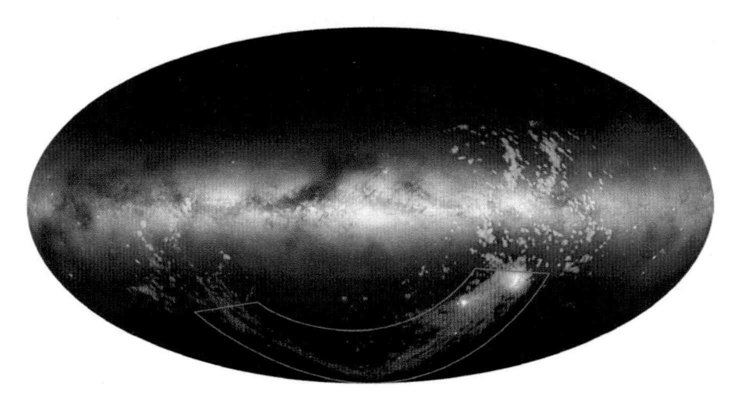

The trail of gas known as the Magellanic Stream has been imaged in neutral hydrogen (HI) and is seen in this combined optical and radio image in shades of pink. The HI line is observed at 21 cm in the radio frequency part of the electromagnetic spectrum. The LMC and SMC can be seen as bright spots in the Magellanic Stream.

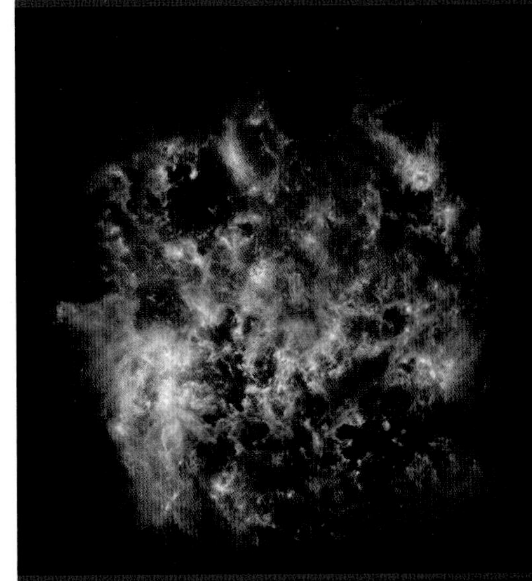

This infrared view of the LMC highlights several well-known areas of star formation, including the Tarantula Nebula (30 Doradus), which is seen in the left-central region. The different colors in this image highlight varying dust temperatures in the LMC. The coolest dust is red and the warmest dust is blue, with the brightest blue-white regions highlighting hot dust found near young, high-mass stars.

A Supernova in the Tarantula

In 1987, a supernova was observed in the LMC near the Tarantula Nebula (see "Supernova 1987A"). This was the closest supernova to have occurred since 1604, when Kepler observed the brightening of the stellar explosion that bears his name. This supernova has been observed almost continuously since it occurred over 25 years ago, and its evolution has increased understanding of supernovae and their impact on their surrounding environment.

Rotation of the Milky Way

The stars, gas, and dust in the Milky Way Galaxy rotate around the Galactic Center. In the inner part of the galaxy, the rotation can be traced in stars, and further out in the disk, clouds of neutral hydrogen can be observed to measure the rotational velocity.

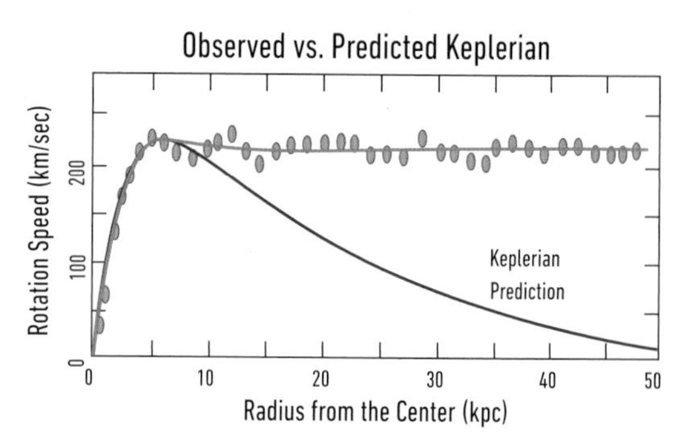

This schematic rotation curve of the Milky Way Galaxy shows what would be expected if the material were in Keplerian rotation around the Galactic Center (red) and the actual velocities of material in the galactic disk (green). Most of the mass in the luminous gas, dust, and stars is located within about 10 kpc from the Galactic Center.

If most of the mass of the Milky Way were concentrated near the center, you would expect to see that the components of the galaxy rotate around following Kepler's laws (see "Planetary Motion")—the inner reaches should rotate more quickly (like Mercury in the solar system), and the outer edges more slowly (like Neptune). But this is not what is observed. At larger and larger distances from the Galactic Center, the velocity of rotation levels off, resulting in what astronomers call a *flat rotation curve*. The fact that the rotation curve does not fall off, but instead is flat, indicates there is much more mass in the Milky Way than you can see in its stars, gas, and dust.

The flat rotation curves of the Milky Way and many other galaxies are well observed. Observations of the stars in the solar neighborhood in the 1930s first indicated that their motions could not be explained by luminous matter alone. Studies of the nearest companion galaxy, Andromeda, made in the late 1930s showed that it, too, seemed to rotate in a way that was not consistent with the amount of matter visible in its stars. Recent studies have not shown gravitational evidence for dark matter in the neighborhood of the Sun. At least in the solar neighborhood, the gravitational effects are consistent with observed luminous matter.

By the 1970s, the work of Vera Rubin and other astronomers made it clear that many galaxies—including the Milky Way—were observed to have "flat" rotation curves, implying that most of their mass was contained far outside the stars and gas visible to normal telescopes.

The explanation for the strange rotation curves of the Milky Way and other galaxies is likely the presence of dark matter, but its exact distribution and nature are unknown.

Early models of the distribution of dark matter showed it was contained in a roughly spherical halo around the Milky Way. More recent work proposes that dark matter is contained in two components—disk and halo—with the disk making up a smaller proportion of the total mass. The halo, taking up a majority of the mass, is said to keep material rotating quickly even far out into the disk.

This artists' conception of the Milky Way Galaxy shows the visible disk of the galaxy and its large halo of dark matter (shown in blue), which is detected in the rotation curve of the Galaxy.

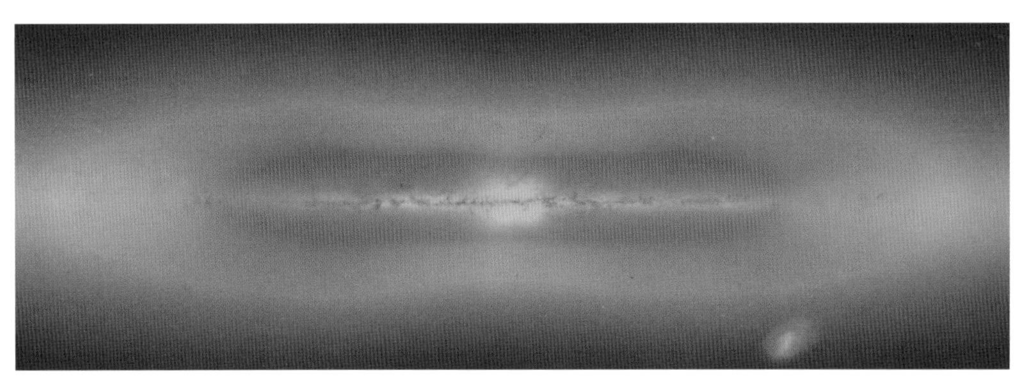

This image superimposes a proposed dark matter disk (in red) over a 2-micron (2MASS) image of the Milky Way.

What Is Dark Matter?

As you read earlier, the rotation of the Milky Way Galaxy indicates that most of the mass of the galaxy consists of something not yet understood; astronomers refer to it as *dark matter*.

While astronomers see its gravitational signature in the rotation curve of the Milky Way, they do not know what form it takes since it emits no photons. Surprisingly, the stars, gas, and dust observed in the Milky Way, or "luminous matter," only make up about 5 percent of its mass.

A similar problem is seen on larger scales, in the realm of galaxy clusters. The problem of the "missing mass" of the universe has been known since the 1930s and is a fundamental quandary. What could this dark matter possibly be?

Baryonic Matter, Antimatter, and MACHOs

If dark matter were made of cold clouds of normal (baryonic) matter, you would be able to observe it by its absorption of radiation.

Dark matter doesn't appear to be antimatter either, which consists of particles like positrons that have all the properties of an electron but an opposite charge. Matter and antimatter annihilate when they meet, forming high-energy photons—gamma rays. There aren't enough diffuse gamma rays observed for antimatter to be dark matter.

Another possibility is that dark matter are very large black holes with the masses of galaxies. But searches for gravitational lensing (the bending of background light by mass) from these types of black holes have come up empty-handed.

It is possible that dark matter could be baryonic if it consists of smaller objects—such as black holes, neutron stars, or brown dwarfs—in the halo of the Milky Way referred to as *Massive Compact Halo Objects (MACHOs)*. A number of MACHO searches have been carried out; however, they have not produced enough discoveries to conclude that MACHOs could solve the dark matter question.

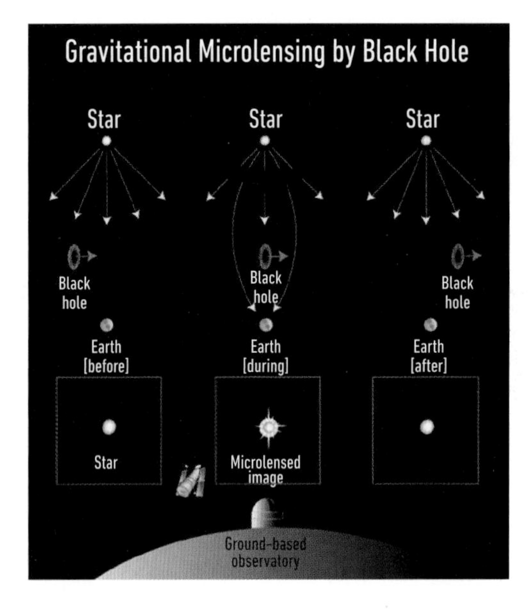

One candidate for dark matter is a collection of black holes either in the disk or the halo of the Milky Way. This diagram shows how the light from a background star is amplified when a MACHO passes between the star and Earth, making the star look brighter while the black hole passes in front of it.

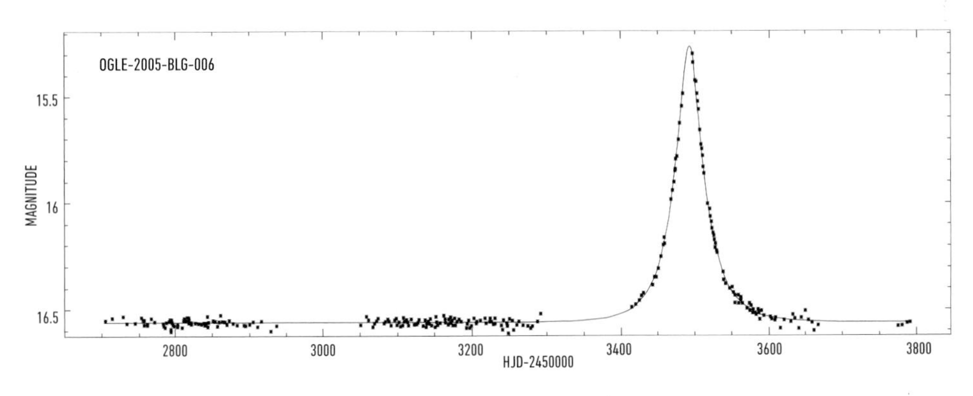

When a MACHO passes between Earth and a distant star, it produces a distinctive "light curve." This plot shows the apparent magnitude of the star (*y*-axis) and time (*x*-axis) based on the Optical Gravitational Lensing Experiment (OGLE) search mission.

WIMPS

Another acronym describes the last class of object that might be dark matter: WIMPs, which stands for Weakly Interacting Massive Particles. "Weakly interacting" means the particles generally can pass through ordinary matter, while "massive" means they must have some mass in order to explain the missing mass in the Milky Way and beyond.

The three possible WIMPs that could be dark matter are neutrinos, axions, and neutralinos. While neutrinos have been detected, they do not appear to be massive enough to be dark matter. The latter two particles (axions and neutralinos) have not been observed.

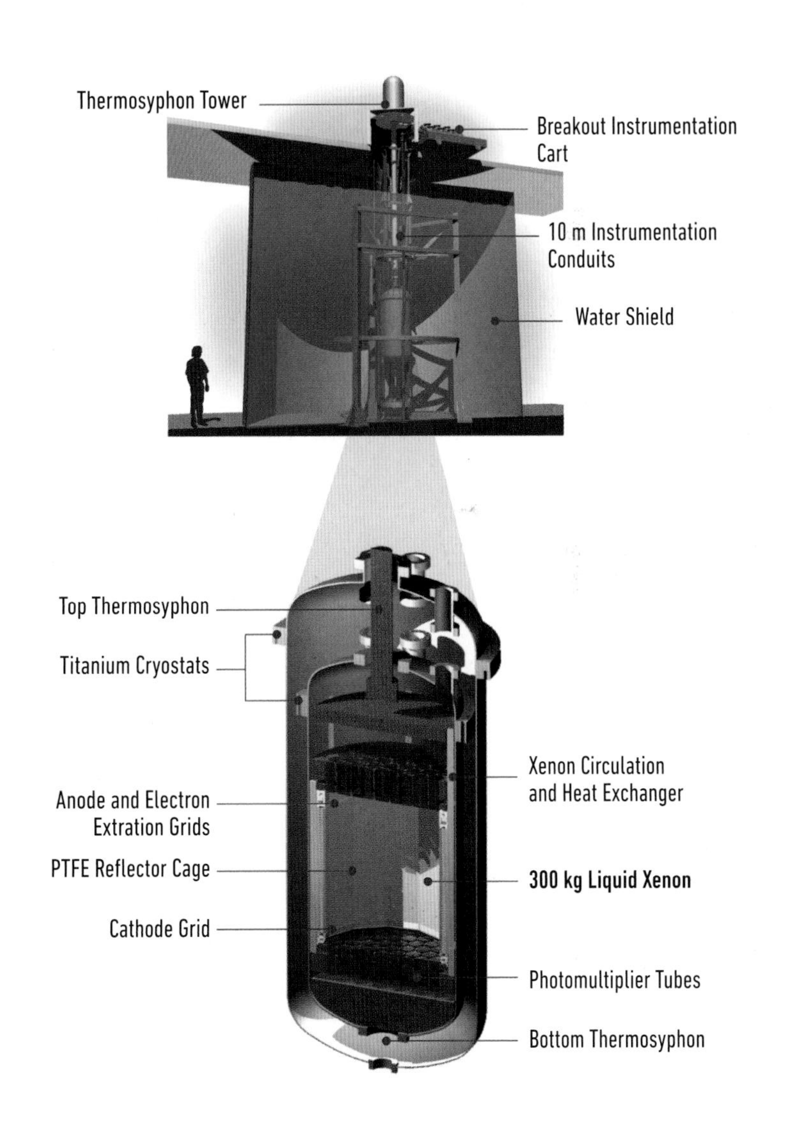

The Large Underground Xenon (LUX) detector was built deep underground to be able to detect the very rare interactions between WIMPs and ordinary matter. The first run in 2013 did not detect any events consistent with WIMPs and may have ruled out the low-mass end of WIMPs as candidates to explain dark matter.

Distances in the Milky Way

One of the fundamental facts that astronomers want to know about an object is its distance. The set of techniques used is sometimes called the *distance ladder*. Here I discuss how astronomers measure distances to objects in the Milky Way and nearby galaxies.

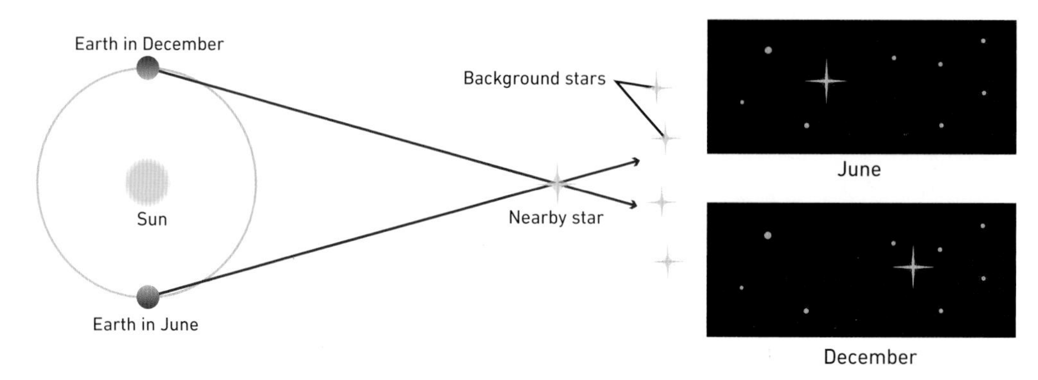

Measurements of the apparent motion of a nearby star relative to more-distant stars can be used to determine the distances to stars out to a few hundred parsecs (or about 500 light-years).

Parallax

Distances to the nearest stars can be measured by their *parallax,* or apparent motion as Earth orbits the Sun. If you hold out your index finger at arm's length and look at it through one eye and then the other, your finger will appear to move back and forth. In this analogy, your finger is the nearby star, your eyes are the positions of Earth separated by six months as it orbits the Sun, and whatever is beyond your finger are the more-distant stars.

Nearby stars are seen to move back and forth with respect to more-distant stars because of Earth's orbit around the Sun giving two different perspectives, an effect known as *stellar parallax.*

The distance to a star (in parsecs) is equal to 1 over the parallax (in arcseconds). As an example, Proxima Centauri is 1.3 parsecs away, and its observed parallax is less than 1 arcsecond.

The problem with stars is that the distances are so large the parallax of even Proxima Centauri, the nearest star, can only be observed with a telescope. So the parallax of stars is a useful method to measure only the distances in the local region of the Milky Way.

Standard Candles

For distances beyond a few hundred parsecs, astronomers have had to find *standard candles,* or astronomical sources that have a known intrinsic brightness. Because the brightness of an object falls off like distance squared, the difference between the intrinsic brightness and the observed brightness can tell you how far away the object is.

Cepheid variable stars (first catalogued by Henrietta Leavitt at Harvard College Observatory) are recognized by the distinctive way their brightness varies, rapidly increasing and then fading more slowly. This type of variable star is thought to occur after core helium burning has begun in a star.

The brightness variations are caused by the regular expansion and contraction of the star's outer layers, with the maximum brightness corresponding to its maximum surface temperature. Cepheids are ideal standard candles because they are very bright stars (many thousands of times more luminous than the Sun), and there is a direct relationship between the period of brightness variation and their luminosity. In order to determine the distance to a Cepheid variable, one only needs to make a careful measurement of its period and its observed brightness.

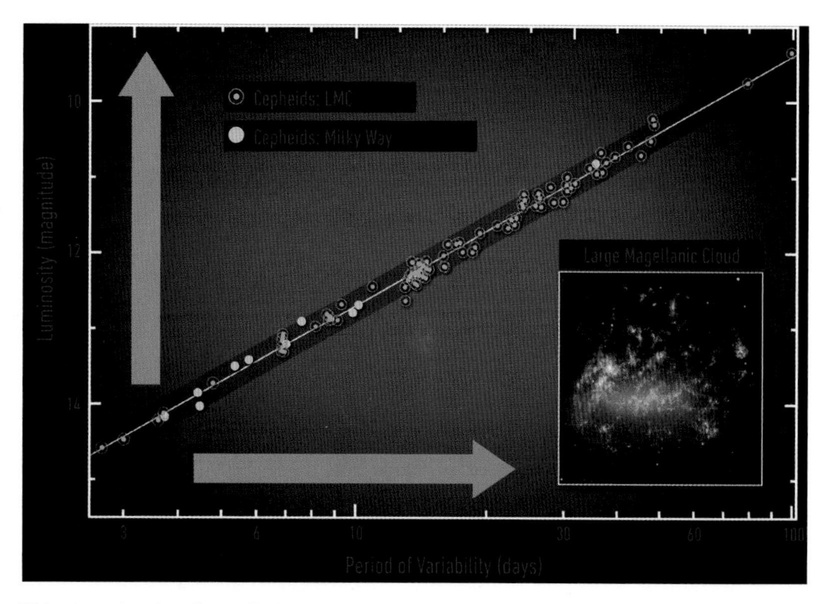

This plot, showing the period-luminosity relationship, is taken from observations of Cepheids in the Milky Way (yellow) and Large Magellanic Cloud (blue) made with Spitzer Space Telescope. Stars that change their brightness more slowly (long period) are more massive and luminous.

The Cepheid variable star V1 in the nearby Andromeda Galaxy changed history, showing that Andromeda was not a nearby nebula, but in fact a galaxy in its own right. This series of inset images show the Cepheid variable star V1 in the disk of the Andromeda Galaxy as observed by the Hubble Space Telescope. This Cepheid was first observed by Edwin Hubble in 1923.

Searching for Life in the Milky Way

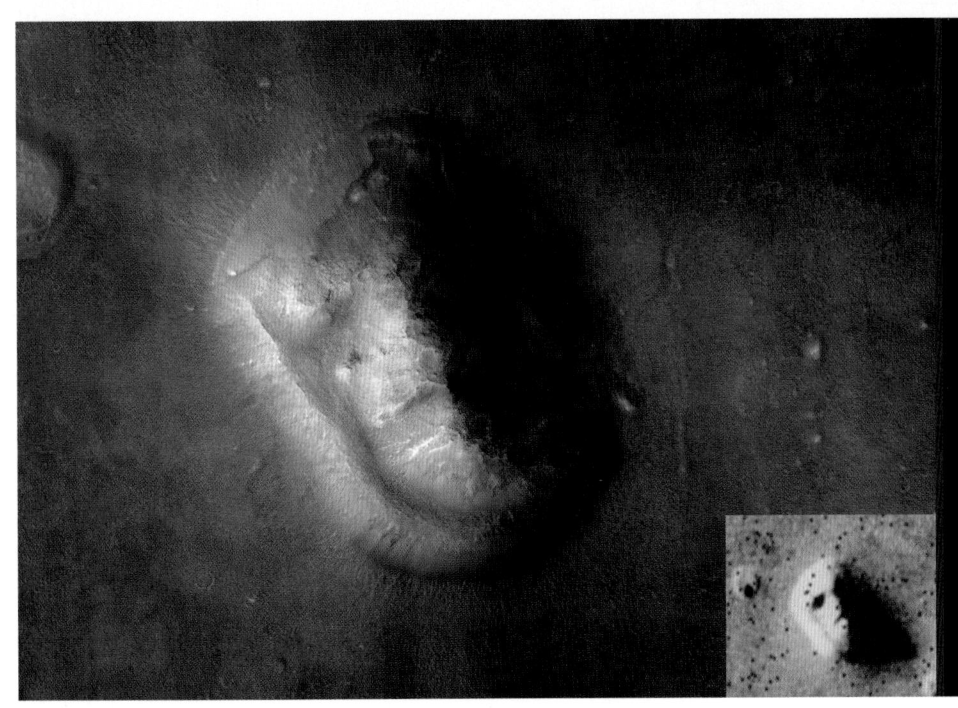

Some people are fascinated by the idea that we are not alone, or that the solar system has been visited by alien life. The "Face on Mars" (inset) is a famous example of a surface feature on Mars that was thought to indicate the possibility of alien visitors. Later high-resolution observations of the feature showed it to be nothing more than a weathered mesa.

Why search for life in the Milky Way? The question "Are we alone?" has one of two interesting answers—either humans have company in the universe and are just one of a number of intelligent, curious beings inhabiting the universe, or humans are in fact alone. If we have company in the universe, it might tell us the chemical and biological processes that led to humans are common. However, if time goes on and we continue to find no evidence for extraterrestrial life, it might tell us that the conditions that led to humanity are rare indeed.

What Are the Odds? The Drake Equation

The Milky Way Galaxy is a very large place, stretching 100,000 light-years across and about 1,000 light-years thick. How can you calculate the odds of extraterrestrial life over such a vast expanse? That's where the Drake equation comes in.

First proposed by Frank Drake in 1961, the Drake equation provides a quick way to estimate how many intelligent, communicating civilizations there might be in the galaxy. It was never intended to provide an exact number of civilizations; instead, it was meant to highlight the factors people should consider when wondering if they're alone.

There are seven terms in the Drake equation. From left to right, they go from well-known (the star formation rate of the Milky Way) to completely unknown (the lifetime of civilizations). The equation is as follows:

$$N = R_* \times f_p \times n_e \times f_l \times f_i \times f_c \times L$$

N: the number of communicating, intelligent civilizations in the Milky Way

R_*: the star formation rate of the Milky Way (stars per year)

f_p: the fraction of stars with planets (number between 0 and 1)

n_e: the number of life-supporting planets per star

f_l: the probability (number between 0 and 1) on an Earthlike planet that life arises

f_i: the probability (number between 0 and 1) that intelligence evolves

f_c: the probability (number between 0 and 1) that communicating civilizations arise

L: the lifetime of a civilization (in years)

Human existence in the Milky Way Galaxy makes it clear that N is not zero. Other than that, though, the final value of N has very few constraints on it. For example, when the equation was first proposed, the number of life-supporting planets was thought to relate only to terrestrial planets at the right distance from the host star for water to exist, but that idea has changed based on observations of the large moons of gas giant planets.

Based on this equation, Frank Drake's original estimates gave a value of around 10,000 civilizations. While the number of civilizations seems sizeable, these civilizations would be (on average) very far apart. Using some rough numbers, and considering the Milky Way to be a very flat cylinder, the 10,000 civilizations would typically be separated by 1,000 light-years.

These great distances between civilizations implies that the search for life in the universe will have to be done not by remote spacecraft or rovers, but by astronomers using the information brought to them by electromagnetic radiation.

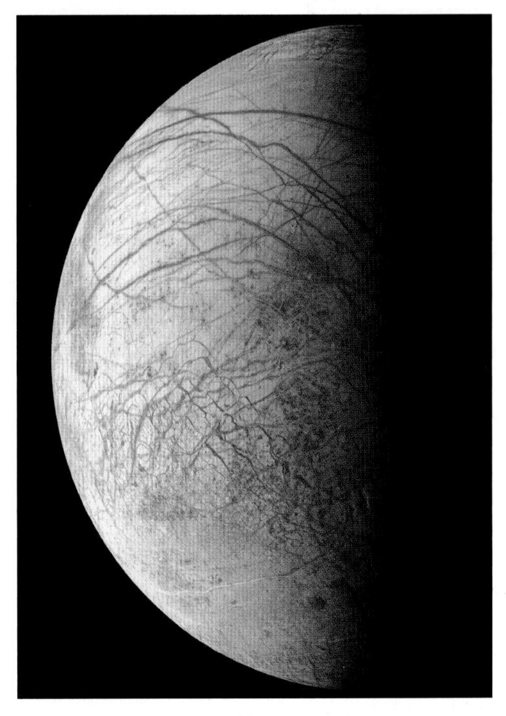

Europa, shown here, may have extensive oceans, warmed by the tidal flexing of the moon. Robotic exploration of the outer solar system and the moons there have made people consider the presence of life on large moons of gas giant planets.

The Allen Telescope Array (ATA) is a privately funded, dedicated set of telescopes that carries out fundamental astronomical research and simultaneously searches for radio signals from extraterrestrial civilizations. The plan is for the ATA to eventually have 350 6.1-m dishes.

Other Planets and Intelligent Communications

While the Drake equation can help people to understand the high probability that other intelligent civilizations exist in the Milky Way, what is the best way to uncover them? People can look for life in the Milky Way in two ways: search for locations that might support life or look for electromagnetic communication from other civilizations in the galaxy.

The efforts to search for intelligent communications are largely carried out by the Search for Extraterrestrial Intelligence (SETI) Institute, a private, nonprofit organization that has a mission to "explore, understand, and explain the origin, nature and prevalence of life in the universe."

Astronomers in the SETI community have focused on detecting radio frequency emissions of the 21-cm line of neutral hydrogen, which is the result of a fundamental property of the most common element in the universe. If other intelligent civilizations existed, the reasoning goes, they would use this frequency as a carrier for any sort of communications. The search for signals is highly technical and requires the use of sophisticated radio frequency detectors. As a result, the Allen Telescope Array (ATA) operated by the SETI Institute is both a dedicated SETI instrument and a highly capable radio interferometer.

Many astronomers are also engaged in the search for locations that might support life. The search for extrasolar planets, or planets orbiting other stars, has been incredibly productive in the past decade. These searches have turned up planets that are similar in mass to Earth and orbit their host star at such a distance that they would have liquid water. While finding planets such as this proves nothing about the existence of other life or civilizations, it provides powerful evidence that the conditions that could support life are widespread in the Milky Way.

NASA's orbiting Kepler mission has identified a number of planets in the "habitable zone" of their host stars. This planet, Kepler-22b, is about 2.4 times the mass of Earth and orbits its star every 289 days.

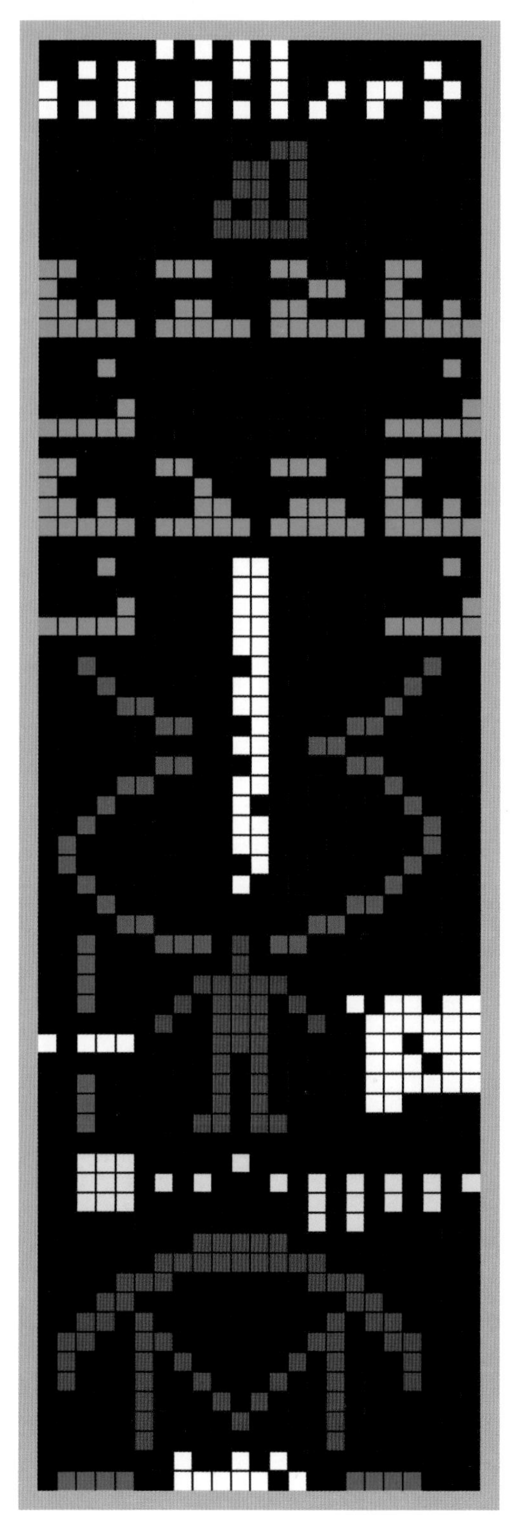

This message was sent in binary form from the Arecibo radio telescope in Puerto Rico toward the M13 globular cluster in 1974. When arranged in a rectangle with sides the length of two prime numbers (23 and 73), the 0s and 1s form a picture that is meant to communicate facts about our civilization.

Exoplanets: Discoveries

Astronomers have explored the planets and moons in the solar system, but for hundreds of years, they have wondered about planets orbiting other stars. Even before the telescope, Italian astronomer Giordano Bruno postulated that the stars were other suns orbited by inhabited worlds.

If astronomers want to observe planets around other stars, how should they go about it? Here are the current techniques:

- **Direct imaging:** The most obvious approach, it involves turning a telescope toward the host star and looking for the planet. This is an appealing idea, because it would provide a picture of the exoplanet.
- **Exoplanetary transits:** This means watching the planet pass in front of the star, in the same way that Venus and Mercury pass in front of the Sun. Transits have been a rich source of new discoveries, especially since the launch of the Kepler spacecraft in 2009.
- **Radial velocity method:** This looks for small velocity changes in the host star as it is tugged back and forth by the orbiting planet. Like exoplanetary transits, this has been extremely productive.
- **Astrometry:** These are high-resolution observations that allow astronomers to watch the host star wobble in the sky.

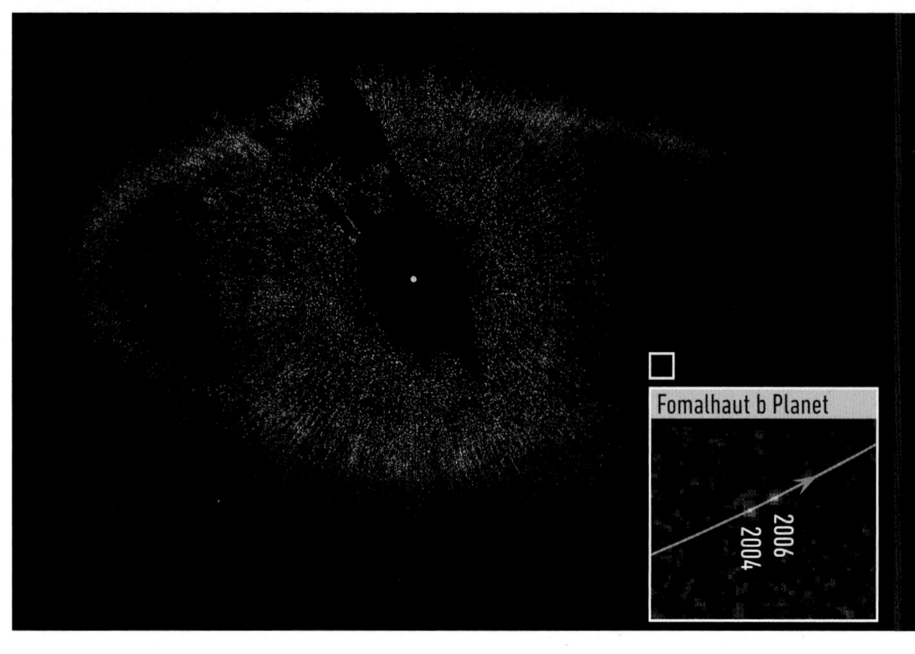

While difficult to accomplish, direct images have been made of planets orbiting other stars. This image of the planet Fomalhaut b was taken by the Hubble Space Telescope. The inset shows two images of the planet in its orbit, 50 AU from the star.

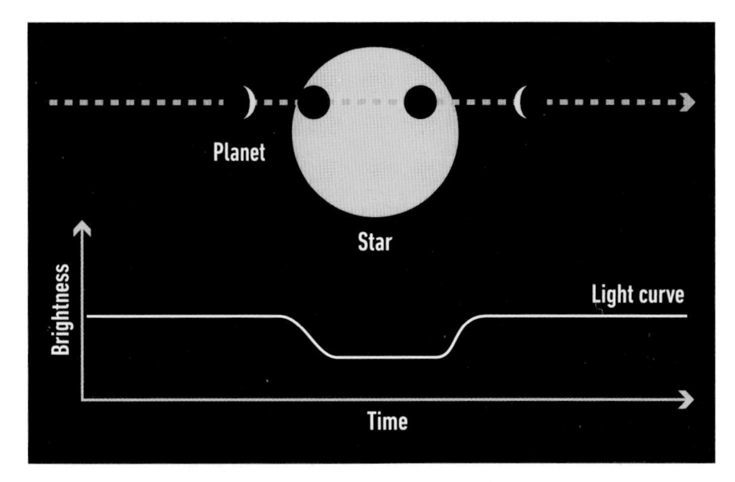

When a planet orbiting another star passes between the star and the observer, the brightness of the star is slightly dimmed—this is known as exoplanetary transit. This plot shows a schematic of an exoplanetary transit (top) and the observed light curve (bottom).

The first confirmed exoplanet around a main sequence star was discovered using the radial velocity method by Mayor and Queloz in 1995. The rate of discovery since then has been rapid and has been made with a variety of methods, with the vast majority of observations by exoplanetary transits and radial velocity changes.

The search for exoplanets passed a milestone in 2013, with the confirmed detection of over 1,000 planets. The vast majority of the detections are Jupiter-sized planets (727), not because they are the most numerous, but because their masses and diameters make them easier to detect.

The star emits blueshifted (shorter-wavelength) light as it moves toward you and redshifted (longer-wavelength) light as it moves away. The first exoplanets discovered were unusual from the perspective of the solar system: Jupiter-sized planets in orbits much smaller than the orbit of Mercury.

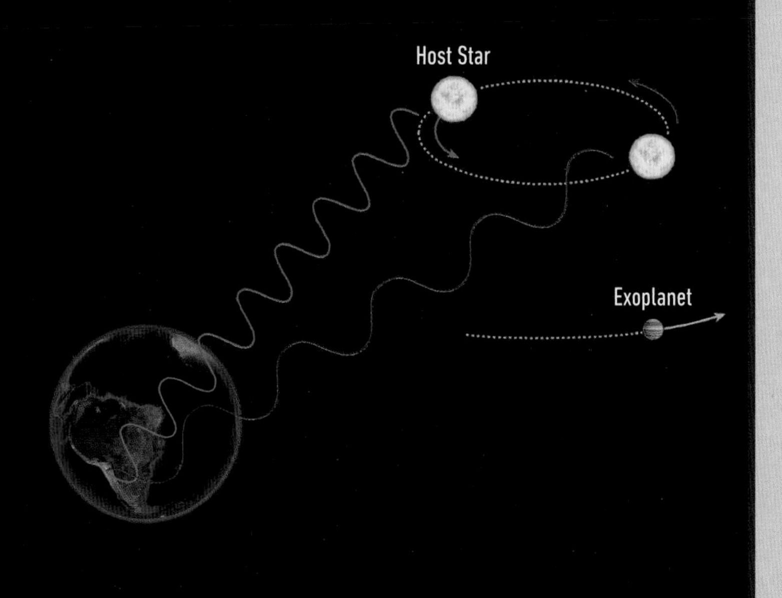

The Doppler Shift

The wavelengths of sound emitted by a moving object are changed by its motion, an effect known as the *Doppler shift*. For example, a car will have a higher pitch (shorter wavelength) when moving toward you with a blaring horn and a lower pitch (longer wavelength) when moving away.

Astronomers have been able to use the Doppler shift of spectral absorption lines from a star to measure its velocity very precisely. These measurements show that some stars wobble back and forth in a predictable way. This wobble is the result of an orbiting planet tugging on the star as it orbits. The star and the exoplanet orbit their common center of mass, causing the star to appear to move back and forth slightly as its planet goes around it.

Exoplanets: Kepler and Future Missions

Many methods have been successful in discovering exoplanets, but the exoplanetary transit method holds particular promise for discovering smaller planets like Earth.

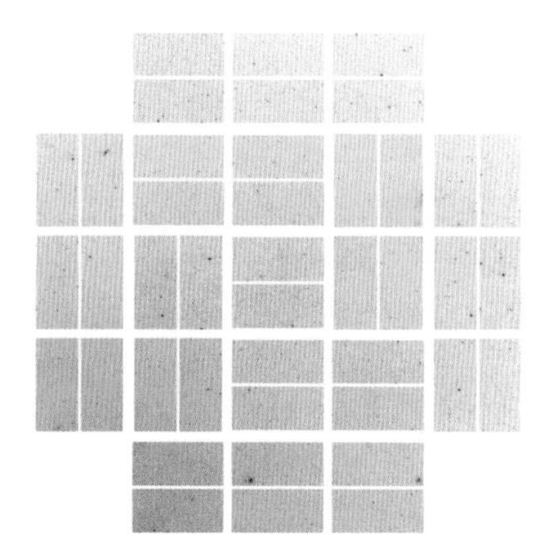

This first-light image shows what data from the Kepler mission looks like.

The Kepler Space Observatory

NASA's Kepler Space Observatory was launched with the primary goal of discovering Earthlike planets in the habitable zones of their host stars. Unlike other telescopes that are made to be able to move around the sky and look at various sources, Kepler was designed to stare at one position in the sky—near the constellation Cygnus, the Swan.

Kepler has an array of digital detectors (known as CCDs, or charge coupled devices) arranged in the focal plane of the telescope. Day and night, Kepler stared at the same place in the sky, observing about 150,000 stars. The brightness of every star in the field of view was monitored, and if any star dipped in brightness in a systematic way, it might host a planet.

By 2013, the Kepler mission had contributed 242 confirmed planets to a list of over 1,000 known. In fact, the number of exoplanets discovered has increased so much that astronomers can now start to think about their general properties. In February 2014, Kepler announced the addition of 715 new planets orbiting 305 stars, almost doubling the number of known exoplanets.

The mechanical failure of gyroscopes in the Kepler spacecraft—essential to fix its gaze on one part of the sky—failed in 2013, ending its ability to identify new candidate planets.

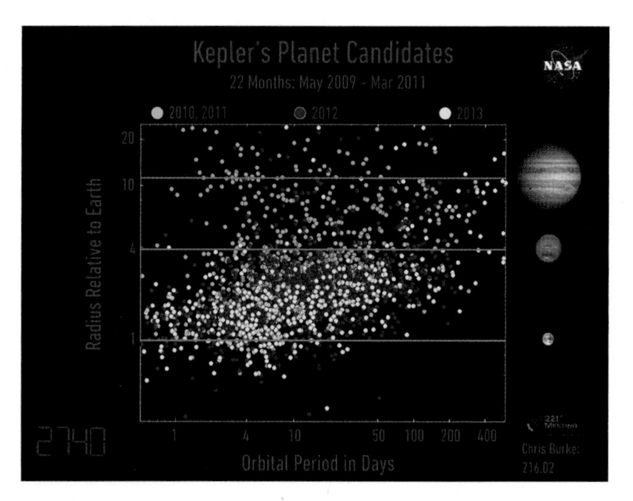

In this plot, Kepler data releases are shown in different colors—blue (2010 and 2011), red (2012), and yellow (2013). The *y*-axis shows the planet's size (relative to Earth) and the *x*-axis shows the time required to orbit the star.

The Future

Several existing space telescopes, including Spitzer and Hubble, have proven to be highly effective ways to observe transits. Competition for observing time on these telescopes is intense, though, driving the need for dedicated transit telescopes in space.

Missions are in the works that will build on the success of the Kepler mission. The proposed TESS (Transiting Exoplanet Survey Satellite) will be able to search the entire sky for Earthlike exoplanets instead of fixing its view on one part of the sky. TESS will focus its efforts on bright, nearby stars, complementing the efforts of Kepler, which surveyed all stars (bright and faint) in a particular part of the sky.

This overlay of the Kepler field of view shows the locations of the Earth-sized planetary candidates (left) and all planetary candidates (right) as of January 2013. Before the Kepler mission was launched, there were three known exoplanets in this part of the sky. As of February 2014, there are over 950 confirmed planets and many thousands of candidates.

The Goldilocks Zone

Planets orbit stars at various distances. In the solar system, planets orbit from the scorching proximity of Mercury to the frigid distances of Neptune. Not all stars have the same temperature as the Sun—some are hotter and others cooler. When a planet orbits a star at a distance where the temperatures allow for liquid water to exist, it is said to be in the *habitable zone.* Some astronomers call this the Goldilocks zone: not too hot, not too cold, but just right.

PART 5

Galaxies

The Milky Way is just one of a mind-boggling number of galaxies in the observable universe. In this part, you learn about other galaxies in the universe.

You begin by exploring the galaxies closest to the Milky Way—the Local Group, and the closest neighbor, the Andromeda Galaxy. You then get to examine examples of each type of galaxy—spirals, barred spirals, ellipticals and irregulars—and how their various structures may have formed. You also learn about the supermassive black holes that lie in the central regions of all galaxies, as well as the active galaxies that derive energy from them (such as Seyferts, quasars, blazars, and radio galaxies). The arrangement of galaxies into clusters and superclusters are some of the largest structures visible in the universe, so I finish with a discussion of these Very-Large-Scale Structures (VLSS).

The Andromeda Galaxy in the Night Sky

The Andromeda Galaxy (or Messier 31) is the nearest spiral galaxy to the Milky Way and part of the Local Group of galaxies. Slightly larger than the Milky Way, the galaxy can be seen to have two companions, M32 and M110, both of which appear to have had interactions with Andromeda. Andromeda and its companions are hurtling toward the Milky Way and will collide with this galaxy in about 4 billion years.

While it is the closest galaxy, it is still 2.5 million light-years away and just looks like a chalky smudge to the naked eye. The galaxy is visible in the night sky without a telescope, though, which earns it the honor of being the most-distant object visible without a telescope by a long shot.

The constellation (and the galaxy) are located at a high declination (about +40 degrees), meaning they are easily visible only in the Northern Hemisphere.

While the galaxy looks indistinct and faint without a telescope, it is quite large in photographic and CCD images, spanning several degrees on the sky, or many times the size of the full moon. (What observers see with the naked eye is just the bright central bulge of the galaxy.)

Observations at other wavelengths highlight star-formation processes in the galaxy. As the closest spiral galaxy to the Milky Way, Andromeda provides a valuable laboratory for understanding the structure and evolution of spiral galaxies in general.

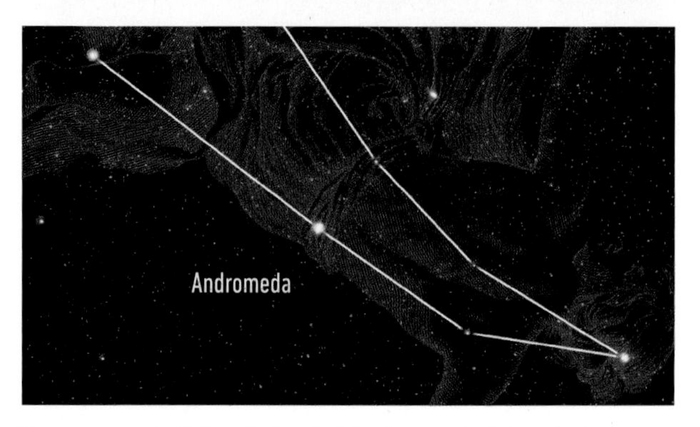

The Andromeda Galaxy is located in the constellation Andromeda. This graphic overlay shows the position of the galaxy in the constellation. The galaxy is the fuzzy, oblong, pink object in the upper right.

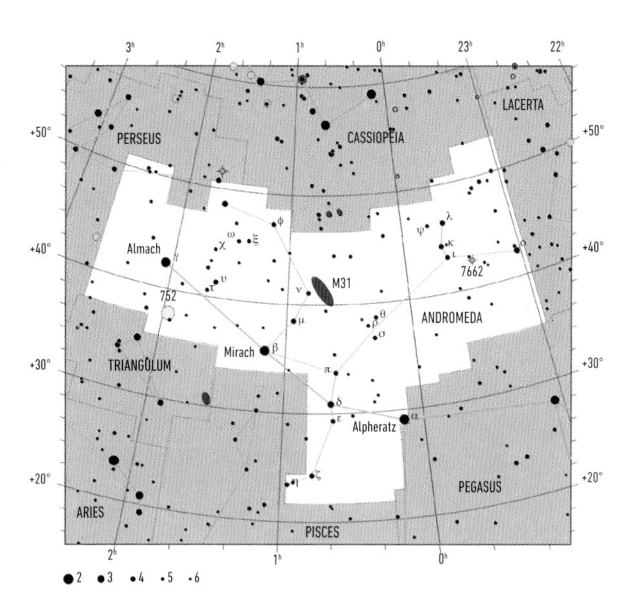

This finding chart shows the position of the Andromeda Galaxy (large red oval) among its neighboring stars and constellations. The star in the head of Andromeda, Alpheratz, is the brightest star in the Andromeda constellation. Two stars, Mirach and μ Andromeda, "point" at the galaxy and make it a bit easier to find in the night sky.

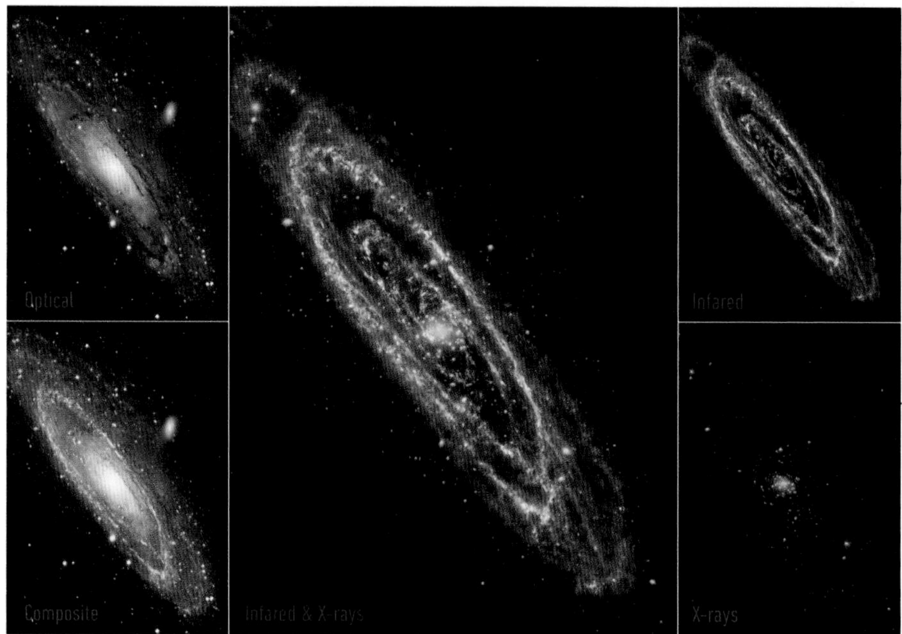

The European Space Agency (ESA) released these images of the Andromeda Galaxy that highlight star formation as seen in the infrared and X-ray. Infrared images show there are several concentric rings of dust in the galaxy, while X-rays highlight the positions of recent supernovae.

This image of the Andromeda Galaxy was taken with the Hyper Suprime-Cam (HSC) on the Subaru Telescope located at Mauna Kea in Hawaii. The Subaru Telescope has an 8.2-m diameter primary mirror, making it very sensitive; this camera is specialized to image large areas of the sky, up to 1.5 degree across. As this image shows, the Andromeda Galaxy is larger than the field of view of even the camera.

The Local Group

As you know, farther out into the universe, stars are collected into a variety of different groupings called *galaxies*. These galaxies are likewise collected into groups, one of which is known as the Local Group (see "Galaxy Clusters" and "Galaxy Superclusters" for more information).

The Local Group is a group of galaxies (or galaxy cluster) held together by gravity that consists of the Milky Way Galaxy, the Andromeda Galaxy, and a variety of other small galaxies—in all, it contains over 50 galaxies. The Milky Way and Andromeda galaxies are by far the most massive.

Galaxies in the Local Group are seen in projection (top) and in perspective (bottom). Each ring represents 2 Mpc (150 km/s in redshift). The galaxies within the central circle are considered members of the Local Group.

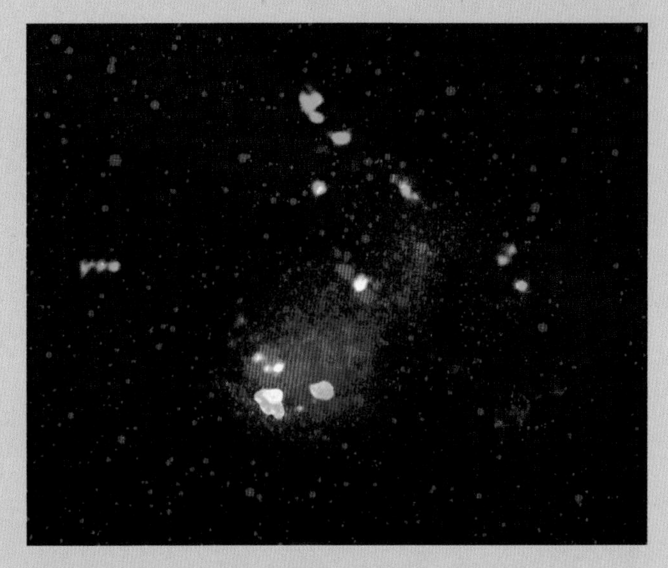

This composite view of the Local Group member IC10 shows the dwarf irregular galaxy at many different wavelengths: optical (blue), H-α (red), and carbon monoxide (green). Astronomers were surprised to find molecular gas scattered in clumps far from the center of the galaxy.

The third-largest galaxy in the Local Group is the spiral galaxy M33, or the Triangulum Galaxy. The pink emission (from the hydrogen alpha line) highlights regions of star formation in the disk of this galaxy. Variable stars (a valuable standard candle) in the Triangulum Galaxy have been used to establish the distance scale (see "Distances in the Milky Way").

As the largest galaxies in the Local Group, both the Milky Way and Andromeda have a host of smaller satellite galaxies, including the Large and Small Magellanic Clouds and M32 and M110 respectively. These satellite galaxies show evidence of interactions with the larger spirals they accompany.

A number of other smaller galaxies are located on the fringes of the Local Group. Most of the members cluster around the larger Andromeda and Milky Way galaxies, but there are some far-flung members. For example, NGC 3109, a small, irregular galaxy, appears to be interacting gravitationally with its neighbor, a dwarf elliptical galaxy called the Antila Dwarf.

This ultraviolet image of the galaxy NGC 3109 shows one of the most distant members of the Local Group. This galaxy is much smaller than the Milky Way, with only about 2 billion solar masses. It is almost twice as far away as Andromeda

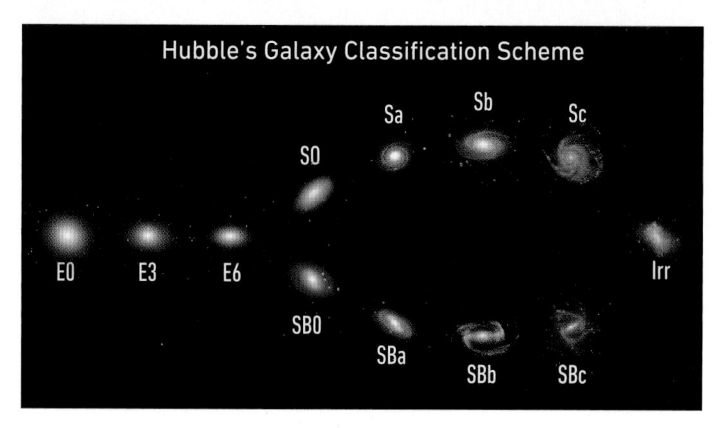

Edwin Hubble is credited for the first systematic classification of galaxies. Galaxies come in four basic varieties: elliptical (E), spiral (S), barred spiral (SB), and irregular (Irr). These galaxy types are often displayed in what is called the "tuning fork" diagram, named so because the diagram itself resembles the shape of a musical tuning fork.

Spiral galaxies are some of the loveliest objects in the night sky. Astronomers can observe them from a number of angles and see them face on, edge on, and everywhere in between.

Once large numbers of galaxies had been observed, it became clear that many of them had these distinctive spiral arm structures and a central bright bulge. The spiral arms themselves highlight the location of massive star formation in spiral galaxies, while recent observations support the idea that the center of all spiral galaxies contains a supermassive black hole.

Spiral galaxies are classified by several prominent features, such as the presence or absence of a bar and the size of the central bulge. Spiral galaxies with a bar—a linear feature near their center—are classified as barred spirals (SB), while those without are simply classified as spirals . Further, the size of the central bulge gives a spiral one of three subclasses: a, b, or c. The largest central bulges are classified as a, medium-size bulges are classified as b, and the smallest bulges are classified as c. For example, NGC 1300 is classified as an SBbc, indicating that it is a barred spiral with a bulge that is intermediate between b and c.

What Is a Spiral Arm?

Spiral arms in other galaxies are thought to be the result of density waves passing through the disk of the galaxy (see "Galaxy Detail: Spiral Arms"). While you might imagine that spiral arms are physical structures spinning around like a fan blade, they are in fact waves. As a wave passes through the disk of a galaxy, those regions are compressed, triggering star formation and briefly lighting up that part of the disk with the energy from newly formed high-mass stars. These density waves can be triggered by near encounters and collisions between galaxies (see "Galaxy Collisions").

If galaxies are somewhere between being clearly elliptical (no apparent disk) or clearly spiral, with a disk but no spiral arms, they are classified as lenticular galaxies (S0). Galaxies that do not fit into the elliptical or spiral classification are irregulars. For example, the Large and Small Magellanic Clouds are classified as irregular galaxies.

This infrared image from the Very Large Telescope shows a piece of the sky about 6.4 arcminutes across that includes a barred spiral galaxy, NGC 1300. As you can see, its arms extend from the end of a clearly delineated bar feature.

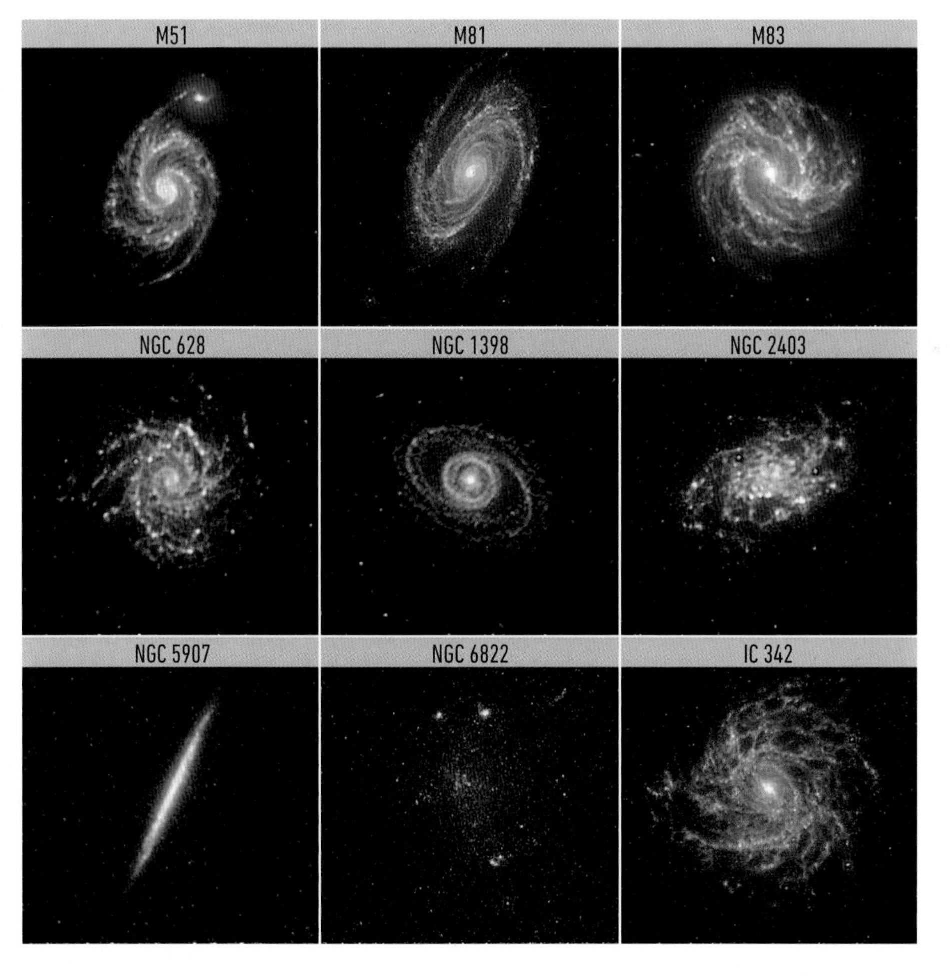

NASA's Wide-field Infrared Survey Explorer (WISE) mission released this showcase of a number of different spiral galaxy types. The false-color image is made up from observations at 3.4-micron (blue), 4.6-micron (cyan), 12-micron (green), and 22-micron (red) wavelength emission. All the galaxies shown are spirals, except NGC 6822, which is irregular.

Elliptical Galaxies

Elliptical galaxies, which comprise both the smallest and the largest known galaxies, are found to be common in the centers of galaxy clusters. In current galaxy evolution models, elliptical galaxies are the end result of galaxy collisions, and as a result are eventually located at the gravitational center of galaxy clusters.

They do not have spiral arms, nor are they irregularly shaped—instead, they vary in shape from circular to highly elongated. Edwin Hubble first classified them based on their shape, with the roundest being E0 galaxies and the most cigar-shaped being E7. However, the apparent shape of an elliptical galaxy may have nothing to do with its actual three-dimensional shape. To visualize this in smaller terms, just consider that a hot dog bun viewed end on might look quite circular and a hamburger bun viewed from the side might look cigar-shaped. Astronomers must often look at other information, including the velocities of stars in an elliptical galaxy, in order to figure out the physical shape of a galaxy.

Unlike spirals, ellipticals appear to have very little of the gas needed to make new stars. The shape of the tuning fork diagram (see "Spiral Galaxies") shows that Hubble originally thought ellipticals might evolve into spiral galaxies. However, their lack of gas and dust and the population of stars they contain indicate that elliptical galaxies are made of older stars. In fact, the stars in elliptical galaxies are some of the oldest stars known and are sometimes used to determine the age of the universe.

NGC 1132 is a giant elliptical galaxy that is 300 million light-years away, or over 100 times farther away than the Andromeda Galaxy. It represents the large end of elliptical galaxies, with a diameter twice that of the Milky Way. The galaxy is surrounded by a large number of globular clusters and background galaxies, but its smooth shape shows no evidence of gas or star formation.

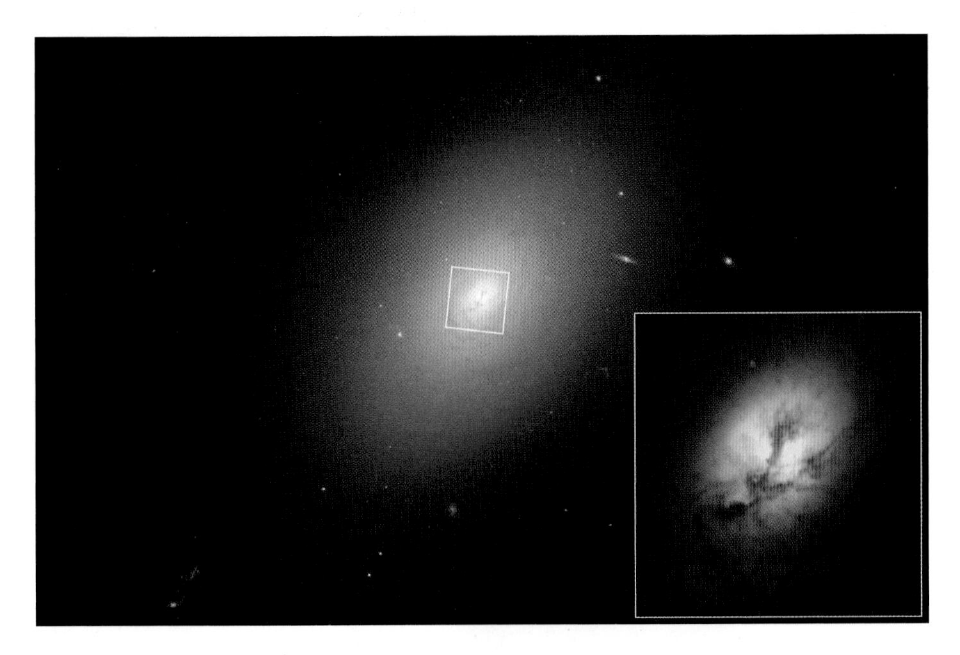

Understanding of elliptical galaxies is still evolving. At first glance, NGC 4150 appears to be a typical elliptical galaxy. However, the inset ultraviolet (blue) and visible (orange) images from the Hubble Space Telescope show that young stars are still being formed near the galaxy's core. This recent activity may be the result of new fuel provided by a merger between galaxies.

Current understanding of elliptical galaxies suggests they are the result of many galactic collisions. When galaxies collide, much of the gas and dust needed to make new stars is flung into intergalactic space; the rest is triggered into vast waves of star formation. In fact, the star formation apparent in elliptical galaxies like NGC 4150 indicates that some of the collisions may have been recent. These collisions may also be related to the presence and activity of the supermassive black holes found in the cores of giant elliptical and spiral galaxies.

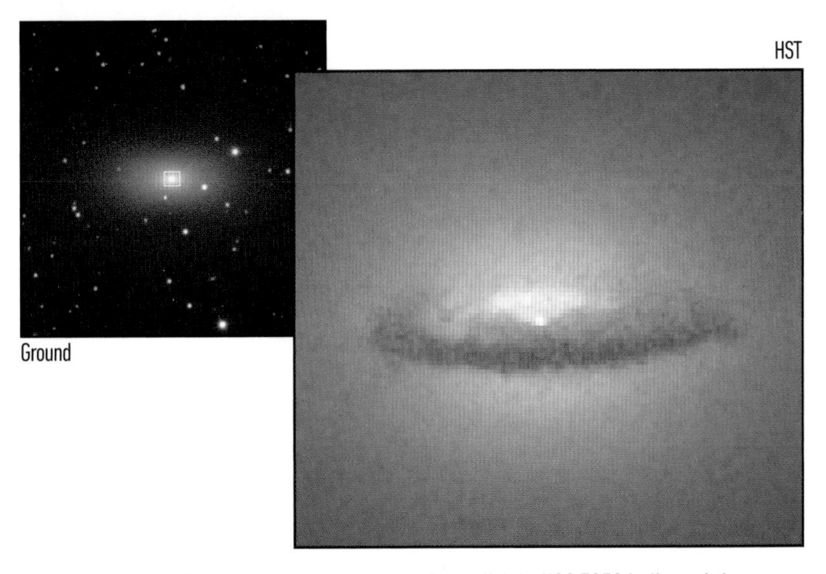

The discovery of a 3,700-light-year-diameter dusty disk in NGC 7052 indicated the presence of a 300-million-solar-mass black hole, much larger than the 4-million-solar-mass black hole located at the center of the Milky Way. Black holes like the one at the center of NGC 7052 are evidence of past mergers.

Irregular Galaxies

One of the four original classifications of Edwin Hubble, irregular (Irr) galaxies have no distinct shape but typically contain a mixture of gas, dust, and stars. Orbiting in crowded environments like the Fornax Cluster, irregular galaxies are torn up within a few billion years as they encounter larger galaxies.

With no discernable regular structure, the irregular galaxy NGC 1427A is in the Fornax Cluster and shows evidence of widespread star formation (blue). A face-on grand design spiral galaxy can be seen in the background of the irregular galaxy to the upper left.

Types of Irregular Galaxies

Irregular galaxies are divided into two structural subtypes: Irr I and Irr II. Irr I galaxies have some hint of a structure that could place them into another Hubble type, while Irr II galaxies do not. Dwarf irregular galaxies (dIrr) have very low metallicity and a high gas content, consistent with them being some of the earliest galaxies in the universe.

Metallicity of Dwarf Irregulars

In the same way that asteroids are rich sources of information about the early solar system, dwarf irregular galaxies tell astronomers much about the early universe and the material from which galaxies like the Milky Way may have formed. This is due to the low metallicity of dwarf irregulars, indicating that they formed from material in the early universe and have not had much processing since then.

The Spitzer Space Telescope took this image of the Small Magellanic Cloud (SMC) in the infrared between 3.6 and 160 microns. Within its chaotic boundaries are old stars (blue) and young stars surrounded by dust (green and red). The SMC is typically classified as a dwarf irregular galaxy.

Metallicity means how much of a star or galaxy's matter is made up of atoms ("metals" to astronomers) heavier than hydrogen and helium. For the most part, these heavier elements are generated by fusion in stars and by their explosive ends. Stars are divided by their age into Population (Pop) I and II stars. Pop I stars have the highest metallicity (and formed more recently), while Pop II stars are lower in metallicity (and older). Low metallicity means that a star or galaxy is made of stuff that has not been processed through as many cycles of star birth and death. The oldest galaxies should be composed of stars containing a larger fraction of hydrogen and helium than later stars.

Because they have very low metallicity, dwarf irregular galaxies are thought to contain very old stars. The low metallicity of many dwarf irregular galaxies also indicates they have not had significant interactions with other galaxies. Dwarf irregular galaxies are some of the earliest galaxies observable.

The dwarf irregular galaxy Leo A as imaged by the Subaru Telescope located at the summit of Mauna Kea in Hawaii. This galaxy is about 10,000 light-years across and about 2.5 million light-years away, the same distance as the Andromeda Galaxy. Theories of galaxy formation indicate that galaxies like Leo A might be the building blocks of larger galaxies.

The dwarf irregular galaxy Henize 2-10 as imaged at optical, radio, and X-ray frequencies contains both high-mass star formation and a massive central black hole. Many galaxies in the early universe probably resembled Henize 2-10.

How Distant Are Galaxies?

Distances form an important part of people's understanding of the universe. For example, one of the early debates in astronomy in the twentieth century was whether Andromeda was a small spiral nebula contained in the Milky Way Galaxy or a more-distant spiral galaxy all its own. The only way to resolve the debate was to accurately measure its distance. Once this was done using the standard candle of Cepheid variable stars in Andromeda (see "Distances in the Milky Way"), its distance was determined to be about 2.5 million light-years, making it a galaxy all its own—and the universe a much bigger place.

But resolving individual stars in ever-more-distant galaxies is difficult for two reasons: the brightness of stars decreases with distance squared, and even the best telescopes don't have the resolution required to pinpoint an individual star in a distant galaxy. The discovery of other standard candles has overcome both of these limitations.

A Type Ia supernova (SN2014J)—the kind used to determine distances to many galaxies—recently occurred in the galaxy M82, an irregular galaxy in Ursa Major. It was discovered by chance in an observing session by a group of students at the University of London Observatory. The position of the supernova is shown with the white arrow.

Units of Distance for Galaxies

On the largest scales, people need big units to talk about distances. Meters, kilometers, and astronomical units won't do. Parsecs and light-years are even relatively small units when discussing the distances to galaxies (see "Stars and Spectroscopy"). Therefore, millions of light-years or millions of parsecs (abbreviated Mpc) are generally used for the distance of galaxies. Using these units, Andromeda is at a distance of 2.5 million light-years or 0.8 Mpc and the Virgo Cluster is at a distance of 56 million light-years or 17 Mpc.

After its discovery, SN2014J was imaged by many other astronomers. These images from the NASA SWIFT telescope show M82 before and after the supernova occurred. This combined image shows ultraviolet (blue and green) and visible (red) light. Several important, nearby supernovae have been detected by accident.

When it comes to successfully measuring large distances in space like those of galaxies, one of the tricks is to figure out something that is easy to measure (like the period of a Cepheid variable star) to tell you something harder (like its distance). With galaxies too distant to observe individual stars, the Tully-Fisher method provides one way to do that. Astronomers determined there was a relationship between the intrinsic brightness of a galaxy and the rotational velocity of the galaxy (determined from observations of the neutral hydrogen [HI] line).

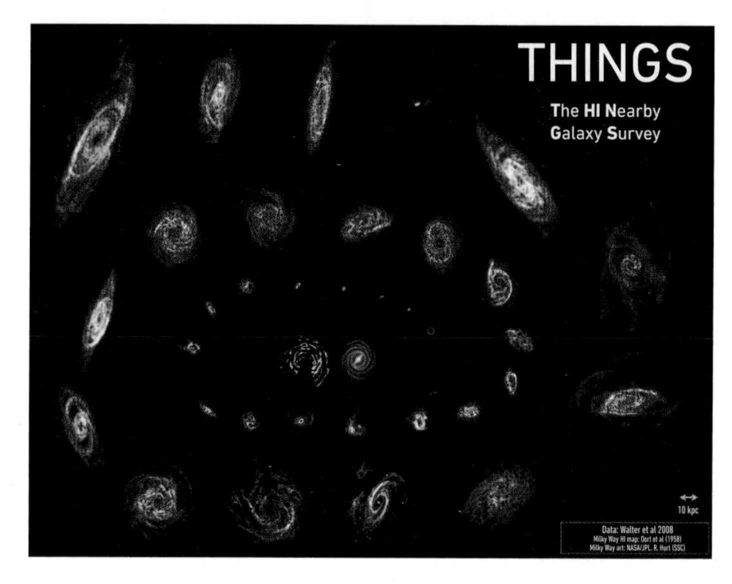

Observations of the HI line in galaxies can be used to determine their distances using the Tully-Fisher method. These images show a group of nearby galaxies as imaged at 21 cm with the Very Large Array. These images and the spectral lines associated with them can be used to determine the rotational velocity of each galaxy.

Another recently developed method to measure the distance of galaxies is to use the peak brightness of a Type Ia supernova. These supernovae result from one star pulling matter from another star to the point where it explodes and have a consistent intrinsic peak brightness.

These two independent methods give similar distances to the nearby Virgo Cluster, giving astronomers confidence that these methods can be used interchangeably.

Galaxy Clusters

In the same way that gravity pulls stars into galaxies, it pulls galaxies into groupings called *galaxy clusters.* Careful mapping of more and more distant galaxies in the twentieth century made it clear that galaxies were not randomly distributed, but rather gathered into distinct gravitational groupings.

Galaxy Clusters Near the Local Group

As you read previously, the Milky Way, Andromeda, and some other smaller galaxies in this region of the universe make up the Local Group. Several large galaxy clusters are found relatively close to the Local Group, including the Eridanus, Fornax, Virgo, and Coma Clusters. These clusters are named by the constellation in which they are found and typically contain hundreds to thousands of galaxies.

While very large, Virgo (about 17 Mpc distant) is considered an irregular cluster, with its galaxies spread over a large region of the sky. The Fornax Cluster is located at a distance of 19 Mpc (60 million light-years) and contains a large number of elliptical and lenticular galaxies. Eridanus is about 23 Mpc away and contains several hundred spiral, elliptical, and irregular galaxies. The Coma Cluster (about 90 Mpc away) is the closest regular cluster.

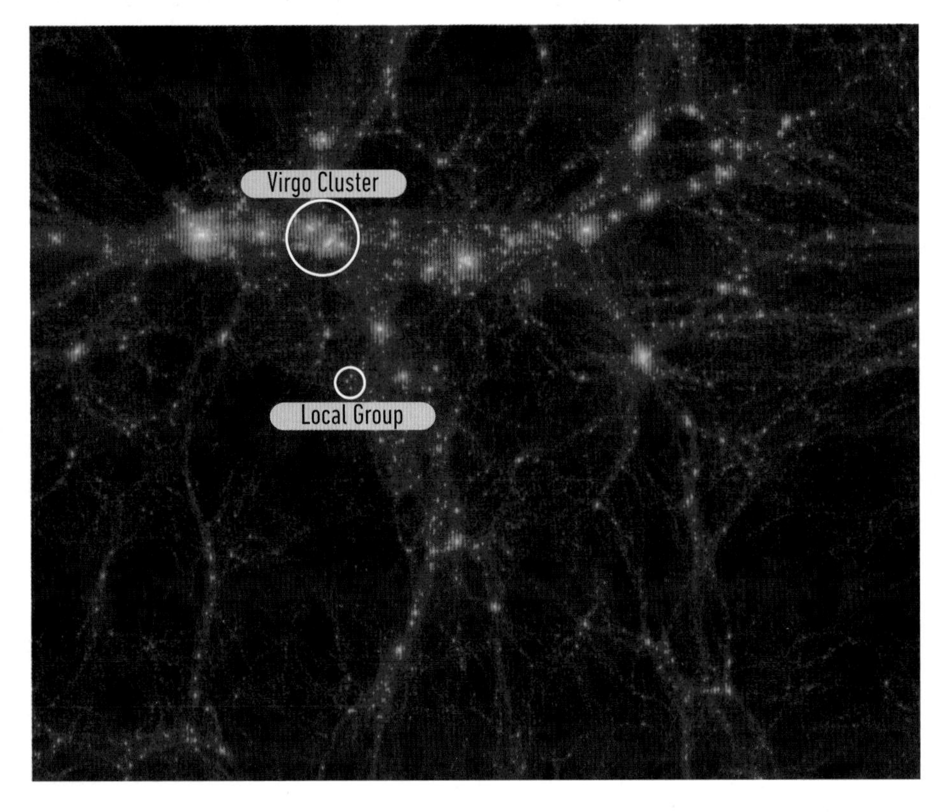

Numerical simulations of the evolution of a dark-matter-dominated universe on supercomputers are able to generate structures similar to those seen in the Virgo supercluster. Virgo and the Local Group are labeled.

Composition of Galaxy Clusters

The clusters themselves, like galaxies on smaller scales, are observed to contain large amounts of dark matter, as seen by its gravitational effects. Surprisingly, the mass of a cluster is contained mostly in this last component—dark matter comprises about 90 percent of the mass of a cluster of galaxies. Accounting for the rest, luminous matter between the galaxies (hot gas mostly visible as X-rays) makes up about 9 percent of the cluster mass, while the material in galaxies themselves only makes up about 1 percent of the cluster mass.

This false-color image shows combined optical (blue) and infrared (red and green) images of the Coma Cluster of galaxies (Abell 1656). Thousands of faint green objects highlighted in the inset image (upper-right corner) are dwarf galaxies in Coma. The two large elliptical galaxies in the center of the cluster are NGC 4889 and NGC 4874.

Less well known but no less impressive is the galaxy cluster MACS J0152-2852. Most of the objects in this image are galaxies, each containing hundreds of billions of stars. As in many galaxy clusters, giant elliptical galaxies (yellow) occupy the gravitational center and contain mostly older stars. The young, hot stars in spiral galaxies near the cluster edge are bluer.

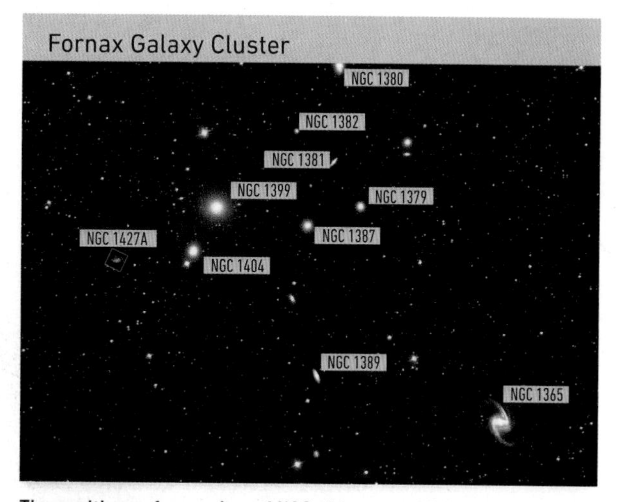

The positions of a number of NGC objects are highlighted in this Hubble Space Telescope image of the Fornax Cluster. The boxed source, NGC 1427A, shows evidence of a recent interaction with the gas contained between the galaxies in the cluster.

The Hubble Law

As you have seen, galaxies are made of stars, and stars have spectral lines, most of which are absorption lines. When a source of electromagnetic radiation is moving relative to the observer, the wavelength of the light will be redshifted if the two are moving apart and blueshifted if they are moving together.

In the early twentieth century, American astronomer Vesto Slipher began to measure the spectra of galaxies and noticed that the spectral lines of galaxies were redshifted. The redshifts measured indicated the Milky Way and its surrounding galaxies were moving apart.

Once the period-luminosity relationship of Cepheid variable stars was established in 1912 (see "Distances in the Milky Way"), astronomers began to use these stars to determine the distances to galaxies. In 1929, Edwin Hubble and his colleague Milton Humason published a remarkable result. Combining their measurements of the distances to nearby galaxies with the redshifts measured by Slipher, Hubble and Humason presented data that indicated the two were related—that is, the more distant the galaxy, the faster it was moving away. What is now called Hubble's Law is illustrated by the following equation:

$$v = H_0 \times d$$

v is relative velocity, H_0 is the Hubble constant (the slope of the line), and d is the distance.

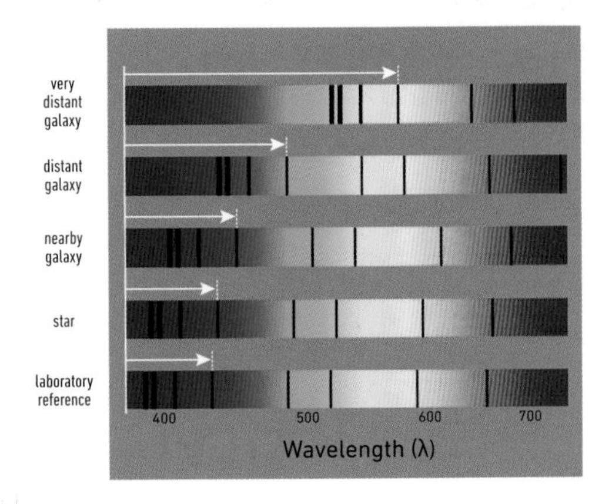

This schematic shows how an absorption line is observed to be shifted to a longer (redder) wavelength the farther away it is. Galaxies are observed to have absorption lines that are shifted to the red end of the spectrum with distance. Notice how the original line is shifted from the blue part of the spectrum all the way to the yellow part of the spectrum.

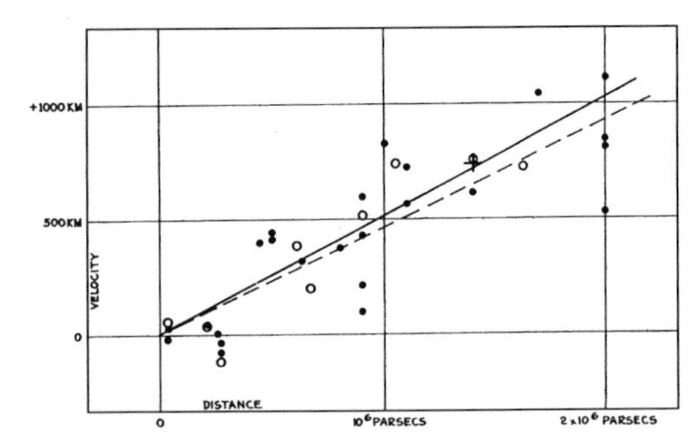

The original Hubble Diagram as published in the *Proceedings of the National Academy of Sciences* in 1929. There is considerable scatter in this original data, but the slope of this line provides a very important constant—the Hubble constant, which can be used to determine the age of the universe.

The simplicity of the relationship between the distance to a galaxy and its redshift is that it is linear. The data from the original observations were fit with a straight line, showing that the slope of the line was a constant. By measuring of a number of velocities (redshifts) and distances (from Cepheid variable stars), Hubble and Humason determined the slope of the line. This slope is called the *Hubble constant* (H_0).

Disagreements about the value of the Hubble constant fueled one of the big debates of twentieth-century astronomy, since its value determines the age of the universe (see "The Hubble Constant and the Age of the Universe"). As astronomers have been able to measure the distance and determine the redshift to more-distant galaxies, they have been able to arrive at a more-precise determination of the value of the Hubble constant. The current accepted value for the Hubble constant is about 70 kilometers per second per megaparsec (70 km/s/Mpc), which is much smaller than Hubble's original value of 500 km/s/Mpc. The overall recession of galaxies is sometimes called the *Hubble flow.*

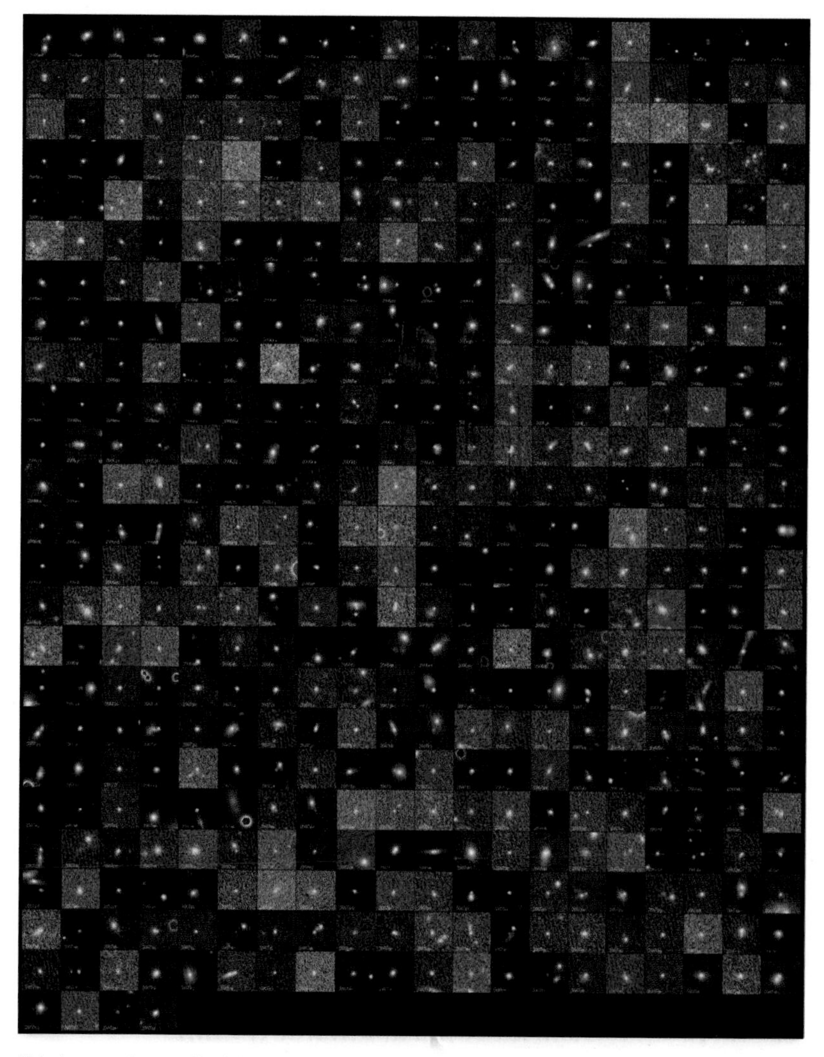

This image shows all of the more than 500 Type Ia supernovae detected by the Sloan Digital Sky Survey (SDSS) supernova project between 2005 and 2007. The discovery that Type Ia supernovae can be used as a standard candle allowed the determination of the distances and redshifts of much-more-distant galaxies.

Galaxy Collisions

As you read previously, the Local Group is a galaxy cluster in which a number of galaxies of different types are orbiting via their mutual gravitational attraction. In situations like this, it is perhaps inevitable that two of the galaxies will collide. For example, there is evidence that the Large and Small Magellanic Clouds have collided with the Milky Way in the past, and Andromeda and the Milky Way appear to be on a collision course for a major encounter 4 billion years from now. Earlier in the history of the universe, collisions were probably more frequent.

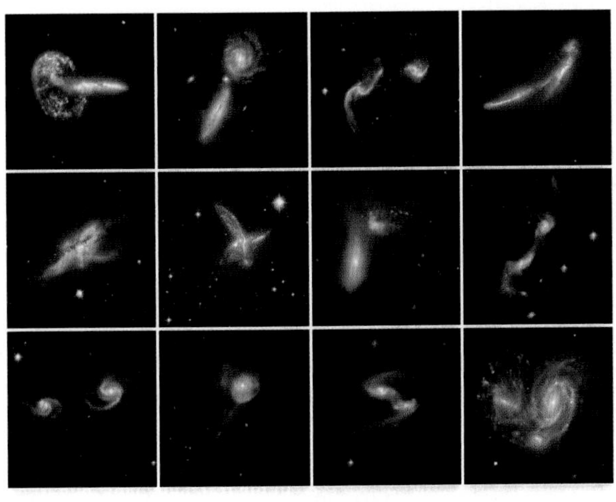

These collision images from the Hubble Space Telescope, Spitzer Space Telescope, and Chandra X-Ray Observatory show a variety of ways in which galaxies can collide. The long "tails" evident in many of the collisions are the result of tidal gravitational forces.

Two spiral galaxies in the cluster ARP 274 are colliding. The bright stars in the foreground are part of the Milky Way Galaxy. Spiral galaxies are enormous—typically hundreds of thousands of light-years across—so collisions take many hundreds of millions of years to occur.

When two galaxies collide, the stars are so far apart that they almost never collide. However, many of the stars may be thrown out into intergalactic space because of the gravitational interactions. Observations of galaxy collisions often show vast tails of gas and stars. The gas clouds in galaxies are large enough that when galaxies collide, they, too, have collisions; these collisions of gas clouds can start gravitational collapse and trigger vast waves of star formation. In colliding galaxies, giant molecular clouds are found close to waves of recent star formation.

These collisions produce complex and fascinating results, and in the crowded environments of galaxy clusters, galaxy collisions are normal events.

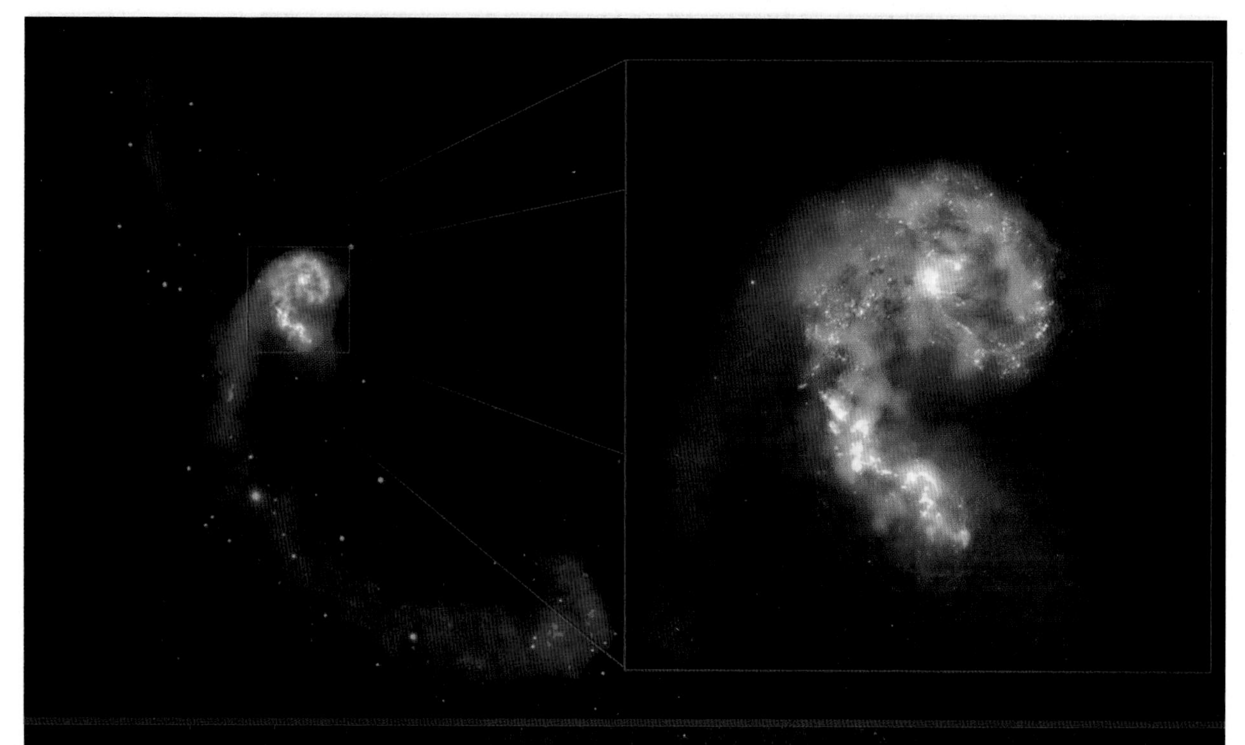

In this image of the Antennae, neutral hydrogen (HI) gas is shown in blue, thrown out in two tidal tails that extend far beyond most of the optical emission from the pair. The inset box to the right shows optical emission overlaid with millimeter and sub-millimeter wavelength emission from the Atacama Large Millimeter Array (ALMA). The large amounts of gas in these tails are often not visible optically, but streamers of gas—similar to those seen connecting the Milky Way to the Magellanic Clouds—are often apparent in radio images of galaxy mergers.

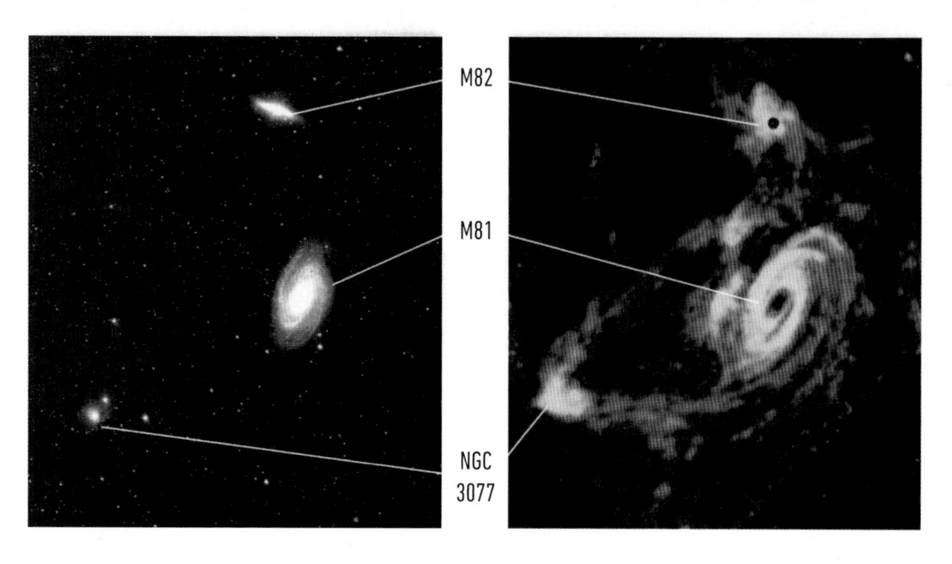

This 1-degree-wide view of the sky near the spiral galaxy M81 shows the importance of multifrequency images for understanding galaxy collisions. To the left, M81, M82, and NGC 3077 are seen in an optical image from the Palomar Sky Survey. To the right, the same field of view shows neutral hydrogen (HI) as imaged by the Very Large Array.

Gravitational Lensing

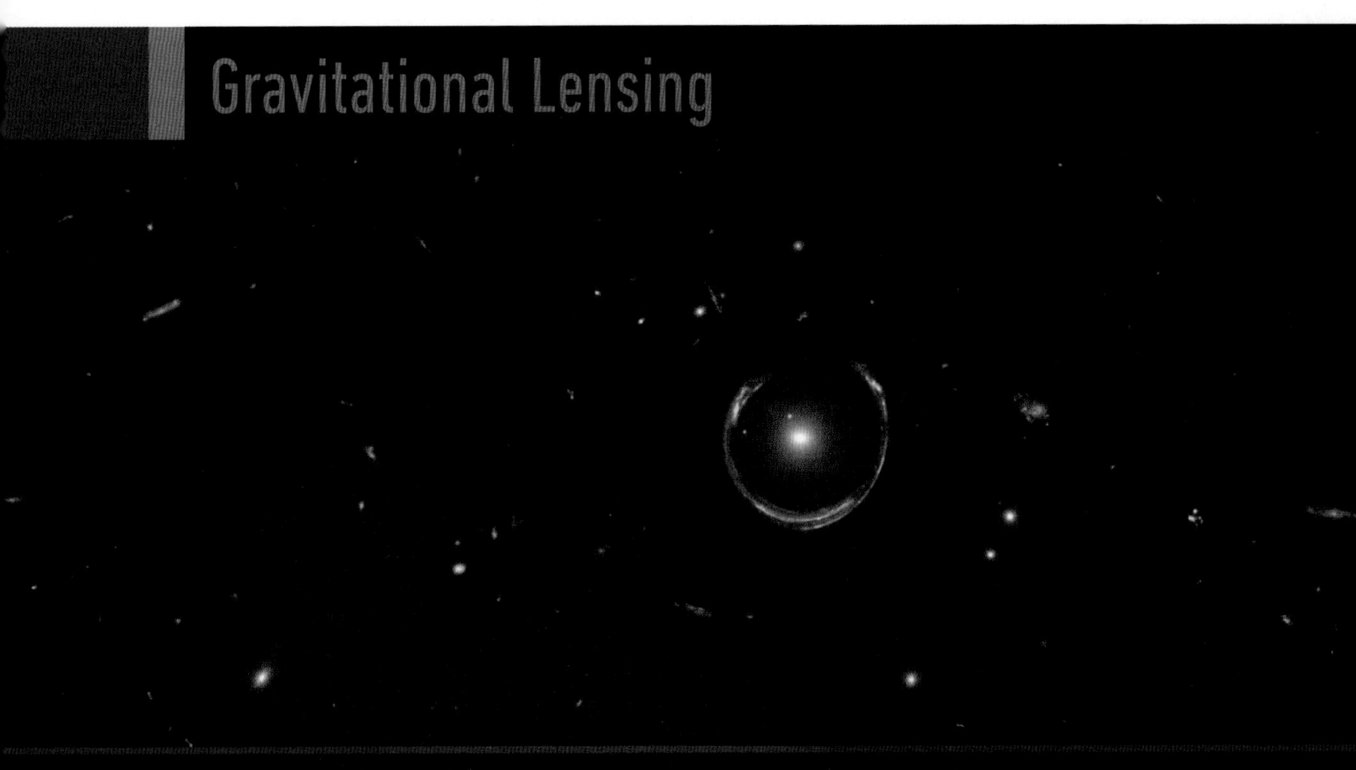

In this Hubble Space Telescope image, you see the image of a distant blue galaxy severely distorted by the mass of a foreground galaxy. A very precise alignment in the position of the two galaxies is required to form a ring image like this.

Clusters of galaxies contain only a small amount (1 percent) of their total mass in visible material. The majority of the mass of a galaxy cluster (90 percent) is contained in dark matter, while about 9 percent is hot, intracluster gas. While astronomers still do not know what dark matter is (see "What Is Dark Matter?"), you can see its gravitational effects in the motions of cluster members.

According to Albert Einstein's general theory of relativity, mass distorts space in such a way that the trajectories of photons are bent. In this way, mass (whether it is luminous or dark) will bend light in a similar manner to glass in a lens. These effects have been seen throughout space in objects that are referred to as *gravitational lenses* and provide clear evidence that dark matter is present between Earth and distant galaxies.

Gravitational lenses provide definitive support of Einstein's general theory of relativity but are remarkable in other ways. Like a glass lens, the dark matter around a galaxy or in a cluster environment can sometimes provide a magnified image of a background source that would be otherwise unobservable.

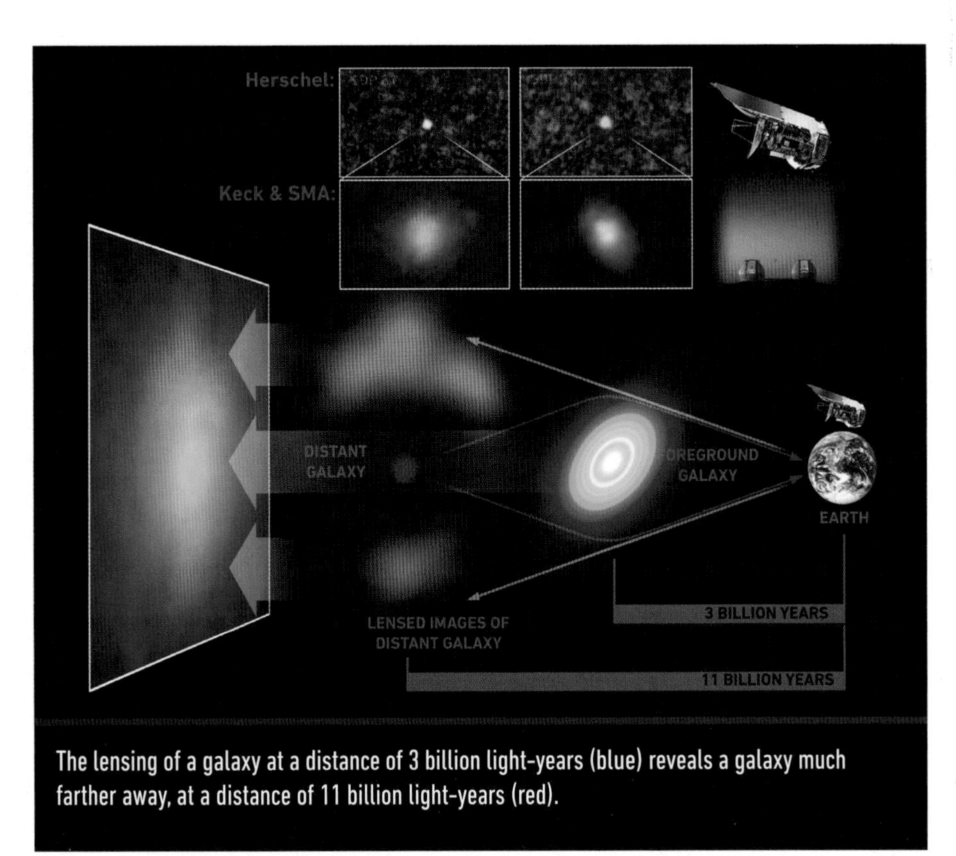

The lensing of a galaxy at a distance of 3 billion light-years (blue) reveals a galaxy much farther away, at a distance of 11 billion light-years (red).

The amount and exact geometry of the distortion allow astronomers to calculate the mass of the dark matter in the foreground associated with the lensing galaxy. These calculations support other independent lines of study that most of the mass of galaxies is contained in its dark matter.

The gravitational field of a foreground galaxy can allow astronomers to view a background source that would be too faint or small to view without the presence of the lensing.

Sometimes background sources are not lensed by a single foreground source but by the mass contained in a cluster of galaxies. Red and blue arcs seen in this image are multiple images of background sources by the dark matter in the cluster Abell 2218 (3 billion light-years away).

Dark Matter in Galaxies

The rotation curves of galaxies indicate that like the Milky Way, most of their mass is contained in dark matter. Galaxies group together gravitationally into clusters, and on these even larger scales, the contribution of dark matter is dominant, making up about 90 percent of the mass of a cluster. Astronomers can use gravitational lensing evident in galaxy clusters to determine the location and mass of dark matter, even if they do not know what dark matter is made of. In images of galaxy clusters, optically visible galaxies can be seen embedded within large concentrations of dark matter. Computers are used to model the distribution of dark matter that would cause the observed lensing.

Gravitational lenses provide a powerful way for astronomers to do something that they could never do before: "image" dark matter. Of course, because dark matter does not emit any radiation, you can't truly image it. However, the distortions it causes because of its gravitational field can be used to construct a distribution of its location in a cluster, pulling back the veil on this mysterious, dominant component of mass in galaxy clusters.

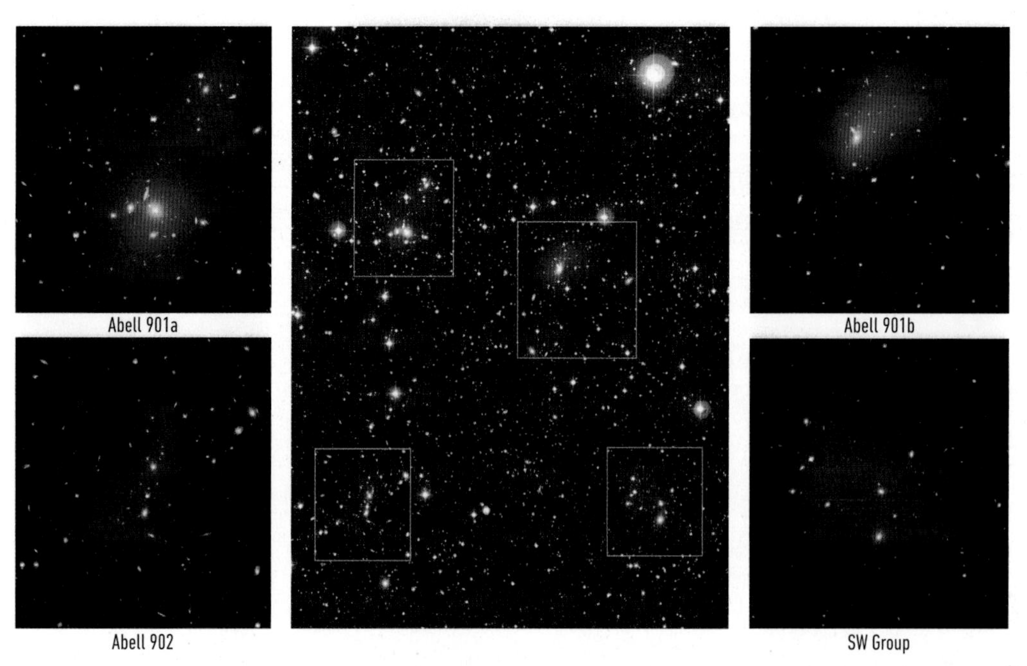

These images show a combination of visible light from galaxies in the Abell 901/902 supercluster of galaxies overlaid with models of the distribution of dark matter in the cluster. The diffuse magenta regions are the location of dark matter in the supercluster, as determined from weak gravitational lensing from the light of over 60,000 background galaxies.

Astronomers have superimposed the distribution of dark matter in the galaxy cluster ZWCl 10024+1652 over an Hubble Space Telescope image of the cluster. The optical image shows complex patterns of gravitational lensing from matter in the cluster. The diffuse blue "haze" shows an unusual ring of dark matter, believed to be the result from the collision between two clusters of galaxies.

The Magellan and Hubble Space Telescope images provide the optical view of the Bullet Cluster (orange and white), showing the visible light from the galaxies in the cluster. The hot gas in the cluster is seen in pink, as imaged by the Chandra X-Ray Observatory. The dark matter distribution is seen in blue.

Colliding Clusters

The Bullet Cluster is a pair of galaxy clusters that have collided. The two clusters have passed through one another, and a combination of images allows observers to see the complex interaction between luminous and dark matter in the Bullet.

The hot, intracluster gas (red) is distorted by the collision of the two clusters. Gravitational lensing of background sources in the Bullet Cluster was used to derive the distribution of the dark matter there (blue). Dark matter contains 10 times as much mass as the hot gas and almost 100 times as much mass as the galaxies themselves. So while your eyes may be drawn to the bright galaxies, their mass tells only a small part of a much larger story.

Galaxy Evolution: How Galaxies Change

Galaxies evolve far too slowly to watch them change in a human lifetime. Even a single star evolves exceedingly slowly, taking millions or billions of years to move through its various stages of life. Astronomers understand the lifetimes of stars by observing many stars at different points in their evolution and connecting these into a complete picture. The best way to understand galaxies is similar: observe a large number of galaxies at different stages of evolution and connect these stages.

A Basic Picture of Formation and Evolution

Based on observations of distant and nearby galaxies, astronomers have come up with a basic picture of galaxy formation and evolution. Careful studies of galaxies in the early universe (2 to 3 billion years after the Big Bang) indicate that many of them were disk shaped, already contained more stars than the Milky Way, and were similar to irregular, gas-rich dwarf galaxies. These galaxies merged into ever-larger collections of gas, dust, and stars and formed central black holes. Galaxies in the early universe were also making stars at a much faster rate than galaxies do in the current universe.

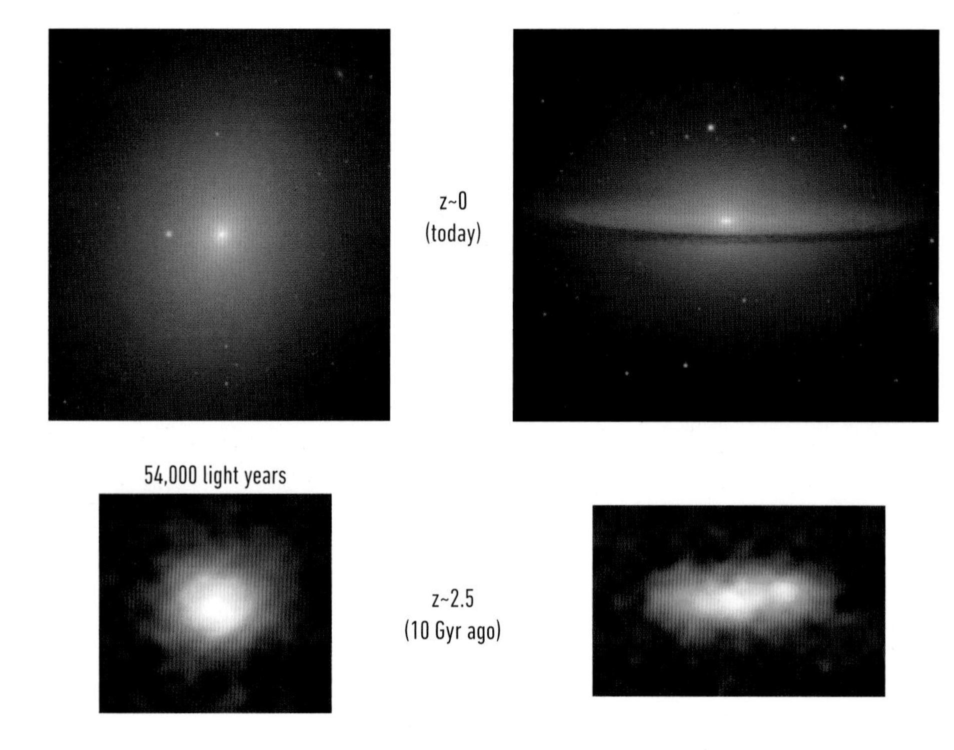

This comparison image shows the difference in size between elliptical and spiral galaxies in the current universe (top) and 10 billion years ago (bottom). The top images show NGC 4472 (an elliptical galaxy) and NGC 4594 (the "Sombrero" Galaxy) and indicate that galaxies today are both more massive and have a larger bulge than galaxies in the early universe.

Collisions and Their Relation to Galaxy Types

Galaxy collisions and mergers are an important part of the evolution of galaxies. In the early universe, collisions were even more common than now, since the universe was filled with a larger number of small galaxies. Understanding the nature and frequency of galaxy collisions is an important part of galaxy evolution models.

The prevalence of giant elliptical galaxies at the core of large galaxy clusters indicates that elliptical galaxies are likely the result of collisions of galaxies in the dense cluster environment. Younger galaxies give off more ultraviolet radiation, generated by hot, young, high-mass stars. When star formation ends, most of the radiation from a galaxy is in the red part of the spectrum, coming from the photospheres of cooler, low-mass stars that live much longer lifetimes. Missions like GALEX have found examples of transitional galaxies that have intermediate properties between young and old galaxies.

The merger of two massive galaxies would have led to the formation of a radio galaxy, a quasar or blazar (see "Active Galaxies and Active Galactic Nuclei") depending on viewing angle, and then an elliptical galaxy. A more unequal merger (with a large galaxy absorbing a smaller one) might lead to a spiral galaxy and feed a central black hole. Such a galaxy would light up as a Seyfert and then become a normal spiral galaxy.

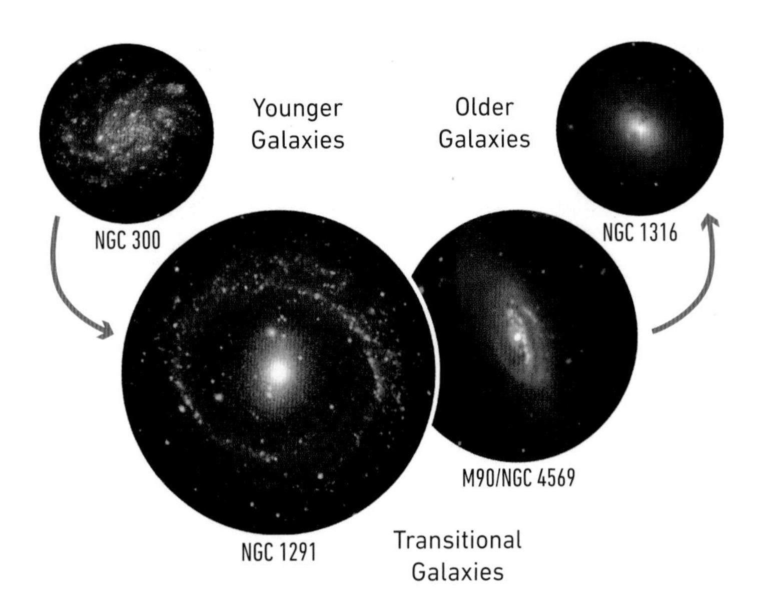

The GALEX mission has identified what astronomers believe are transitional galaxies, with properties that locate them between the younger galaxies filled with star formation to the older elliptical galaxies where the raw material for star formation has been effectively used up.

These two interacting galaxies ("The Mice") are also catalogued as NGC 4676. Both are spiral galaxies, and the tails of material are typical of spiral galaxy mergers, where gravitational tidal forces throw gas and dust out to many times the size of the original galaxies.

What Are Quasars?

An artist's impression of the quasar 3C 279 shows the accretion disk and jet around a distant supermassive black hole.

Quasars are a type of active galactic nucleus; the emission from a quasar is thought to originate from the region near a supermassive black hole at the center of a distant galaxy.

Quasars, like a number of interesting objects in the universe, were discovered quite by accident. But in order for them to be discovered, radio telescopes needed to have high-enough resolution to pinpoint the source of radio emission, and spectroscopy had to have been used to discover the redshift of galaxies. Once these advances were in place, it was only a matter of time and curiosity.

Radio Telescopes and the Discovery of Quasars

In 1936, Grote Reber built the first functional radio telescope and discovered several bright sources of radio emission. One bright radio source, Sagittarius A, turned out to be the Galactic Center region; another bright radio source, Cassiopeia A, was later determined to be a supernova remnant in the Milky Way. The nature of a third radio source, Cygnus A, was a mystery until sensitive optical images detected an optical emission at the position of Cygnus A. Cygnus A was also detected to have an unusual and unexplained set of spectral emission lines (most galaxies were known to have absorption lines).

When first observed with optical telescopes, objects like Cygnus A had a very starlike appearance. They were first called *quasi-stellar radio sources,* later shortened to *quasars.*

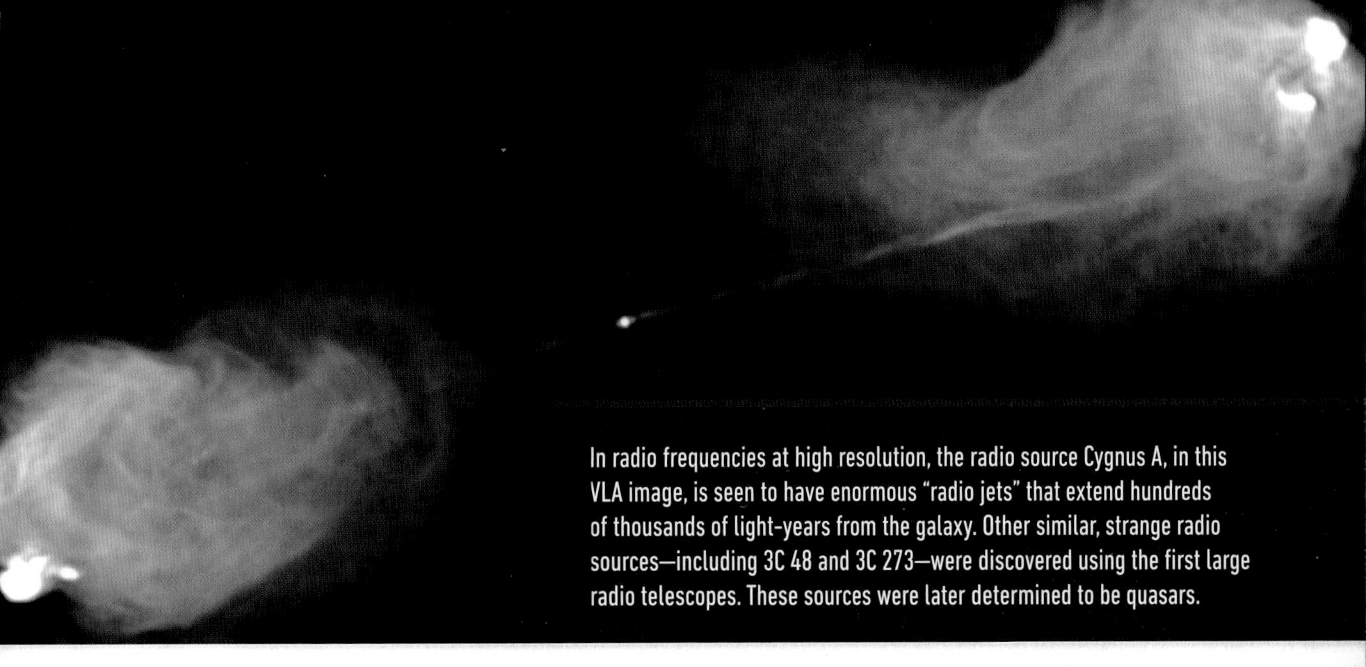

In radio frequencies at high resolution, the radio source Cygnus A, in this VLA image, is seen to have enormous "radio jets" that extend hundreds of thousands of light-years from the galaxy. Other similar, strange radio sources—including 3C 48 and 3C 273—were discovered using the first large radio telescopes. These sources were later determined to be quasars.

How Spectroscopy Determined Their Distance

The mystery was finally resolved by Maarten Schmidt in the early 1960s. He noticed that the spectral lines in these strange radio sources were redshifted significantly, so the red emission line of hydrogen (the H-α line detected toward many sources) was in the infrared part of the spectrum. This shift corresponded to a velocity shift of 44,000 km/s, far too high to be associated with anything in the Milky Way. These sources must be galaxies, and their velocities placed them at very great distances, according to the Hubble Law.

Using the velocities from these emission lines and the Hubble Law, astronomers determined that the galaxy Cygnus A was very distant, at 230 Mpc (740 million light-years) away. This confirmed that quasars are some of the most distant objects observed.

What Is Redshift?

For very-distant galaxies, astronomers refer to the redshift of the galaxy. This term (abbreviated with the letter *z*) is a measurement of the shift in spectral lines from their known wavelength. Objects moving away have their wavelengths shifted (to the red part of the spectrum) proportional to their velocity. So according to the Hubble Law (see "The Hubble Law"), the higher the velocity, the greater the distance—and therefore, the greater the redshift. More precisely, the redshift is given by the following:

$$z = \frac{\text{change in wavelength}}{\text{original wavelength}}$$

If astronomers measure the redshift for spectral lines in a quasar or galaxy, they can get a measurement of its distance from the Milky Way.

The many quasars that have been catalogued have large redshifts, ranging from 0.06 to 5.8. For comparison, an object with a redshift of 5 is at a distance of 12.5 billion light-years.

Radio Galaxies

As you read previously, early images of Cygnus A showed it to have an unusual radio structure: jets that extended far beyond the small galaxy itself, which originated at a bright point source near the galaxy's center. This was the first radio galaxy discovered; the fast evolution of radio telescopes after World War II brought a host of related discoveries.

A *radio galaxy* is a type of active galaxy that has bright radio frequency emission. Careful analysis of the radio emission from these galaxies shows the emission comes from high-velocity electrons spiraling around magnetic fields lines, also called *synchrotron radiation*.

Synchrotron radiation results when charged particles moving close to the speed of light are forced to move in a curved path by a strong magnetic field. While synchrotron radiation is often detected at radio frequencies, it is emitted across the electromagnetic spectrum.

The galaxies that accompany these radio structures are typically giant elliptical galaxies and are common enough that they have their own "anatomy." Sources like Hercules A exhibit all of the typical characteristics, including jets (bright, linear features emerging from the galactic center), lobes (diffuse features located along and at the end of the jets), and hotspots (bright spots within the lobes).

Hercules A, a giant elliptical galaxy (shown in the center of this image), appears ordinary at optical wavelengths but is a monster in mass: over 1,000 times the mass of the Milky Way. The radio emission from the galaxy (shown in red) comes from its core and from two oppositely directed radio jets and lobes.

Radio Astronomy and Radar

World War II led to a number of advances in radio frequency detection. Radar (radio detection and ranging) was used to great advantage by the British in the Battle of Britain to detect German aircraft. Radar systems send out a pulse of radio-frequency electromagnetic radiation and then detect the reflected pulse.

Detectors and skills first used in radar systems during WWII were put to use in the following decades by radio astronomers. The sensitive radar dishes that could detect incoming aircraft could also detect the faint radio emission from distant galaxies.

This composite image shows the giant elliptical galaxy M87 in visible (yellow/green), radio (red), and X-ray (blue) emission, highlighting very-large-scale emission around the galaxy. X-ray and radio emission show the remains of a series of outflows from the supermassive black hole at the center of the galaxy.

VLA Radio Galaxy Images

Some of the best images of radio galaxies have been made with the Very Large Array (VLA), dedicated in 1980 and recommissioned in 2012. This radio interferometer is made up of 27 individual 25-m diameter telescopes that can be moved to have sensitivity to small-scale structures, large-scale structures, or structures in between. Radio astronomers often combine images from all four configurations in order to make a complete image of a source.

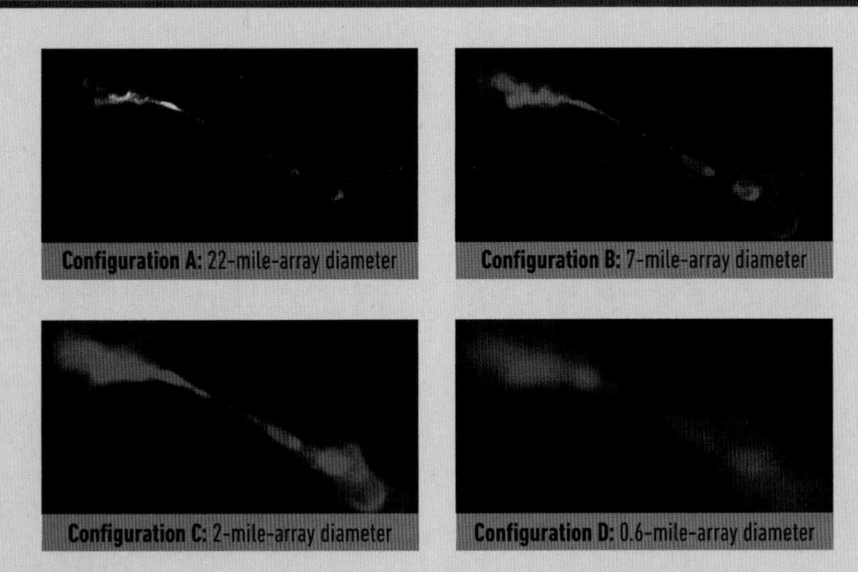

Configuration A: 22-mile-array diameter
Configuration B: 7-mile-array diameter
Configuration C: 2-mile-array diameter
Configuration D: 0.6-mile-array diameter

These images show the data from different configurations of the Very Large Array (VLA) that were combined to generate the final image of the radio galaxy Hercules A. The 27 telescopes in the VLA have to be moved on a custom-made set of double railroad tracks in order to make images in different configurations. The VLA is in each configuration for a period of 3 to 4 months.

Active Galaxies and Active Galactic Nuclei

The discovery of quasars and radio galaxies was actually preceded by the detection of another strange energetic type of source. In 1929, the source BL Lacertae was discovered and classified as a variable star. It was not until almost 50 years later that further observations revealed this source was located in an elliptical galaxy. When more sources of this type were discovered, they were classified as *blazars*. Radio frequency observations led to the conclusion that blazars were simply radio galaxies seen "end on"—looking down the axis of the radio jet.

In the following decades, a unified model was proposed that explained quasars, radio galaxies, and blazars as different views of the same object—the nucleus of a galaxy harboring a supermassive black hole. Sources viewed from the side are seen as radio galaxies, sources viewed down the axis of the radio jet are blazars, and sources viewed from an angle in between are radio-loud quasars.

The galaxy NGC 1097 is a face-on barred spiral in the Fornax constellation, classified as a Seyfert galaxy. The nucleus of NGC 1097, like that of a quasar, has strong emission lines and contains a black hole with 100 million solar masses. The central ring contains star formation triggered by the infall of material toward the black hole.

December 2, 2009

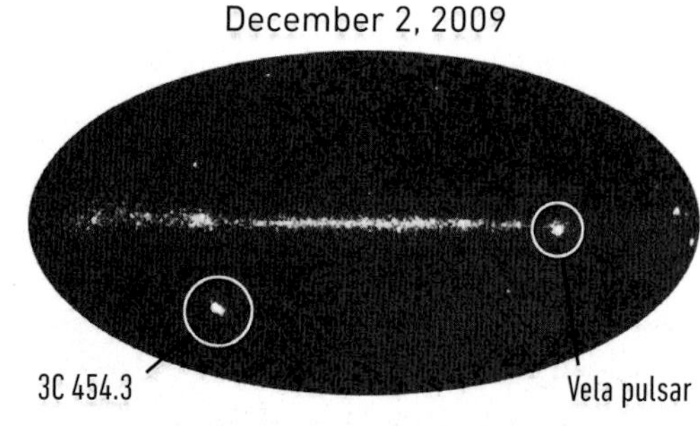

November 3, 2009

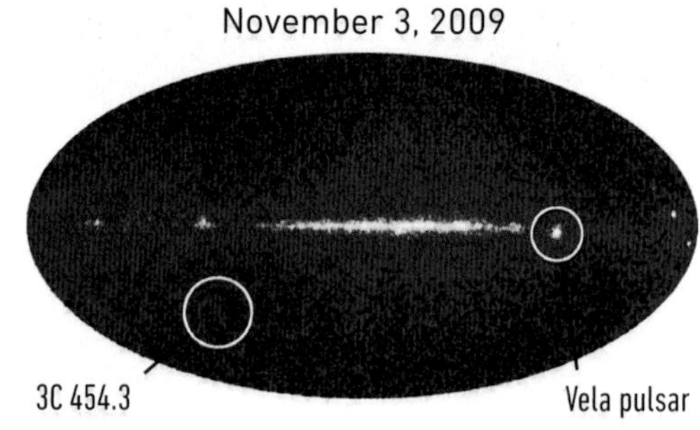

The flaring blazar 3C 454.3, located 7.2 billion light-years away, is shown in these two all-sky gamma-ray images from the Fermi Large Area Telescope. Flaring is related to activities very close to the galaxy's supermassive black hole, and the same flare was seen at radio and visible wavelengths.

This artist's impression of an AGN shows the components of a unified model that explains many different sources, from quasars to radio galaxies. The black hole (here hidden from view by a torus of material) is fed by a smaller disk that is the source of the jets of material that move out in both directions.

These objects, along with Seyfert galaxies, are collectively referred to as *active galaxies,* and the central regions of these galaxies are called *active galactic nuclei (AGNs).* Seyfert galaxies are a subclass of spiral galaxies that have an unusually bright, compact core. Like quasars, Seyfert galaxy nuclei have strong emission lines that are thought to arise from a central supermassive black hole.

As active galaxies, blazars, quasars, Seyfert galaxies, and radio galaxies have a common bond: each one of them hosts a supermassive black hole in its nucleus. In active galaxies, this black hole is currently accreting material and powering energetic jets of material. Galaxies are thought to go through periods of activity when material is "fed" to the region around the black hole. For this reason, merging galaxies can become "active" as material from the merger is fed into the central regions of the galaxies.

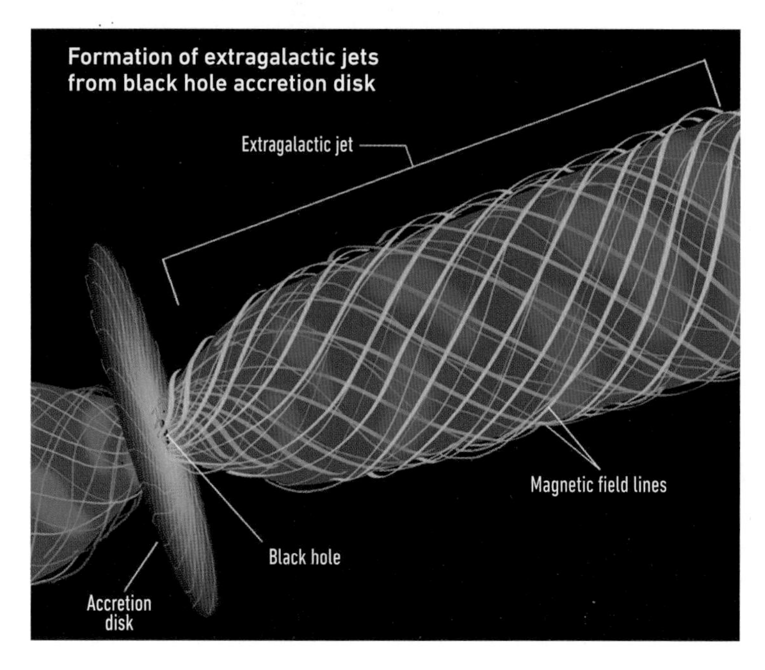

In this schematic, you see the region close to an AGN. Magnetic fields that are threaded through the accretion disk direct charged particles outward along the jet. The powerful magnetic fields in the jet are often detected as synchrotron radiation at radio frequencies.

Supermassive Black Holes in Other Galaxies

As you learned in Part 3, stellar mass black holes are the remnants of exploded high-mass stars that began their lives with more than 20 solar masses. These objects are scattered throughout the disk of the Milky Way and other galaxies. But a much larger black hole—with the mass of 4 million Suns—is located at the Galactic Center toward the constellation Sagittarius.

Observations of active galaxies and galaxy rotation curves have made it clear that the Milky Way is in no way special. Supermassive black holes lurk in the centers of all galaxies. The black holes in these galaxies may or may not be active (see "Active Galaxies and Active Galactic Nuclei"), but it seems clear they are there. In many galactic centers, infrared and radio emission peak at the position of the active region around the black hole.

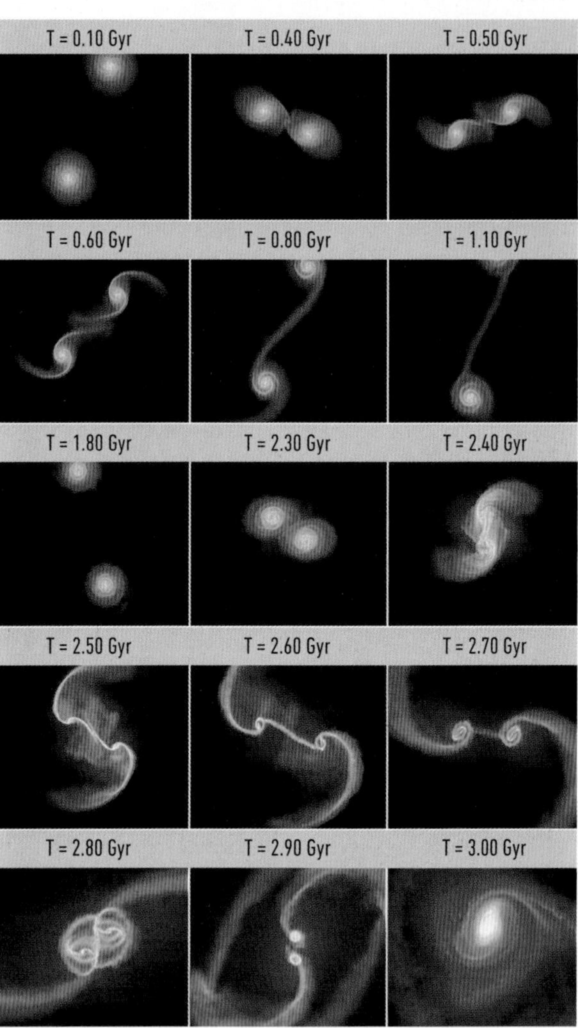

These panels—freeze frames from a numerical simulation—show how the collision of two spiral galaxies can lead to the formation of central supermassive black holes.

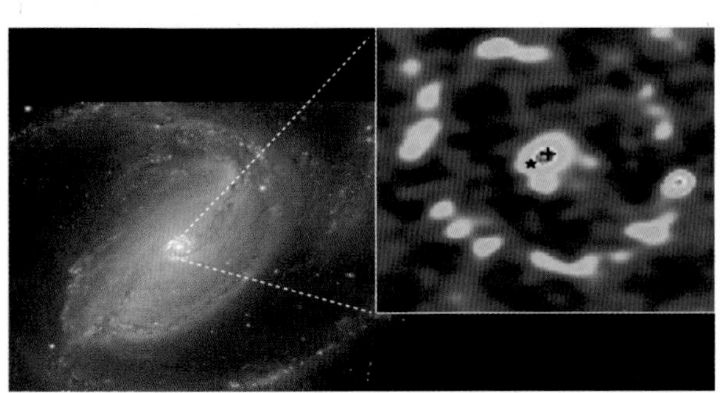

The central 2100 light-year radius region obtained by ALMA

This image shows the optical image of the Seyfert galaxy NGC 1097 with an inset showing a detail of the central 2,100 light-years as seen by Atacama Large Millimeter Array (ALMA). The false-color ALMA image shows a central star-forming ring and emission from dust near the central supermassive black hole.

In any galaxy, the mass of its black hole is thought to be proportional to the mass of its galactic bulge, which is thought to grow with each galactic collision. How supermassive black holes form in the centers of galaxies is still debated, but one leading model is that smaller black holes form as the galaxy forms and then merge with one another to build up to the millions of solar masses seen in most galactic center black holes.

Another possibility is that a single black hole grows in size by consuming nearby gas, dust, and stars. There is evidence from Hubble Deep Field images that the most distant and therefore oldest galaxies experienced many collision events, with their black holes merging in the process. Only about 1 percent of supermassive black holes appear to be "active," coincident with blazars, quasars, and radio galaxies (see "Active Galaxies and Active Galactic Nuclei"). So mergers appear to be one way to funnel material into the very center of a galaxy and "turn on" the black hole that's there.

There is strong observational evidence that supermassive black holes in galaxies can become active during merger and collision events. An X-ray survey by the Swift Telescope has detected many AGNs (circled in the image), and optical observations of these sources show nuclei to be at the centers of galaxies involved in mergers.

The Environments of Black Holes

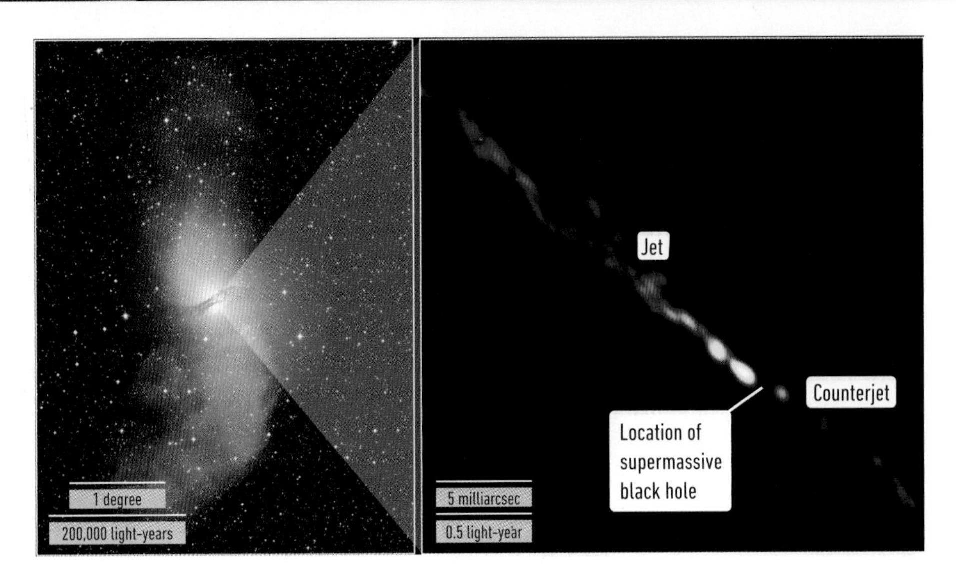

The giant elliptical galaxy NGC 5128 is a well-known radio galaxy (Centaurus A), which is seen at optical and radio wavelengths in the wide field image on the left. To the right is the most detailed radio image ever made of the jet that originates from the region around the galaxy's central supermassive black hole.

The environments of black holes must be strange places indeed. Whether it's around a stellar mass black hole or a black hole with the mass of many millions or even billions of Suns, the intense gravitational field around a black hole distorts space locally, and when fuel is available, drives highly energetic outflows.

Many galaxies are observed to have large-scale radio jets (see "Radio Galaxies"). These jets can be tracked down to very small scales, into the environment around the central black hole. For example, the jet emanating from Centaurus A's 50-million-solar-mass black hole has been imaged down to the few central light-years of the galaxy using a radio interferometer in Australia called TANAMI.

Properties and Composition of Black Holes

Black holes are theorized to have only three physical properties: mass, charge, and angular momentum. The *event horizon* of a black hole defines the separation between the black hole and the rest of the universe. Matter and light can pass through the event horizon toward the center of the black hole, but neither can pass out. Nonrotating black holes have spherical event horizons, while rotating black holes have more "flattened" geometries. The event horizon is far too small to be imaged, even for the closest supermassive black hole, located at the center of the Milky Way.

Even the supermassive black hole at the center of the Milky Way Galaxy is too distant to image the event horizon itself. This artist's conception shows what the environment of a galactic supermassive black hole might look like.

The black hole is believed to be the source of jets of high-energy particles. The base of the jet and the accretion disk around the black hole are the sources of high-energy X-rays. In radio galaxies, this same jet continues out for hundreds of thousands of light-years.

Material near a black hole (outside the event horizon) will form into a flattened disk called an *accretion disk,* and as material migrates inward toward the black hole, it releases energy detectable as X-rays and radio waves. Some of the primary targets for new X-ray telescopes like Chandra are the cores of galaxies thought to harbor such accretion disks.

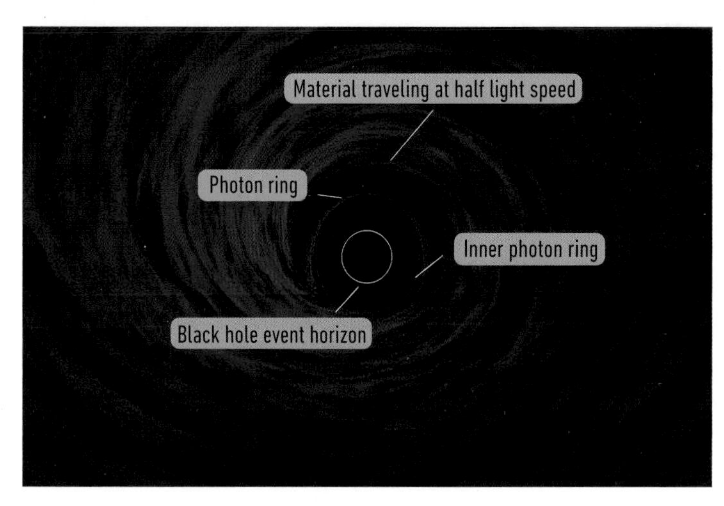

A simulated view of the accretion disk of a black hole as seen with the top of the accretion disk inclined toward the observer. Soft (red) and hard (blue) X-ray emission is shown, as well as the position of the event horizon of the black hole. Gravity is so strong that it bends light from the far side of the accretion disk to make the photon ring.

Hard and Soft X-Rays

Recent studies of black hole environments have tried to explain the detection of hard (higher-energy) X-rays and soft lower-energy X-rays from black hole environments. Simulations show lower-energy X-rays may be generated in the accretion disk of the black hole, while higher-energy X-rays may come from a corona that is above and below the accretion disk. Photon rings, which result from light that is severely bent by the gravity of the supermassive black hole, are observed in simulations of black hole environments.

The Hubble Deep Field Images

Typically, observations of the night sky are dedicated to particularly interesting sources. But arguably one of the most fascinating images in astronomy, the Hubble Deep Field (HDF), was the result of pointing for a long time at one of the most uninteresting positions in the sky possible.

The HDF image, made in 1995, resulted from a decision by Director of the Space Telescope Science Institute (STScI) Robert Williams to use discretionary time that he was awarded as director for a deep image of a typical patch of the sky, far out of the plane of the Milky Way. The patch of sky chosen was free of any bright sources at any wavelength, from radio out to the ultraviolet. The final pointing selected was in the constellation Ursa Major and covers a region only about 2.5 arcminutes on a side. (For comparison, the full moon is about 30 arcminutes across.) Similar observing strategies were used to make images of the HDF (1995), HDF South (1998), the Hubble Ultra Deep Field (HUDF; 2004), and the Hubble Extreme Deep Field (XDF; 2012).

The original HDF image contains thousands of galaxies in a variety of evolutionary stages. Only a few of the points in this image are stars; the rest are galaxies of stars. Notice the variety of galaxy types visible: ellipticals, spirals, and irregulars.

This image shows the HUDF made in 2004. This pointing was in the constellation Fornax and was chosen with criteria similar to the HDF. The field of view of the HUDF is slightly larger than the HDF at about 3 arcminutes on a side. There are about 10,000 galaxies shown here at various redshifts and stages of evolution.

Each of the HDF images has inspired follow-up observations at other wavelengths. These images have given a view into the distant universe. The fact that so many galaxies are visible in such a small patch of the sky emphasizes the true enormity of the visible universe. The study of the nature of galaxies in the early universe has advanced because of these sensitive, unbiased images.

By combining these images with measurements of the redshifts of galaxies, astronomers have been able to make 3D renderings of the Deep Field images and examine the ways in which galaxies evolve from the most distant (oldest) galaxies to those more nearby (youngest).

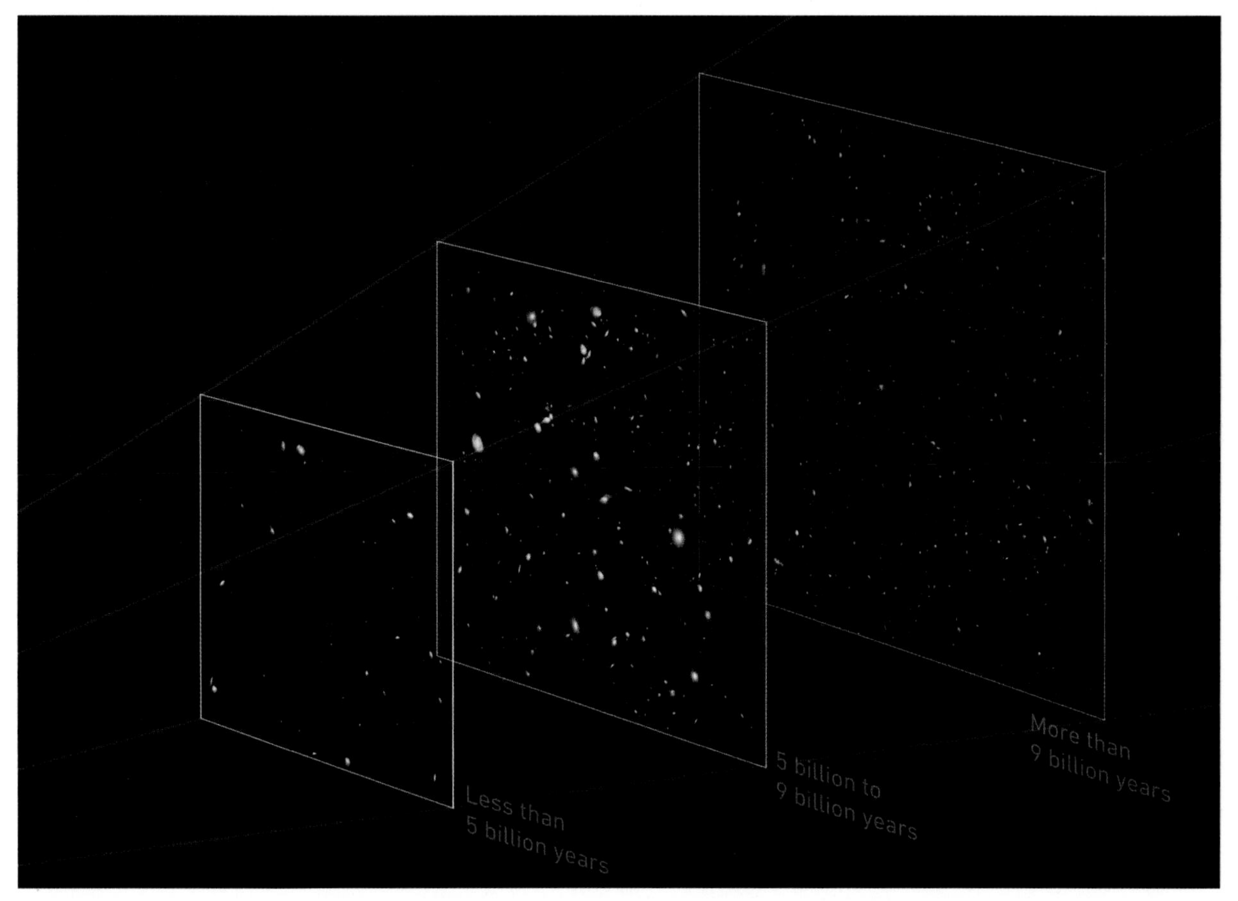

This schematic pulls the HXDF image into planes separated by the approximate distances to the galaxies. The distance (in light-years) is indicated as "lookback time." Traveling at the speed of light, that is how long it has taken photons from these galaxies to reach Earth. Many of the galaxies in the HXDF have lookback times of more than 9 billion years.

Galaxy Superclusters

Matter is collected into ever-larger structures by the forces of gravity. The same force that holds the planets in their orbits around the Sun keeps the Sun and its companions orbiting the center of the Milky Way Galaxy. The Milky Way is attracted to the nearby Andromeda Galaxy, and these two along with other smaller galaxies form the Local Group, which is considered a galaxy cluster (see "Galaxy Clusters"). Groupings of galaxy clusters comprise even larger *superclusters;* in the case of the Local Group, it is part of the Virgo Supercluster.

Makeup and Distribution of Superclusters

A typical supercluster contains 10 to 20 clusters of galaxies and can span hundreds of millions of light-years. Superclusters are the largest-known structures in the visible universe, and on these scales, clusters of galaxies are not held together by gravity, but instead are moving apart from one another as part of the Hubble flow (see "The Hubble Law").

The distribution of superclusters in space has been studied carefully in the past few decades and only becomes apparent when you look at the positions of hundreds of thousands or millions of galaxies in three-dimensional space. In order to make these three-dimensional images, you need to know two things: where the galaxy is in the sky and how far away it is (its redshift).

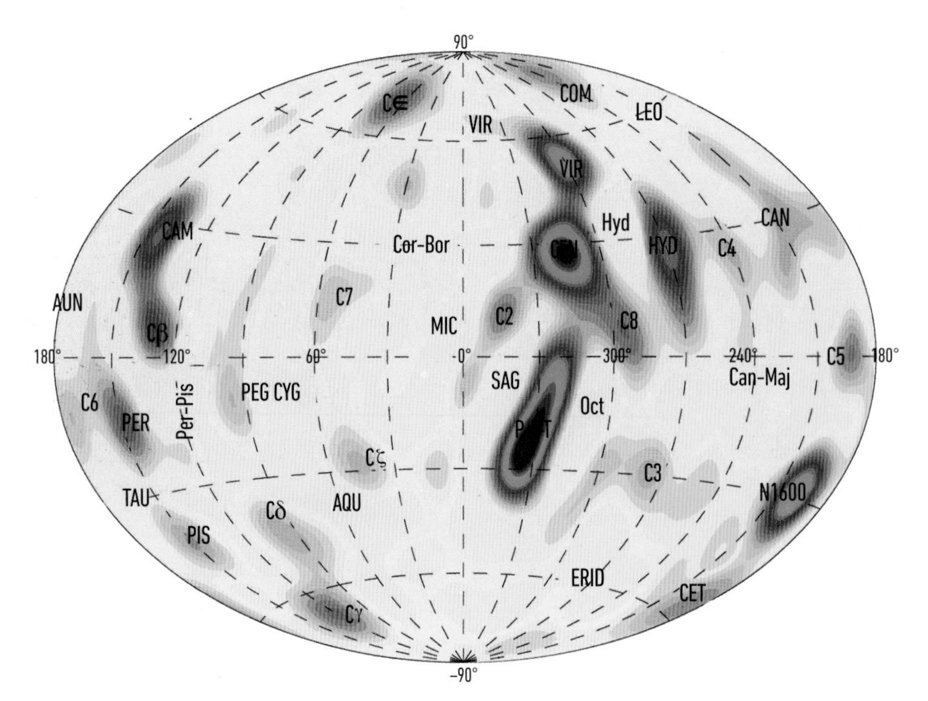

This image, based on 2MASS data, shows the distribution of matter (luminous and dark) in the local universe out to a distance of 130 million light-years. Yellow is relatively low mass density, while blue is relatively high. The largest supercluster in the nearby universe is the "Great Attractor," centered near the mass concentrations labeled CEN (Centaurus), HYD (Hydra), and Pavo-Indus-Telescopium (P-I-T).

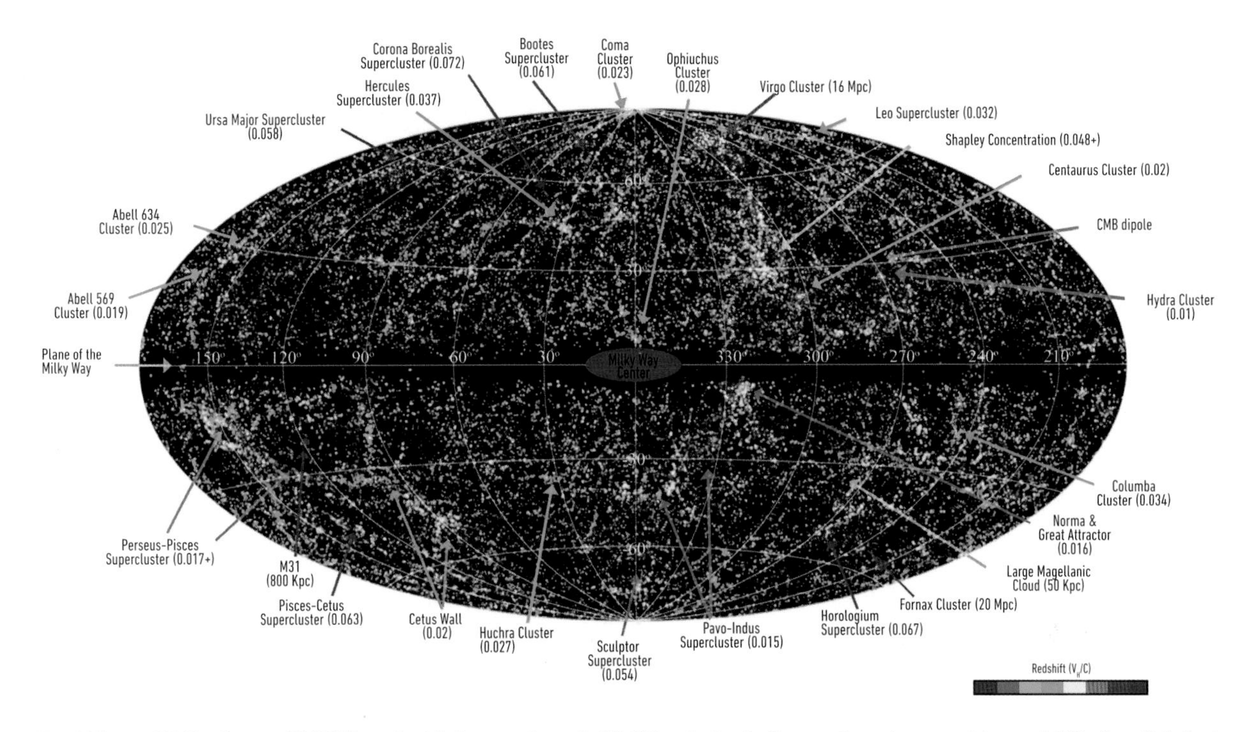

The 2 Micron All Sky Survey (2MASS) made this image of nearly 50,000 galaxies in the nearby universe out to a redshift of z = 0.1. Each dot on this image is a galaxy, with the blue dots the closest and the red dots the most distant.

The Great Attractor

In the local universe, the most massive cluster is called *The Great Attractor*. The Great Attractor is a supercluster of galaxies exerting gravitational forces on all of the local galaxies. The Great Attractor is the gravitational center of the Virgo Supercluster; the Milky Way Galaxy, as well as all of the galaxies in the local supercluster, are moving toward the Great Attractor and will eventually converge there.

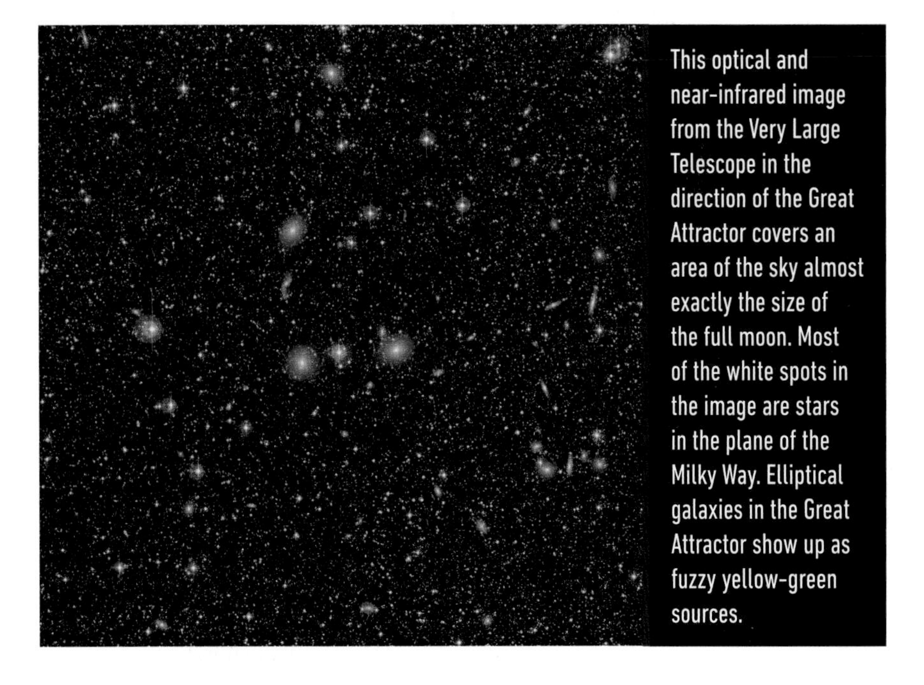

This optical and near-infrared image from the Very Large Telescope in the direction of the Great Attractor covers an area of the sky almost exactly the size of the full moon. Most of the white spots in the image are stars in the plane of the Milky Way. Elliptical galaxies in the Great Attractor show up as fuzzy yellow-green sources.

Very-Large-Scale Structures

When the location of galaxy superclusters in the sky and their distances are combined, you can begin to build up a picture of the structure of the universe on the very largest scales observable. Structures on these scales are referred to as *Very-Large-Scale Structures (VLSS),* and only in the past few decades have astronomers been able to see that the universe is not uniformly filled with matter, but is filled with filaments and voids of luminous and dark matter.

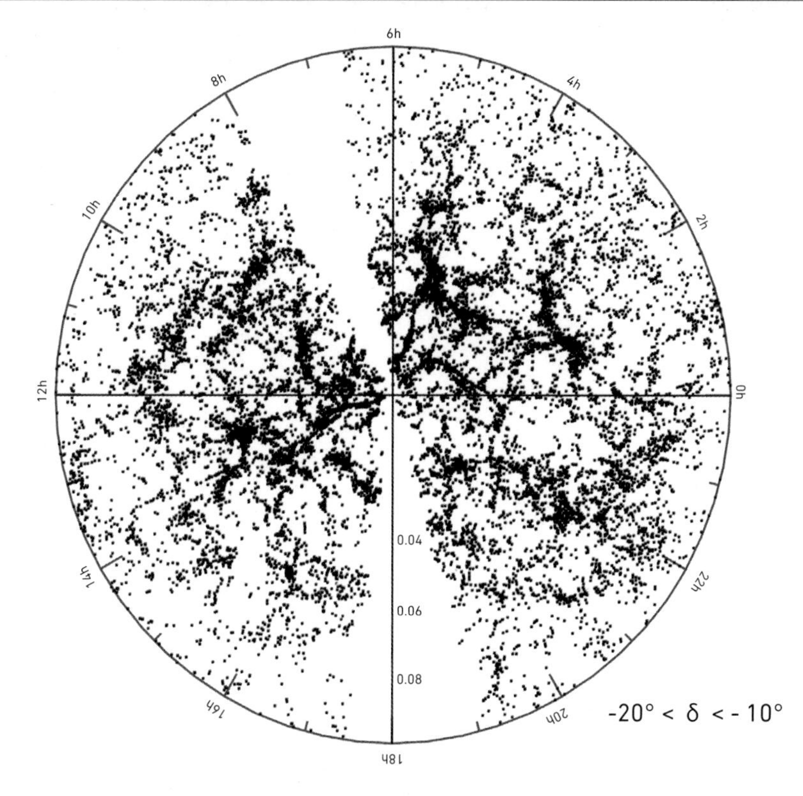

This image shows thousands of galaxies in a thin "slice" of declination below the Celestial Equator between -10 and -20 degrees. Notice how superclusters of galaxies are arranged in thin filaments and bubbles surrounding areas where there are no galaxies (voids). The distances to galaxies (measured in redshift – z) are indicated on the scale that runs from 0.04 to 0.08.

The 6df Galaxy Survey

A survey of galaxies called the *6df Galaxy Survey (6dfGS)* has mapped the southern sky distribution of over 100,000 galaxies in the nearby universe using galaxy spectra. The 6-degree field instrument (which is where the "6df" comes from) of the Anglo-Australian Observatory allows astronomers to make spectra over a large area of the sky quickly. Spectra measure a redshift, which gives a distance to each galaxy. The redshift combined with a sky position can be used to generate a three-dimensional image of galaxy distributions, helping to map out VLSS.

What the 6dfGS and other surveys have found is that superclusters are not distributed evenly throughout the local universe. Instead, they appear to be concentrated in filaments and bubbles separated by large voids. The local universe—out to hundreds of millions of light-years—looks a bit like foam.

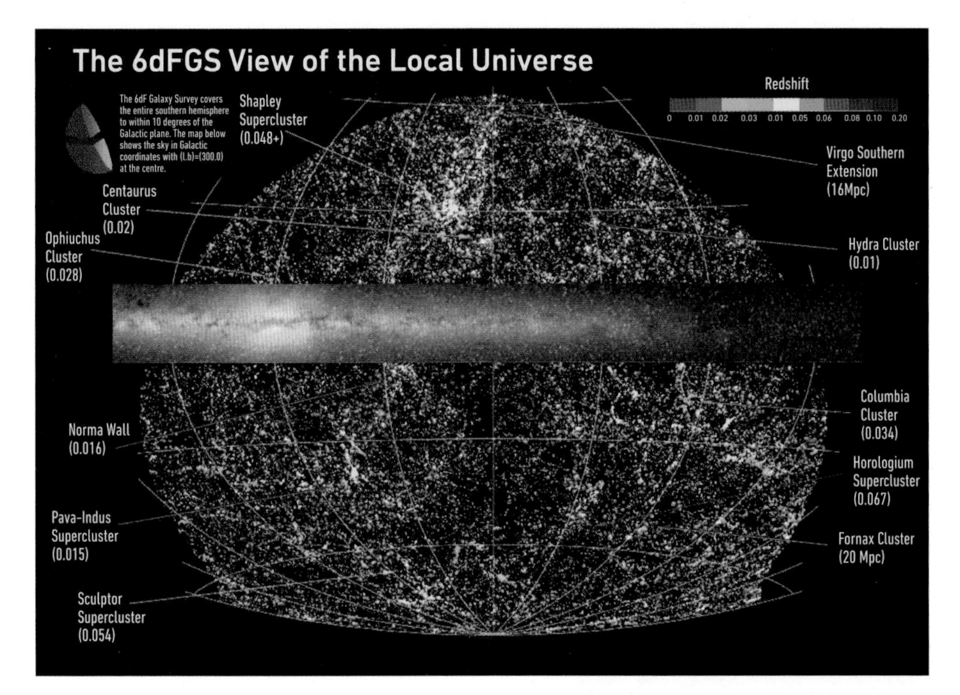

This view of the galaxies imaged by the 6dFGS shows galaxies colored by their redshifts, with the red dots representing the most distant galaxies at a redshift of z = 0.2. The positions of a number of known superclusters and the Galactic Center are indicated. Because this image shows galaxies at all distances, the three-dimensional structure is not obvious.

What Caused the Distribution of VLSS?

Surveys like the 6dfGS have shown us how galaxies are distributed. But how things got to be this way—what caused this distribution—is what cosmologists think about, and that question brings you to the final part of your journey into the universe.

Because it makes up 90 percent of the mass of clusters and superclusters, dark matter shapes the early evolution of the universe as modeled by supercomputers. These simulations, which allow gravity to act on dark and luminous matter, naturally lead to the formation of filaments and voids, in agreement with observations. The luminous matter in galaxies shows up like the foam on ocean waves.

The presence of dark matter in the universe causes galaxy superclusters to collect in bubble- and filamentlike structures surrounding voids in numerical simulations. The scale of the simulation is indicated by the bar. The unusual distance unit Mpc/h reflects the uncertainty in the value of the Hubble constant.

PART 6

The Origin and Fate of the Universe

Because of the finite speed of light, it takes time to see more-distant objects. As a result, the most-distant objects you can see in the universe formed long ago, when the universe was young. In this part, I help you explore these distant realms of the universe and begin to understand the earliest times in the universe.

Strangely, in order to understand the largest structures in the universe, you must have a grasp of the universe on its smallest scales; therefore, I discuss the fundamental forces and particles present in the universe. I also show you the cosmic microwave background, a doorway to the early universe. I then turn to a chronology of the early universe, from its fiery beginnings in the Big Bang to its current accelerating expansion. I end with a discussion of why the universe might be just so (known as the *anthropic principle*) and a review of some remaining unsolved questions in astronomy.

Ancient Views of the Origin and Fate of the Universe

Thinking about our origins is a distinctly human trait. Generally, when people speak of origins, they are talking about the origin of human life, or perhaps the origin of life itself. These events happened long ago, with human origins stretching back over millions of years and the origin of life on Earth into billions of years. But what about the origin of everything? Where did the universe come from? Or if that question is too difficult to answer, then how did the universe become what it is today? Has it always been the same, or was it different in the past?

In the past, these questions were out of the realm of science and were usually explained by creation myths and religious beliefs. But today, these fundamental questions about the universe are pondered by cosmologists and particle physicists, based on our observations of the universe on its largest and smallest scales.

How Do Creation Myths Apply to the Cosmos?

Many world cultures tell of a universe that is cyclic, being reborn over and over again. Others describe a linear history of the world, starting with its creation from nothingness. Many origin stories begin with darkness, formlessness, and chaos, with light bringing order to the world. These ancient ideas speak to a fundamental curiosity about our origins, and many of them are mirrored in our scientific exploration of the origin and evolution of the universe.

This page from the Codex Borgia shows part of the Aztec creation myth, which involved four worlds that existed before the present one.

Chinese and Indian Creation Myths

Ancient Chinese creation stories tell of a time of chaos, and that light created order from this chaos. In Chinese mythology, the Yin and Yang (representing male and female, Sun and Moon) balanced one another to make the universe a harmonious place. Hindu beliefs say that the universe is just one of many that have existed in succession, and that each new universe is born from the remains of the last one.

Michelangelo's The Creation of Adam (1512) captures part of the creation story of Judaism, Christianity, and Islam.

Western Creation Myths

The Book of Genesis in the Old Testament provides the seven-day creation story for Christianity, Islam, and Judaism. In order to bring the universe into being, God said, "Let there be light." And there was light. In Islam, Allah merely said, "Be" and thereby created the heavens and Earth. In the account of Genesis, light and dark were separated first, followed by water and land (which was covered in vegetation); the Sun, Moon, stars, and planets were then placed in the sky. Animals and humans only appear in the final part of the story.

A universe that emerges from this kind of nothingness in a Big Bang is the starting point of our current cosmological models. Astronomers can describe the universe from 10^{-42} seconds after the Big Bang, but not before.

African Creation Myths

Like stories from other parts of the world, many African creation myths involve water. The Boshongo in Central Africa tell of a beginning of only darkness, water, and their god Bumba, who created the Sun, Moon, stars, and planets from vomit. Another Bantu tribe in Africa, the *Fans,* tell of their god Nzame creating the universe and everything in it from nothingness.

Hydrogen, the most common element in the universe, is a fundamental part of the water molecule, and the presence of water is essential for life as people know it.

A Steady State Universe and Einstein

While many origin myths had a beginning, most astronomers were convinced, until the twentieth century, that the universe had always existed. The leading model for the universe was a *steady state universe*—one in which its components (like galaxies and stars) changed with time but was unchanging and eternal on the largest scales.

Newton and Einstein's Views on Gravity in the Universe

The force that powered the dynamism in the universe was gravity, and from the time of Newton in the seventeenth century to the early twentieth century, there was a single description of how gravity worked. According to Newton, any two objects with mass were attracted to one another; this attraction was directly proportional to the product of the masses and inversely proportional to their distance squared. This idea, known as Newton's law of universal gravitation, is written as follows (where *FG* is the gravitational constant, m_1 and m_2 are the masses, *G* is the gravitational constant, and *r* is the distance)

$$F_G = \frac{(G \times m_1 \times m_2)}{r^2}$$

However, a young physicist in the early twentieth century, Albert Einstein, gave the world a different view of gravity that fundamentally changed our understanding of how the universe works on its largest scales and under extreme conditions. Instead of gravity being an attractive force between two objects with mass, Einstein proposed something very different—that mass distorted space and time (something he called *spacetime;* see "Spacetime") in such a way that anything moving near that mass would move in a curved path. Like any good scientific hypothesis, Einstein's model, known as *general relativity,* explained the known behavior of matter and predicted behavior that would only be true if his model was correct.

For example, general relativity predicted the Sun's gravity should deflect the light from a background star, and that this effect should be observable during a solar eclipse. This effect was observed by Sir Arthur Eddington in 1919, in the first confirmation of Einstein's general relativity. In the Newtonian description of gravity, photons (which have no mass) should not be deflected by gravity. However, the gravitational lensing effects commonly observed in galaxy clusters have provided further abundant confirmation of Einstein's prediction.

The original observations of the position of background stars during the eclipse of May 29, 1919, that confirmed one of the predictions of general relativity—that photons are deflected by gravitational forces.

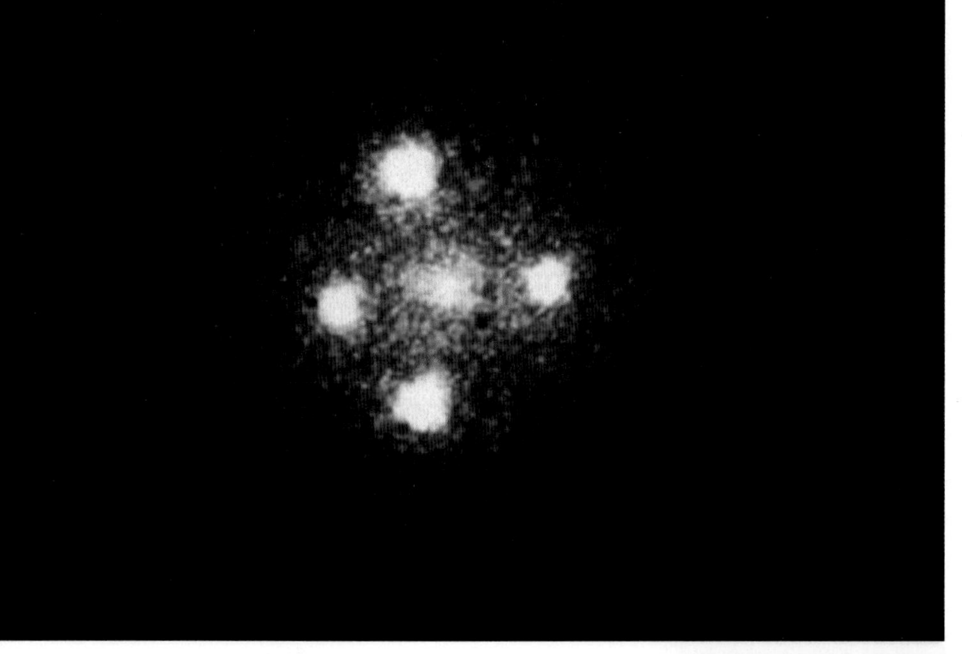

Observations of gravitational lenses like G2237+0305, also called the *Einstein Cross*, provide confirmation of Einstein's general relativity. A foreground galaxy has distorted a single background object into the four objects seen.

The Cosmological Constant

In order to keep the steady state universe from collapsing because of the force of gravity, Einstein introduced a constant called the *cosmological constant* (λ) that provided his equations an outward force that countered gravity. Without such a force, the universe would collapse back in on itself. When Hubble and Humason published their result on the recession of galaxies and an expanding universe (see "Expansion of the Universe"), Einstein realized the constant was unnecessary and abandoned it. Interestingly, this constant has been revived in recent years for a very different reason: to explain the accelerating expansion of the universe (see "The Runaway Universe").

One of Einstein's predictions was only confirmed a few years ago, in 2011, by an orbiting spacecraft called Gravity Probe B. This mission measured that Earth's gravitational field distorts spacetime locally, and confirmed an even more subtle effect: the rotation of Earth distorts spacetime through a process called *frame dragging.*

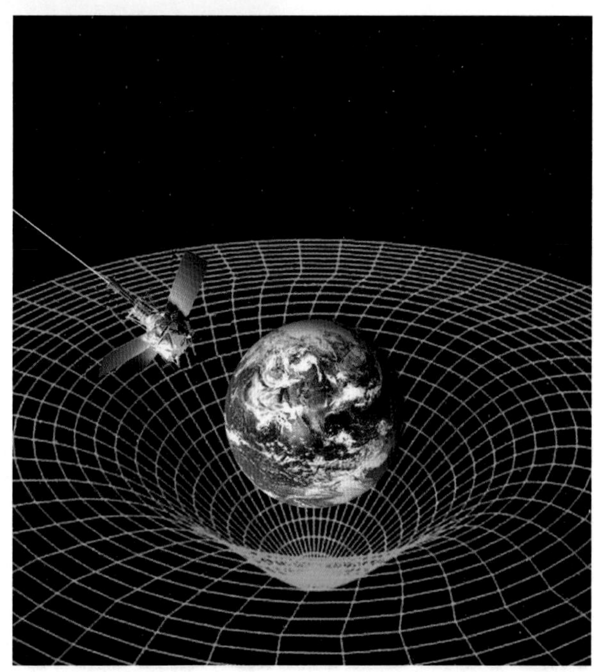

This artist's conception shows Gravity Probe B in polar orbit around Earth. Gyroscopes aboard the satellite confirmed two fundamental predictions of general relativity: the distortion of spacetime by mass, and frame dragging caused by Earth's rotation.

Spacetime

Spacetime is the name for a model of the universe that combines the three dimensions of space with one dimension in time, making a four-dimensional space. In general, this mathematical model emphasizes the geometry of the universe and fits well with the ways in which cosmologists think about the origin and evolution of the universe.

Spacetime is bit a different from the common perception of moving "through" time, because it puts space and time on a more equal footing. Because the four dimensions of spacetime always exist, what people experience as change and evolution can be thought of as motion through this spacetime continuum.

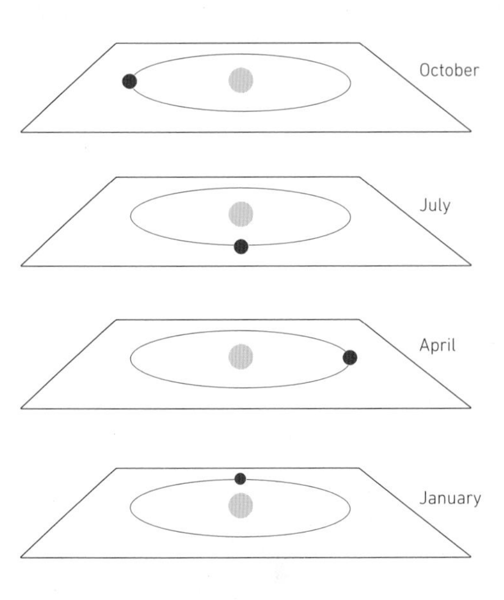

Special Relativity

One of the fundamental tenets of Newtonian mechanics is that the universe has an absolute clock by which the passage of time can be measured. Einstein's theory of special relativity in 1905 changed that by proposing the speed of light is constant for all observers, meaning that observers moving at different velocities would not observe events to be simultaneous—the speed of light, not time, is the absolute. The constant speed of light meant that space and time must be more pliable in order to keep the speed of light constant. Two effects of special relativity, confirmed by experiments on Earth, are called *time dilation* and *length contraction*. Because of time dilation, an object moving at high velocity relative to another object will experience the passage of less time. Length contraction describes how the high-velocity object perceives moving through a shorter distance.

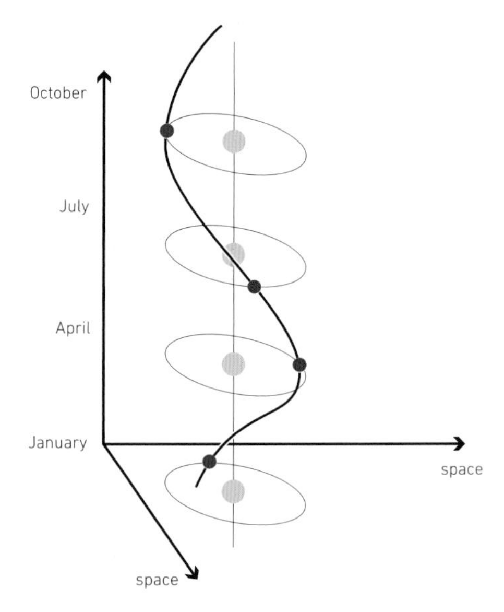

In order to understand spacetime, think of the position of Earth as it orbits the Sun, as shown on top. If you were to stack up these images and trace the position of Earth, you would get a plot like that seen on the bottom. That spiraling pattern is a spacetime diagram.

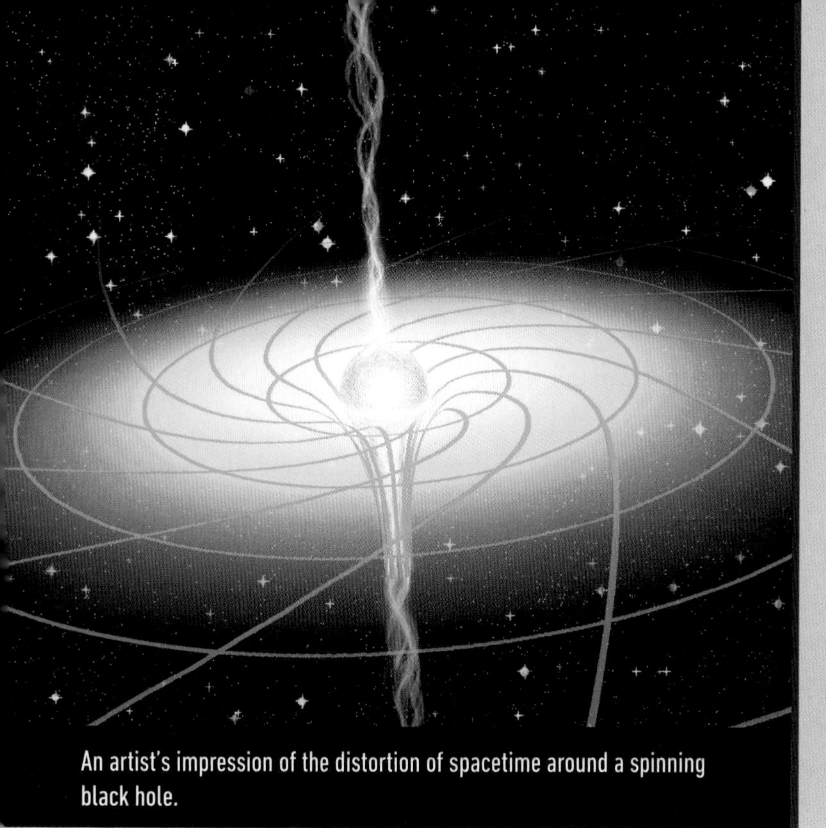
An artist's impression of the distortion of spacetime around a spinning black hole.

Black Holes and Spacetime

Black holes distort spacetime in such a way that radiation and matter cannot escape them. In essence, anything that passes into this sphere (called the *event horizon*) around a black hole is lost to the universe. The distortion of spacetime caused by a black hole is so severe that it results in a singularity—a point in space with finite mass but no volume.

Because photons can be affected by gravity, general relativity predicts an effect known as *gravitational redshift*, in which a gravitational field shifts the wavelength of light moving away from a black hole to longer (redder) wavelengths.

General Relativity

As you read previously, Einstein described how gravity worked by its effects on spacetime via general relativity. In general relativity, spacetime is distorted by the presence of matter. Einstein predicted the mass of the Sun would distort spacetime in such a way that photons from a nearby star would be deflected by its mass. This was not an effect predicted by Newtonian gravity, because photons are massless (see "Fundamental Particles").

The reason this distortion had not been noticed before is that Newton's laws work remarkably well for low-mass, low-velocity objects, and most of what people deal with in the ordinary world has both low mass and low velocity. However, Einstein's ideas work for velocities close to the speed of light and very massive objects (like stars), which is what he and other physicists were focusing their attention on at the time.

Expansion of the Universe

Astronomers Edwin Hubble and Mitchell Humason made observations in the 1920s that confirmed galaxies were moving apart from one another at velocities that were proportional to their distance from the Milky Way. This observed expansion of the universe causes a shift in the wavelength of spectral lines in a galaxy and can be used to determine the distance to a galaxy (see "The Hubble Law").

Understanding Expansion

You might be tempted to think of galaxies as two ships moving apart on the ocean. But the redshift of galaxies is more fundamental than that. The redshifts of galaxies are the result of the expansion of the universe itself; it is as if the ocean between the ships is expanding. So the expansion of the universe apparent in redshifts is fundamentally different than other ways in which two objects move apart.

To better understand how the expansion of the universe works, imagine that galaxies are dots on the surface of a balloon. As you inflate the balloon, each of the dots moves away from every other dot. No matter which galaxy (dot) you observe from, you will see all of the other galaxies (dots) moving away from you as the balloon expands. Likewise, from the perspective of any galaxy, the universe is expanding, and all the other galaxies are moving away.

The balloon analogy is shown with galaxies represented as dots on the surface of an expanding balloon. Like the universe itself, the surface of the balloon has no center and no preferred vantage point. Every galaxy will observe the same expansion, with more-distant galaxies receding more quickly.

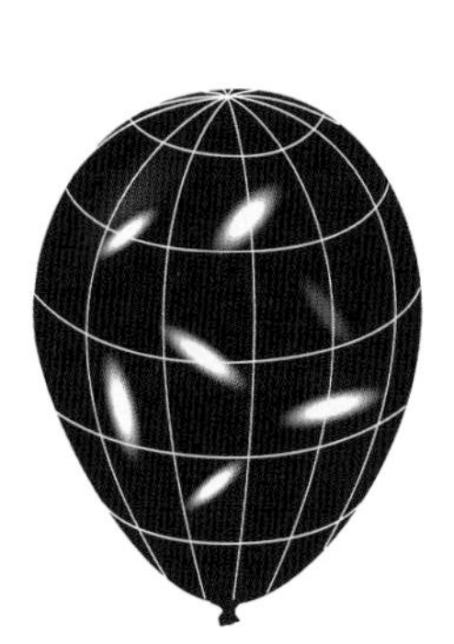

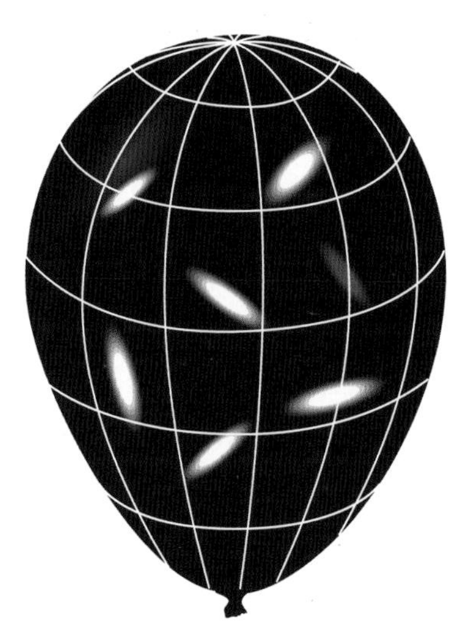

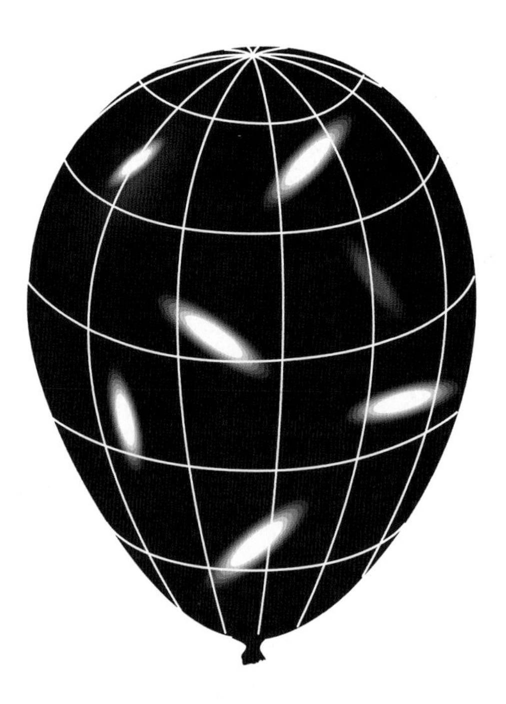

The balloon analogy can also be used to understand the redshift of galaxies. The expansion of the universe itself stretches photons to have longer (redder) wavelengths.

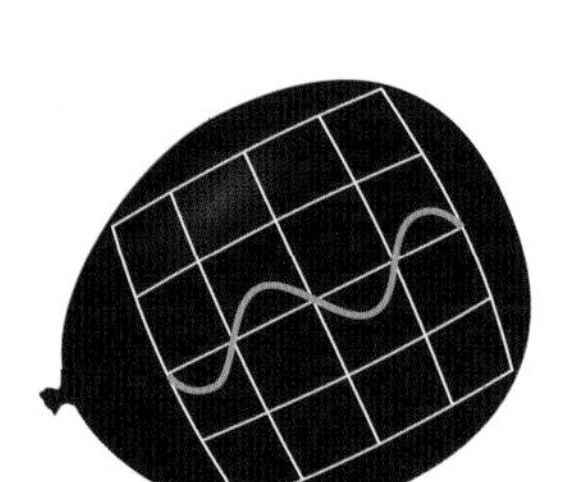

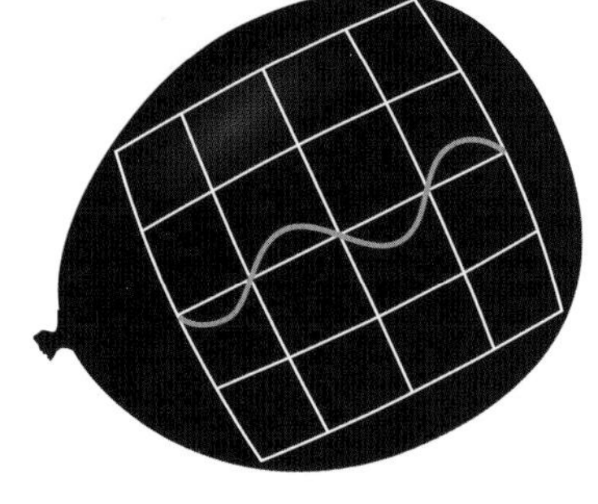

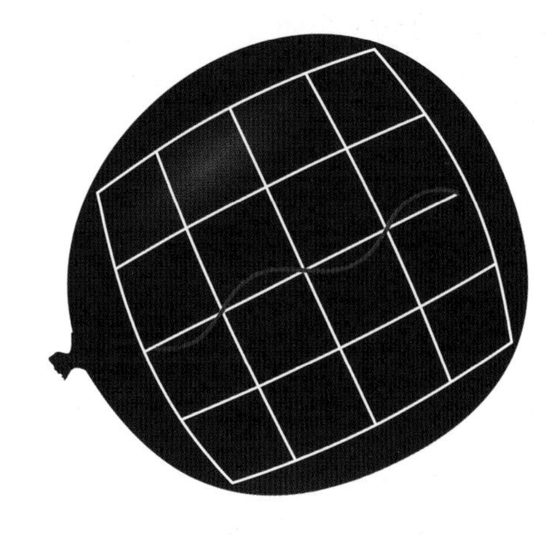

But how do you explain the redshifts of galaxies in relation to the expansion of the universe? Using the balloon example again, a wavelength (which represents a photon) gets longer as the balloon expands. Similarly, when a galaxy emits a photon, that photon is part of spacetime. As the universe expands, that photon is "stretched" out to longer and longer wavelengths. The more distant the galaxy, the greater the "stretch" and the larger the redshift.

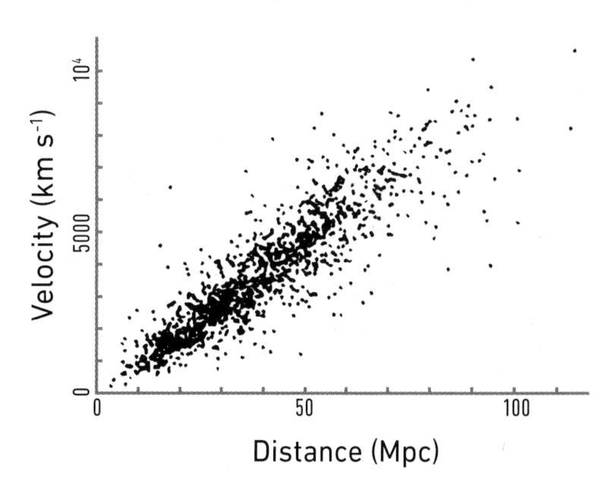

As an example of a more-recent study of the expansion of the universe, this data shows a sample of over 1,300 galaxies out to a distance of over 100 Mpc. There is scatter in the data, but the relationship holds out to great distances.

Studying the Universe's Expansion

As telescopes and detectors have improved, astronomers have been able to explore the expansion of the universe out to greater and greater distances. Hubble's early plot from 1929 only traced the expansion of the universe out to about 2 Mpc, or about 6.5 million light-years. The Andromeda Galaxy is about 2.5 million light-years away, so this first result was limited in its scope.

However, later observations of more-distant galaxies have confirmed the relationship holds true to very great distances—more-distant galaxies are moving away at greater and greater velocities, according to the Hubble Law.

The discovery of standard candles that can be used out to very-great distances (such as Type Ia supernovae) has allowed astronomers to trace the expansion of the universe to even greater distances and to discover that the expansion of the universe has not been constant, but is accelerating (see "The Runaway Universe").

The Hubble Constant and the Age of the Universe

As you read in "The Hubble Law," the exact value of the Hubble constant (H_0), or rate of expansion of the universe, was debated for many decades. The problem was that different observational tests seemed to give different results, with one set of values around 50 km/s/Mpc and the other closer to 100 km/s/Mpc. In years past, the exact value of the Hubble constant might have been considered an unsolved problem in astrophysics. But multiple, independent, observational results have now given astronomers a value of the Hubble constant with a relatively small error bar, very close to 70 km/s/Mpc. In fact, Type Ia supernovae in the past two decades have provided some of the best estimates of the Hubble constant, currently 71 km/s/Mpc +/- 5%. Making large numbers of observations of the same constant reduce the size of the errors.

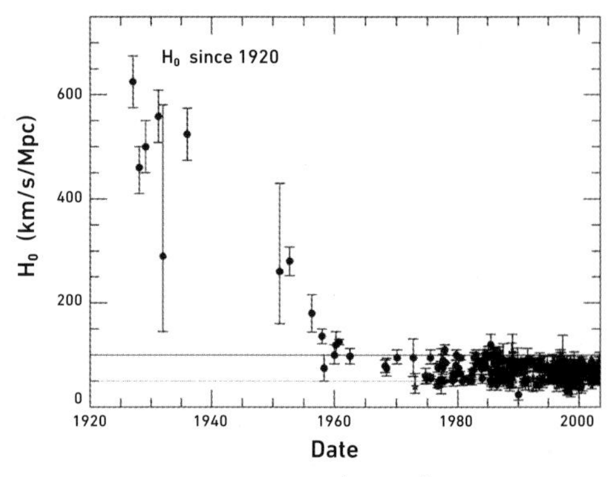

Values of the Hubble constant (circles) with error bars showing its best current value from 1920 until now. Notice how the value has fallen, the error bars have gotten smaller, and the number of measurements have greatly increased.

Why did astronomers debate the value of the Hubble constant so fiercely? Because the Hubble constant provides a very fundamental quantity: the age of the universe.

The early values for the Hubble constant were about 500 km/s/Mpc, giving the universe an age of only about 2 billion years. That age ran into immediate difficulty with radioactive dating of Earth, which showed the planet was at least 3 billion years old. Because the distance to galaxies is such a fundamental part of determining the Hubble constant, a recalibration of the Cepheid variable distance scale lowered the Hubble constant significantly (see "Distances in the Milky Way").

Using Redshifts

If the slope of the Hubble Law is well known, the law itself can be used to determine the distance to galaxies from the redshift of their spectral lines. Redshifts are measured directly, and do not depend on a specific value for the Hubble constant. In order to convert a redshift into an accurate distance, astronomers must know the exact value of the Hubble constant. Because there is some remaining uncertainty in the value of the Hubble constant, astronomers often refer to the redshift of a galaxy instead of its distance in Mpc.

However, the debate was not resolved until the Hubble Space Telescope was developed and launched in the 1980s. One of its primary goals was to accurately determine the distance scale out to hundreds of Mpc and thus resolve the dispute over the value of the constant. Determining distances in the nearby universe (out to about 100 million light-years) depends on the careful measurement of the variability of Cepheid variable stars. There is a very well-known relationship between the period of variability and the luminosity of the star. The Hubble Space Telescope was able to measure the variability of individual stars out to much greater distances than before because of its resolution.

If you look closely at the Hubble constant, you will notice that it has units of kilometers per second per megaparsec (km/s/Mpc). With the right unit conversions, the kilometer and the megaparsec units (both measuring distance) cancel out, leaving just 1/s, or inverse time. The age of the universe can be determined from the Hubble constant (currently) to be about 13.8 billion years.

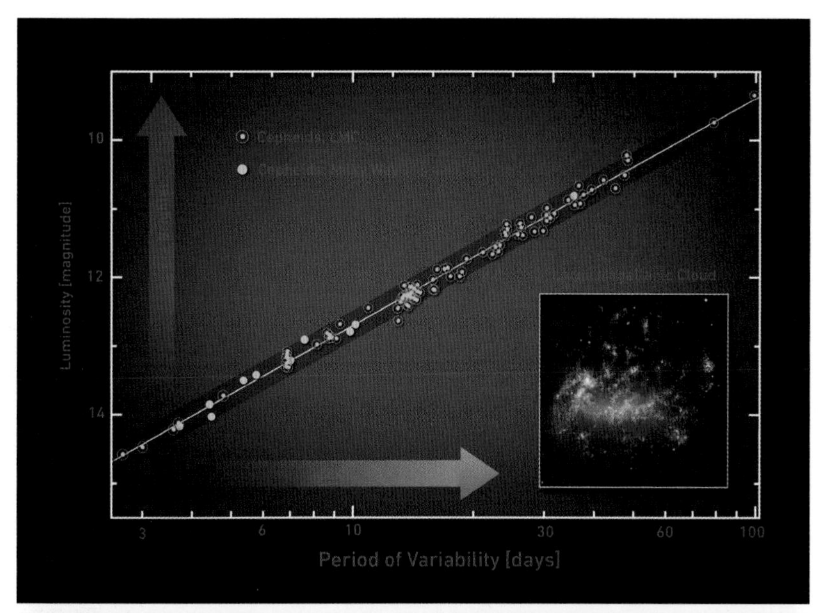

This plot of the period-luminosity relationship in the Milky Way and the Large Magellanic Cloud shows how precisely that relationship holds.

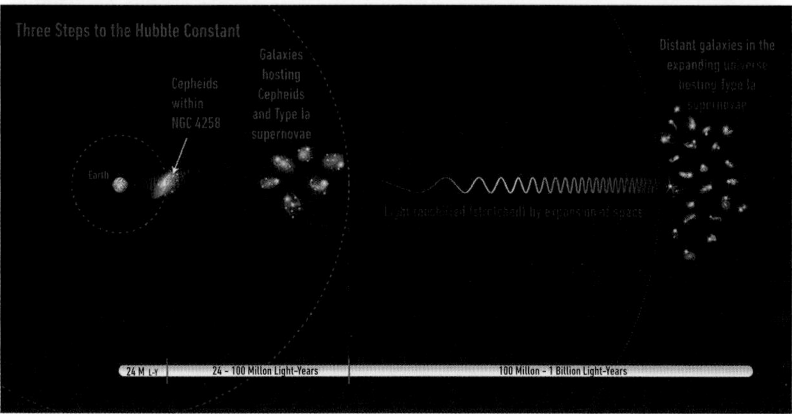

This schematic shows the type of observations that can be used to determine the value of the Hubble constant—Cepheid variables in the closest galaxies, Cepheids and Type Ia supernovae out to about 100 million light-years, and Type Ia supernovae in the most-distant galaxies.

The Cosmic Microwave Background: The Afterglow of the Big Bang

Penzias and Wilson discovered the cosmic microwave background with the 15-m Holmdel horn antenna, which was built for use with communication satellites. They received the Nobel Prize in Physics in 1978 for their discovery.

Many times in the history of astronomy, great discoveries have been made quite by accident. Perhaps the greatest accidental discovery in the history of astronomy is the detection of the cosmic microwave background, which provided conclusive evidence that the universe started as a mighty explosion: the Big Bang.

Discovering the Cosmic Microwave Background

The universe's origin as a spectacular explosion was not a new idea at the time of the cosmic microwave background's discovery in the 1960s. In the 1940s, theoretical physicist George Gamow first proposed this model as an explanation of the recession of galaxies discovered by Hubble and Humason decades earlier. But evidence other than the redshift of galaxies had not been detected. Astronomers at the time proposed that if the universe did have a beginning—and if that beginning was extraordinarily hot and dense—there should be a detectable afterglow from that explosion.

In the mid-1960s, in Holmdel, New Jersey, two engineers with Bell Telephone Labs—Arno Penzias and Robert Wilson—were part of a project to transmit signals via satellite across the Atlantic Ocean. Much to their annoyance, they found the sky had low levels of radio interference, no matter what part of the sky was facing their telescope. Careful cleaning of the "horn" antenna did not reduce the noise—it had a constant strength and came from all areas of the sky, day and night.

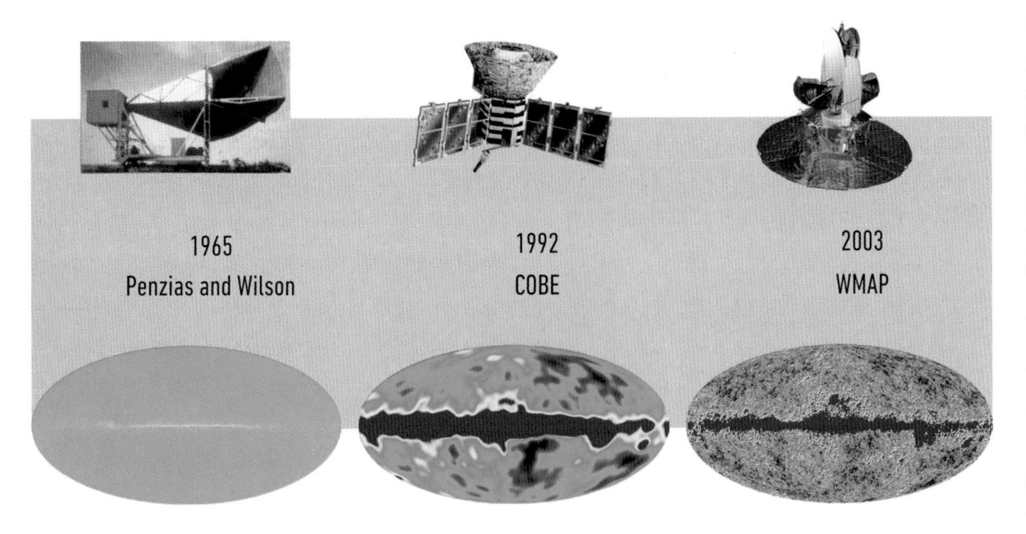

This image shows the sky as imaged by Penzias and Wilson in 1965, and by later orbiting telescopes Cosmic Background Explorer (COBE) and Wilkinson Microwave Anisotropy Probe (WMAP). In all three images, the horizontal plane of the Milky Way is the brightest emission in the microwave view of the sky.

At the same time, in nearby Princeton, New Jersey, a group of astrophysicists was working on the theory of the cosmic microwave background—the thermal radiation left over from the Big Bang—and how it might be detected. The two groups were connected by a common physicist friend, and the Princeton group quickly understood the Bell Labs engineers had discovered the radiation they had hoped to find.

Since then, observations of the cosmic microwave background have been made by a number of ground- and space-based missions, which have confirmed this initial discovery.

Black-Body Radiation

So what was the radiation detected by Penzias and Wilson in 1965? It came from the universe—not from one object in the universe, but from the whole universe.

If you recall, the temperature of an object is reflected in its color—for example, a metal rod glowing blue is hotter than one glowing red. In fact, a metal rod can be extremely hot and not radiate in the visible part of the spectrum at all. The plot of brightness as a function of wavelength is called a *black-body curve,* and its peak brightness tells the temperature of the emitting object.

The cosmic microwave background comes from the black-body radiation of the early universe. While it started incredibly hot and dense, the universe (on average) has expanded and cooled to a very low temperature: 2.73 K, or just a few degrees above absolute zero. The photons have been "stretched" by the expansion of the universe (see "Expansion of the Universe").

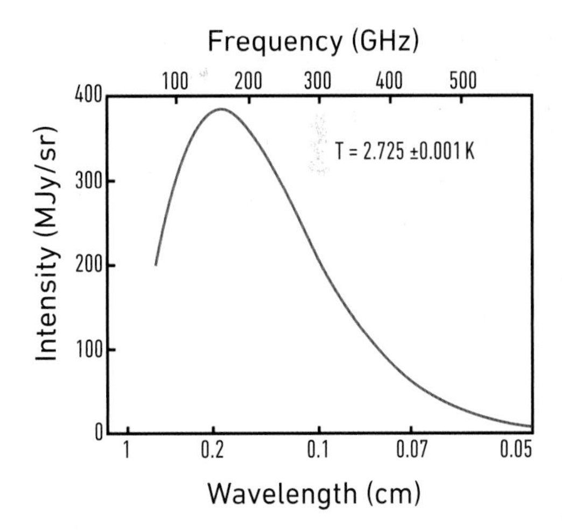

The black-body curve of the cosmic microwave background. Its peak wavelength (in the microwave part of the spectrum) indicates the universe has a temperature of just above absolute zero.

Mapping the Cosmic Microwave Background

The discovery of the cosmic microwave background gave strong support to a universe with a beginning in time: a Big Bang. This radiation is the earliest radiation visible from the universe, allowing observers to see photons that were released only about 380,000 years after the Big Bang (see "Early Universe Timeline").

Once the cosmic microwave background had been discovered, the race was on to begin to map it or to make images of its distribution on the sky.

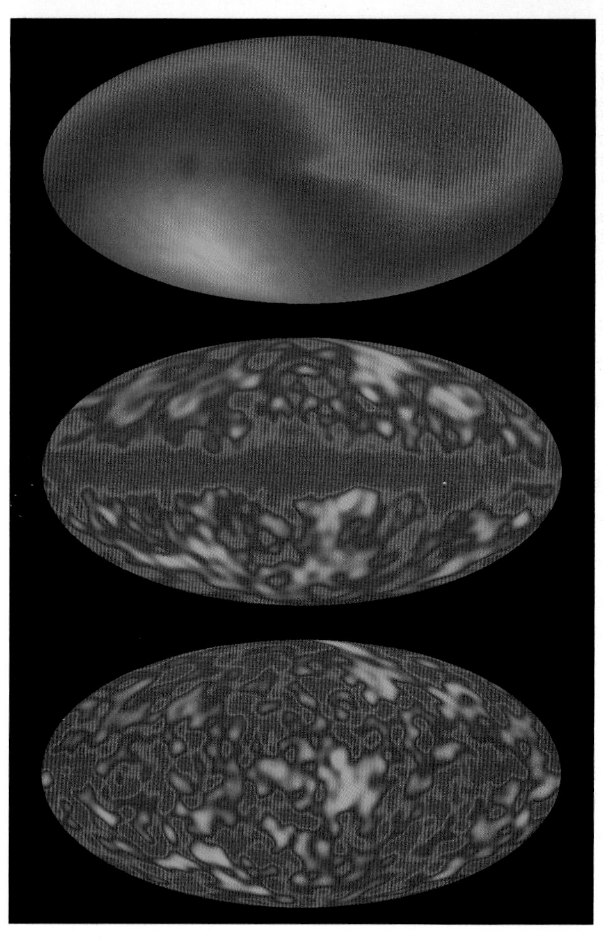

Variations around 2.73 K are shown in this false-color image of the cosmic microwave background. The change introduced by the motion of Local Group and emission from the Milky Way are seen in the top and middle images. Once these effects are removed, the tiniest variations (only a few parts in a million) are visible (bottom image).

Studies of the Cosmic Microwave Background

One of the first missions was the Cosmic Background Explorer (COBE). Launched in 1989, the mission scientists for COBE received the 2006 Nobel Prize in Physics for their observations in support of Big Bang theory. These early missions first found that while the cosmic microwave background was evenly distributed on the sky, there were small, measurable fluctuations in the brightness of the radiation corresponding to tiny variations in the temperature of the early universe.

The angular size of these tiny fluctuations became more apparent during later missions. In 2001, NASA launched the Wilkinson Microwave Anisotropy Probe (WMAP), which mapped the cosmic microwave background with much higher resolution. The European Space Agency's Planck mission, launched in 2009, has provided the sharpest view of the cosmic microwave background to date.

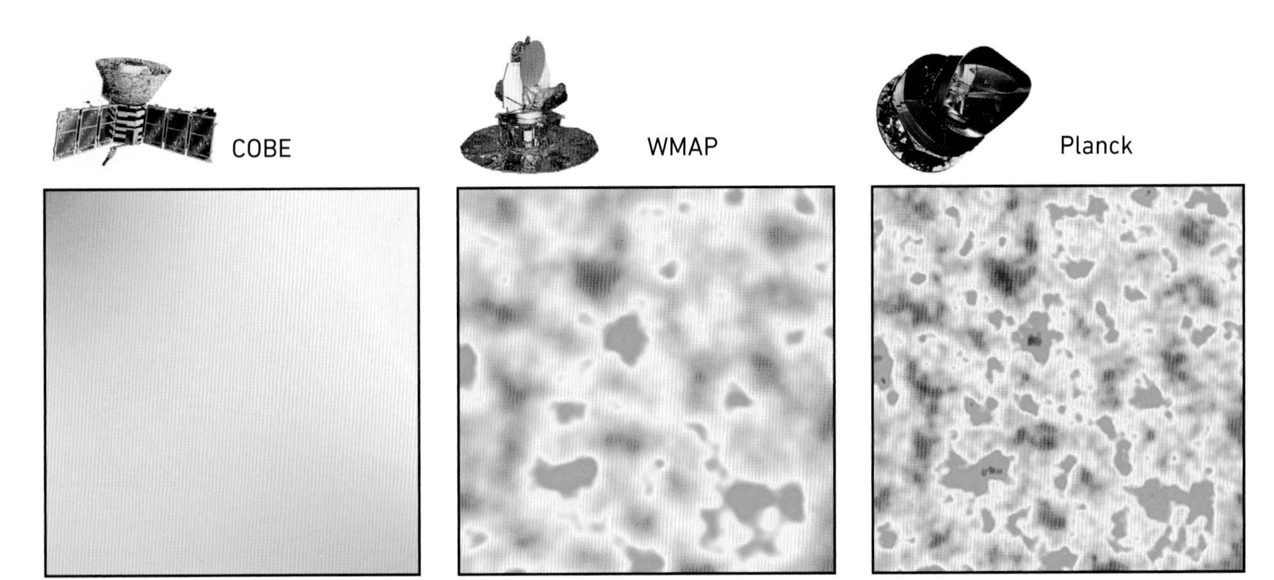

Images of the cosmic microwave background showing a box 10 degrees on a side (or a field about 400 times the size of the Moon), with the spacecraft that captured the images shown above them. As you can see, the most current image from Plank provides the sharpest view of the cosmic microwave background.

What's Known About It

The series of missions to map and study the cosmic microwave background have provided ever-more-sensitive and sharp views of the distribution of the fluctuations first observed by COBE.

All of the missions have confirmed the radiation detected in the cosmic microwave background is not associated with any known structures in the universe. This is because the light detected in the cosmic microwave background was released before the universe had made any stars or galaxies, making it the oldest picture astronomers have of the universe.

This all-sky picture of the cosmic microwave background was made from nine years of data from the WMAP. The false colors show variations in temperature 200 microKelvin above and below 2.73 K. The variations in temperature (and therefore density) are thought to have led to the formation of the first stars and galaxies.

The distribution and size of the fluctuations in the cosmic microwave background are windows into some fundamental properties of the universe. The scale of the fluctuations can be used by cosmologists to determine whether the universe is flat, open, or closed (see "The Geometry of the Universe"), as well as the balance between matter and energy in the early universe.

Early Universe Timeline

The cosmic microwave background gives us a snapshot of the very early universe, when it was only about 380,000 years old. It turns out, there are ways to understand the evolution of the universe at even earlier times. The fact that the universe apparently had a beginning (the Big Bang) and was changing with time naturally led to this question: What was the universe like in the past?

In the coming sections, you will learn about the origin and evolution of the universe from its very earliest moments. To begin with, though, it is important to see how some of the images of the observable universe I have discussed so far fit in to this larger picture.

Remember, the more distant an object, the older the light we are receiving from it; therefore, the faintest, most distant objects show the early universe. The earliest possible view of the universe comes from 380,000 years ago (the cosmic microwave background). Before that, photons were not free to move through the universe, but were constantly scattered by electrons.

The cosmic microwave background was emitted by the first photons to emerge from the haze of hot material that filled the early universe when it had cooled to about 3,000 K (see "Recombination"). Once stars and galaxies formed, they emitted photons that astronomers now observe in some of the most sensitive (deep) images.

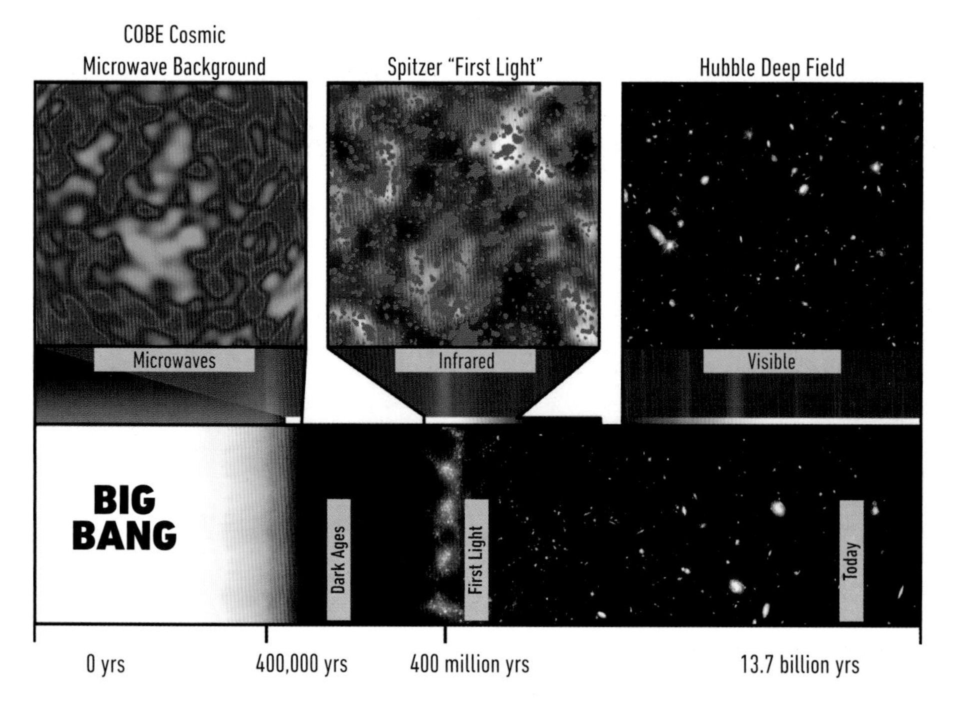

The universe begins with a Big Bang, and the cosmic microwave background is indicated about 380,000 years later. The first light from stars was not emitted until 400 million years later, as ultraviolet and visible light stretched into infrared light by the expansion of the universe. Galaxies in the Hubble Deep Field are from the nearby (and more recent) universe.

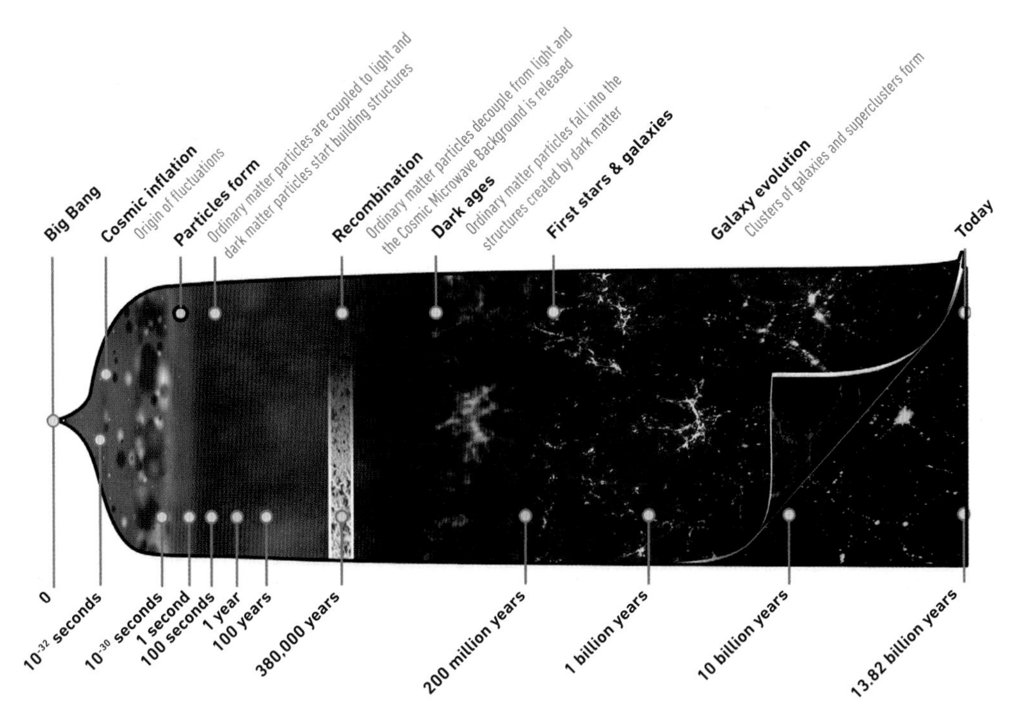

This diagram shows stages in the evolution of the universe, starting with the first moments and coming through to the present day.

The following is an outline of the evolution of the universe and the times in the history of the universe they correspond to:

Big Bang (0 seconds): The start of it all. What was there before the Big Bang? No one knows or perhaps can know.

Cosmic Inflation (10^{-32} seconds): The scale of the universe expands suddenly by 50 orders of magnitude (or 1 with 50 zeroes after it).

Formation of Matter (10^{-12} to 1 seconds): The stuff of the universe; both matter and antimatter take shape.

Nucleosynthesis (3 minutes to 20 minutes): The entire universe briefly acts like the core of a star and fuses some helium and traces of heavier elements (lithium and beryllium).

Recombination (380,000 years): The universe is cool enough for protons and electrons to hold on to one another without breaking apart. This change allows photons to escape for the first time, creating the cosmic microwave background.

Dark Ages (150 million to 800 million years): During this time, dark matter begins to pull ordinary matter into the first structures.

First Stars and Galaxies (150 million to 1 billion years): Guided by dark matter, ordinary matter becomes visible and is distributed along arcs and filaments.

Galaxy Evolution (500 million years to present): Gravity drives the mergers of galaxies, leading to the formation of clusters and superclusters of galaxies.

Fundamental Particles

In order to provide you with a deeper understanding of the early universe, I'd like to discuss the fundamental particles and forces that make up the universe. You have read about some of these forces (like gravity) and particles (like photons) in previous sections, but here I lay them out in one place and in later sections talk about their relationship to the evolution of the early universe.

What Is the Smallest Particle?

Ancient Greek philosophers, who were some of the first people recorded to have pondered the nature of matter, proposed that all things in the universe were made of a smallest indivisible particle called an *atom*.

Scientific inquiry in the intervening millennia verified that basic idea, with a number of complications—there was not just one fundamental particle, but many. An atom has parts, including a nucleus (which holds neutrons and protons) inside a haze of negatively charged electrons. Electrons do not appear to be divisible into smaller particles; they are fundamental. Protons and neutrons, however, are made up of even more fundamental particles called *quarks*.

The current Standard Model of particle physics divides the universe into matter particles (fermions) and force carrier particles (bosons). All particles have a property known as spin; particles with integer spin are bosons, while those with half-integer spin are fermions. Fermions and bosons are considered the most fundamental types of particles in the universe.

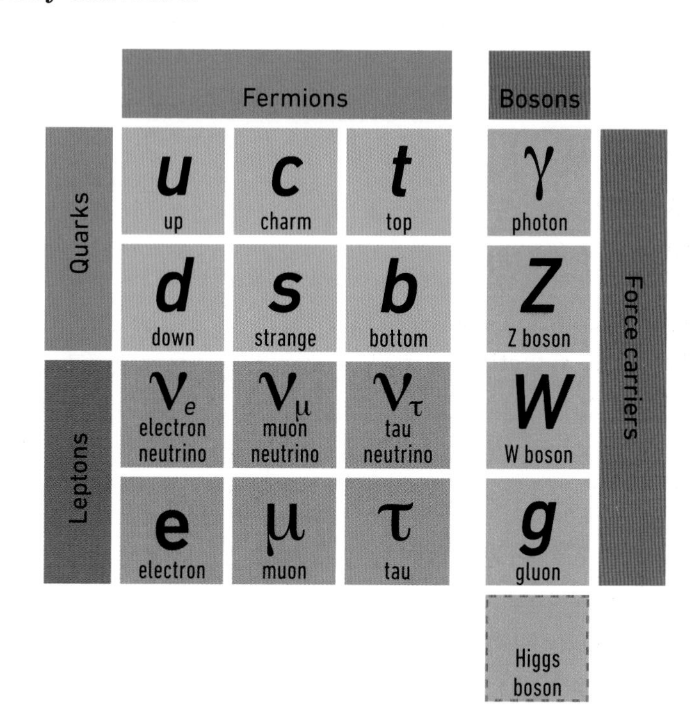

Fundamental particles, shown here, are divided into fermions and bosons. Combinations of fermions make up all of ordinary matter. Bosons are the carriers of the four fundamental forces—gravity, the weak nuclear force, the strong nuclear force, and the electromagnetic force.

Fermions

Fermions are particles that make up the matter of the universe. They are further divided into quarks and leptons. Quarks and leptons are matter particles, while antiquarks and antileptions are antimatter particles. There are 12 fermions (six quarks and six leptons) and 12 antifermions (antiquarks and antileptons).

Quarks and leptons are divided into three generations—divisions in which the particles' interactions are identical, but the particles are more massive. Quarks can combine with one another to form hadrons, the most familiar of which are protons and neutrons. Up and down quarks have the lowest mass and are the most stable, while the heavier quarks (strange, charm, bottom, and top) are unstable and quickly decay into up and down quarks. As an example, a proton is made of three quarks (up, up, down), and a neutron is made of three different quarks (up, down, down).

There are also six types of leptons in three generations. The most familiar are the electronic leptons—the electron and the electron neutrino. The muon, muon neutrino, tau, and tau neutrino are heavier generations of leptons that can be generated in high-energy accelerators.

Bosons

The four fundamental forces—gravity, the weak nuclear force, the strong nuclear force, and the electromagnetic force (see "Fundamental Forces")—are said to have force carriers; these are known as bosons. Bosons, of which there are five, exchange the fundamental forces between particles. Photons carry the electromagnetic interaction, W and Z bosons carry the weak interaction, and gluons carry the strong interaction. A sixth proposed boson that would carry the gravitational force is known as a graviton, though physicists have not yet detected it.

In the Standard Model, the universe is permeated by the Higgs Field, and interactions between some particles and the Higgs Field gives them mass. The Higgs boson, an elementary particle that was proposed and in the 1960s and discovered in 2012 by the Large Hadron Collider (LHC), is thought to explain why some particles have mass and others do not.

The Large Hadron Collider (LHC), built by CERN in Europe, is the world's most powerful particle accelerator; it's able to simulate the conditions of the early universe. In 2012, the LHC announced the discovery of the Higgs boson, consistent with predictions of the Standard Model. One of the LHC's detectors (ATLAS) is shown here.

Fundamental Forces

Four fundamental forces, some of which may be more familiar to you than others, govern all interactions between particles in the universe. In the distant past—when the universe was a hotter, more energetic place—all of these forces were one (see "Inflation and the Very Early Universe").

The Four Forces

The strongest of the four forces is called, appropriately, the *strong nuclear force*. This is the force that holds the nucleus of an atom together. This force counteracts (and is stronger) than the repulsion between the positively charged protons in the nucleus of an atom. While it is the strongest force, its range is very small (10^{-15} m), only covering the diameter of an atom's nucleus.

The next-strongest force is the *electromagnetic force*. With infinite range, this force causes the attraction and repulsion between charged objects, as well as magnetism. Like gravity, the attraction between particles with opposite charge gets weaker with the square of distance.

The *weak nuclear force* is a million times weaker than the strong nuclear force and acts on even smaller scales (10^{-18} m). This force describes phenomena like the beta decay of particles.

The most familiar of the forces, and the one I have spent the most time describing, is the *gravitational force*. While it has infinite range, it is 10^{38} times weaker than the strong nuclear force.

Strong	Electro-magnetic	Weak	Gravity
Force which holds nucleus together		Neutrino interaction induces beta decay	
Strength			
1	$\frac{1}{137}$	10^{-6}	6×10^{-39}
Range (m)			
10^{-15} (diameter of a medium-sized nucleus)	Infinite	10^{-18} (0.1% of the diameter of a proton)	Infinite
Particle			
gluons, π(nucleons)	photon	Intermediate vector bosons W^+, W^-, Z_0,	graviton ?

The four fundamental forces in the universe are the strong nuclear force, weak nuclear force, electromagnetic force, and gravitational force (or gravity). Gravity is by far the weakest of the forces. The strong force, which holds the nucleus of atoms together, is the strongest but only acts over very small distances.

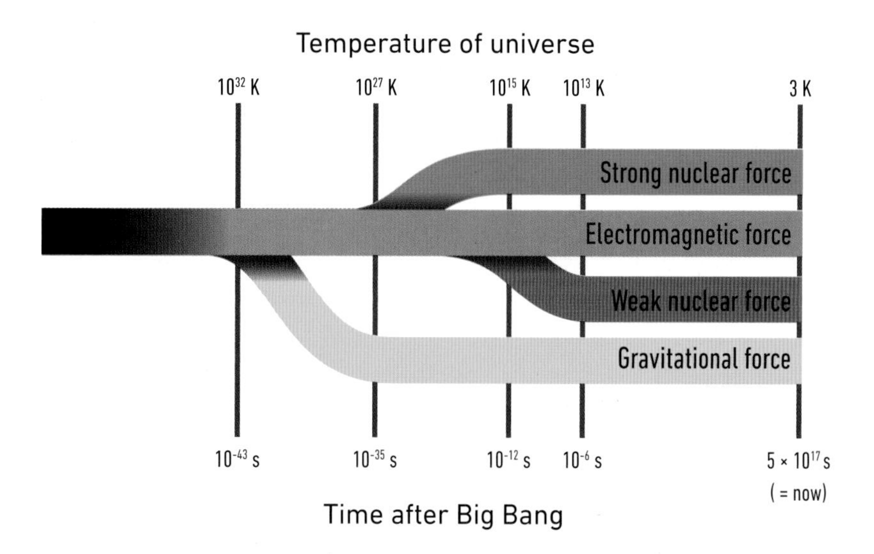

In the current universe, the four forces act differently. In the past, when the universe was much hotter and denser, the forces were not differentiated. Experiments in particle accelerators have confirmed this "unification" of forces.

Fundamental Forces and the Universe

The recession of galaxies and the cosmic microwave background imply that the universe began as a much hotter, more energetic place. If you wind the film of the universe backward, you come to times when the entire universe was at the temperature of the surface of the Sun, the temperature of an HII region, the core of a star, or even hotter.

These predictions about the state of the early universe can be tested in particle accelerators, which can simulate (in a tiny volume) the conditions of the early universe. At energies above 100 GeV or temperatures above 10^{15} K, the electromagnetic and weak nuclear forces are indistinguishable, and are referred to as the *electroweak force*. The unification of these two forces has been verified experimentally. At energies above 10^{14} GeV ($T > 10^{27}$ K), the electroweak and strong nuclear forces should become indistinguishable. In the very early universe, the universe is thought to have been so hot and dense that all four known forces were unified into a single force. Scientists do not yet have the physics to describe all four forces with one mathematical model (or Theory of Everything). In current models of the universe, gravity became a separate force from the other three forces when the universe was a temperature of 10^{32} K. At this time, the universe would have been only 10^{-41} seconds old.

What Is Beta Decay?

Beta decay describes the conversion of a neutron into a proton, an electron, and an antineutrino (see "Fundamental Particles"). Many of these decays happen within the nucleus of an atom containing many neutrons and protons, but a free neutron's beta decay looks like this:

$$n \rightarrow p + e^- + \bar{\nu}_e$$

The Very Early Universe and Inflation

The present state of the universe and the detection of the cosmic microwave background imply that the universe was much hotter, energetic and dense in the past. If you rewind the film of the universe to its first few frames, you will see a universe unimaginably hot and dense. While the last stages of this early phase of the expansion of the universe can be simulated in particle accelerators, most of these early times are beyond experimental confirmation.

Fortunately, cosmologists can use fundamental laws physical laws to postulate how the early universe evolved and compare the observed universe with those models. In this way, it is possible to understand the first picosecond (10^{-12} seconds) of the universe.

The history of the universe has been one of expansion and cooling, but at its origins, the universe contained energies and temperatures almost beyond imagining—and mostly beyond the reach of any particle accelerator.

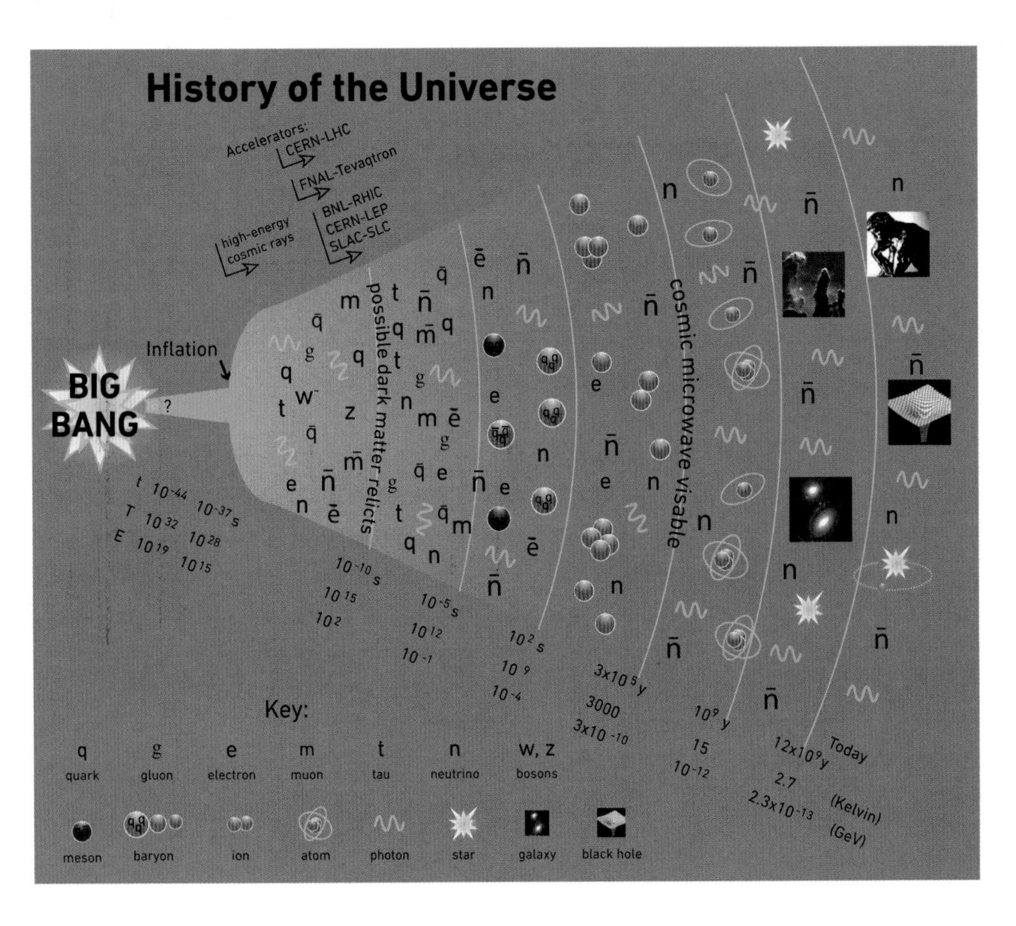

The history of the universe, shown graphically. The changing time, temperature, and energy scales are shown along the bottom. The key indicates the particles that were present at different times in the history of the universe. The accelerators (including the LHC) that can probe early times in the universe are shown.

From $t = 0$ to $t = 10^{-43}$ seconds (10^{19} GeV, 10^{32} K): At these early times, all four of the fundamental forces were unified in a single force. We do not yet have a Theory of Everything (ToE) that explains the universe at these early times, and it is unlikely that we will ever build accelerators that can simulate these conditions. When the energy of an average particle in the early universe fell below 10^{19} GeV, gravity is thought to have "frozen out" or differentiated from the other three forces.

From $t = 10^{-43}$ s to $t = 10^{-35}$ s (10^{14} GeV, 10^{27} K): As the universe continued to expand and cool, temperatures fell to 10^{27} K, and the strong nuclear force "froze out" and was differentiated from the electroweak force. At this moment, the universe is thought to have undergone an amazing expansion or inflation.

From $t = 10^{-35}$ s to $t = 10^{-32}$ s (10^{14} GeV, 10^{27} K): During the inflationary epoch, the universe expanded in scale by a factor of 10^{50}. Energy released from the vacuum is proposed to have kept the temperature of the universe the same, at about 10^{27} K.

From $t = 10^{-32}$ s to $t = 10^{-12}$ s (100GeV, 10^{15} K): After inflation, the universe continued to cool and expand, and at 10^{15} K, the electroweak force "freezes out" and separates into the electromagnetic force and the weak force.

The hot, energetic conditions thought to be present after about 10^{-32} seconds in the early universe can be tested in particle accelerators. Times before that are out of the reach of any current facility.

This image from a bubble chamber (filled with liquid hydrogen) shows the paths of particles created in a particle accelerator. The curved paths result when positively and negatively charged particles interact with a magnetic field applied to the chamber. The straight paths are from neutral particles.

What Is Inflation?

Inflation is essential to answer a number of thorny cosmological questions. First, why does the universe appear to be flat, isotropic (the same in all directions), and homogeneous (the same in all volumes)? Inflationary theory explains this by postulating that the entire visible universe arose from a tiny region that was suddenly expanded to an enormous volume. Otherwise, it is difficult to explain how parts of the universe that have never interacted with one another are at the same temperature (see "Mapping the Cosmic Microwave Background"). Inflation can also explain the origin of Very-Large-Scale Structure (VLSS), in which tiny temperature fluctuations are magnified to the scale of the visible universe and seed the growth of stars and galaxies. In March 2014, the first observational evidence of inflation was confirmed in sensitive images of the cosmic microwave background.

Formation of Matter and Antimatter

On its tiniest scales, the universe is very different than the world you inhabit. Quantum mechanics explores the behavior of matter on these atomic scales, and on these scales, you have to think of matter in different ways.

The Heisenberg Uncertainty Principle

One of the fundamental ideas in quantum mechanics is the Heisenberg uncertainty principle. Simply stated, it says you can't know both the exact position and the momentum of a particle. The product of the uncertainty in where something is and its momentum (related to its velocity) must equal a constant. In measuring the position of a particle, you change its momentum. The act of measurement affects the precision of the measurement.

Energy and time have a similar relationship as position and momentum, meaning the product of the uncertainty in the energy of a system and the time of measurement must also equal a constant. Combining this uncertainty principle with Einstein's mass-energy equation ($E = mc^2$) means at any location in the universe, over brief periods of time, you can't say with absolute certainty how much mass is there. There is a fundamental uncertainty in how much mass a region of space contains.

Stated another way, the uncertainty inherent in quantum mechanics means the universe allows mass to be created on small scales for tiny periods of time. The limitation to this creation is that particles must be created in pairs—particle and antiparticle. In the normal scheme of things, these particles and antiparticles almost immediately meet one another. When a particle and an antiparticle meet, they disappear (annihilate), and their energy is carried away by photons. The universe, on its tiniest scales, is seething with the creation and destruction of these virtual pairs of particles.

$$\Delta p \, \Delta x \geq \frac{1}{2} \hbar$$

$$\Delta E \, \Delta t \geq \frac{1}{2} \hbar$$

The Heisenberg uncertainty principle as expressed for position and momentum (top) and energy and time (bottom). The h-bar symbol is the Planck constant divided by 2π.

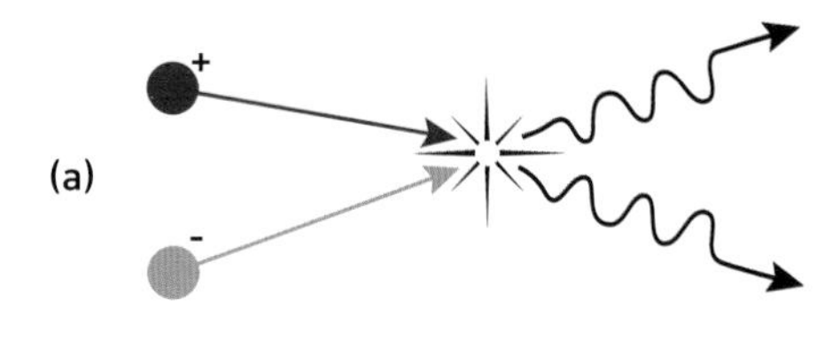

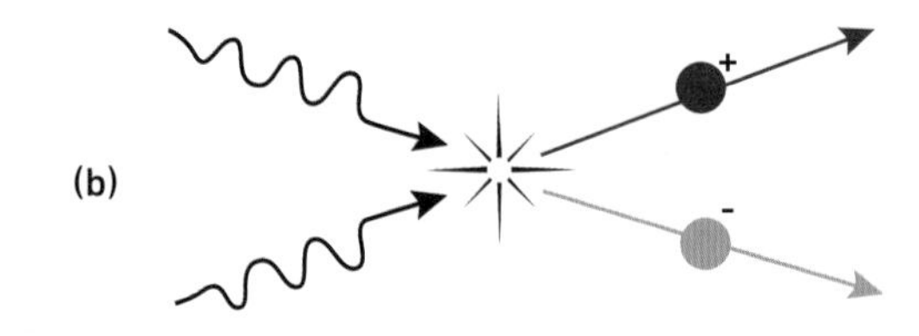

This image shows the process of (a) annihilation and (b) pair production, which is thought to have occurred in the early universe. These processes are regularly detected in particle accelerators.

Explaining Matter and Antimatter in the Universe

This annihilation and creation, paired with the universe's sudden expansion, may be the explanation for why there is matter in the universe.

When the universe underwent inflation, pairs of particles that would generally annihilate almost immediately were suddenly separated, leading to a flood of matter and antimatter.

In the inflated universe, particles and antiparticles collided (annihilated), releasing a wave of gamma rays, which in turn could collide to create pairs of particles and antiparticles. For a brief time, annihilation and pair production were in balance.

But the expansion of the universe continued to lower the average temperature of particles. When particles in the universe reached an average temperature of 10^{13} K (at a time of $t = 10^{-6}$ seconds), a period called *quark confinement* began. What this means is that earlier than this time, there was too much ambient energy for groups of quarks to settle into combination of quarks like neutrons and protons, because they would immediately be broken up by collisions with other energetic particles. However, at this time, when the universe was about one millionth of a second old, the first nuclei were created from combinations of quarks. For reasons that are not yet understood, matter outnumbered antimatter by about 1 billion to 1. A wave of annihilation occurred, and the 1 part in a billion that remained formed the observable universe.

As the universe cooled below 10^{13} K, no more neutrons or protons or their antiparticles could be created. Then, at an age of 1 second, the temperature of the universe fell below 10^{10} K, allowing electrons and positrons to survive.

So at 1 second old, the universe was filled with neutrons, protons, electrons, neutrinos, and high-energy photons. The universe had differentiated into a familiar soup of particles, ready to build atoms, molecules, planets, stars, and galaxies.

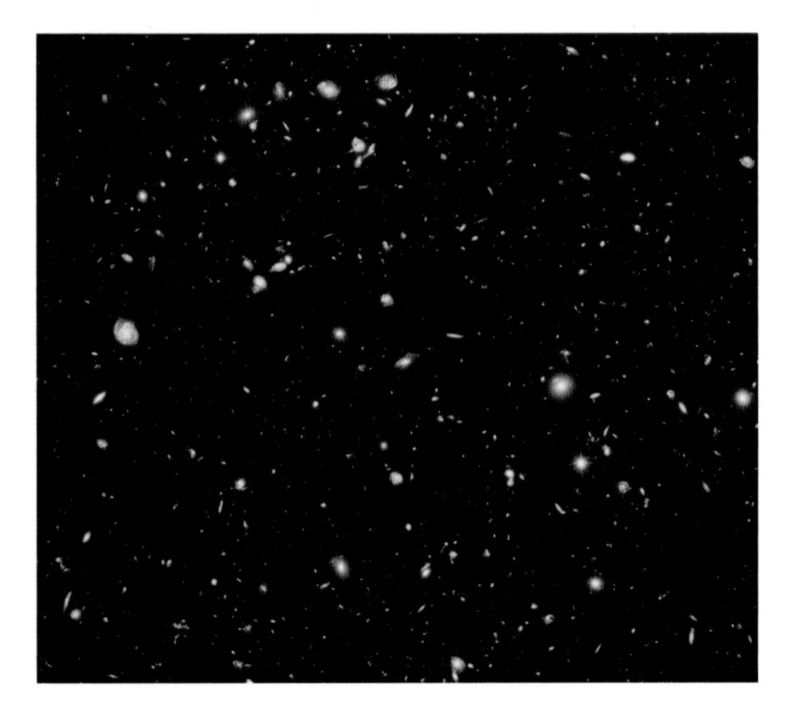

This 2012 image of over 5,000 galaxies (the Hubble Extreme Deep Field) shows the universe in the direction of the Fornax constellation. The visible universe—made up of things like protons, neutrons, and electrons—is the 1 part in a billion left after the annihilation that occurred in the early universe.

Big Bang Nucleosynthesis

When the universe was only a few seconds old, it was as hot as the core of a star and filled with neutrons, protons, electrons, positrons, neutrinos, antineutrinos, and photons. As a result, the conditions were ripe for the formation of elements. This nuclear fusion that created light elements in the early universe is known as *Big Bang nucleosynthesis.*

The Big Bang Nucleosynthesis Process

When it was a second old, the universe had a temperature of about 10 billion K. At this time, the neutrons that had formed from quarks either decayed into protons and electrons (via beta decay) or combined with protons and formed deuterium (^2H). Once deuterium formed, a sudden burst of element formation occurred.

The temperatures of the universe stayed hot enough to fuse elements for only about 15 minutes, so there was only a brief window when deuterium could fuse into helium and a small amount of lithium and beryllium. As in the cores of stars, the process of fusing hydrogen into helium has a few steps, with a proton and a neutron first coming together to form deuterium, and then helium, lithium, and beryllium.

It turns out that the amount of ordinary matter present in the early universe (relative to the photons) will result in a specific relative abundance of hydrogen, helium, lithium, and their isotopes. These predicted abundances are consistent with their observed abundances in the universe, indicating that Big Bang nucleosynthesis is understood quite well. This era of the Big Bang ended with a universe in which about 1 in 10 atoms were helium, with traces of lithium and beryllium.

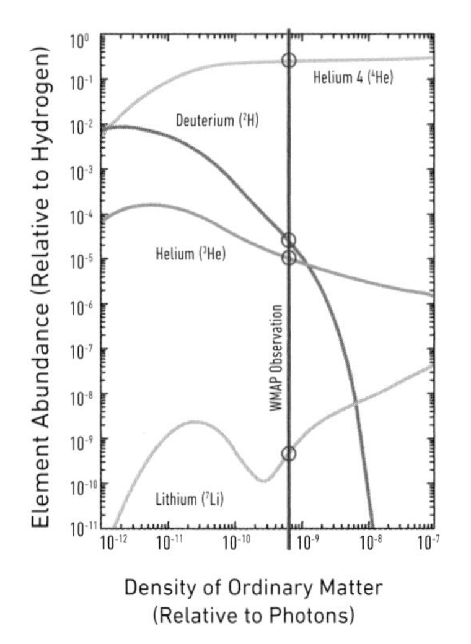

The abundance of isotopes of hydrogen, helium, and lithium in the early universe varies with the density of ordinary matter relative to photons. The red vertical line indicates the predicted densities of these isotopes.

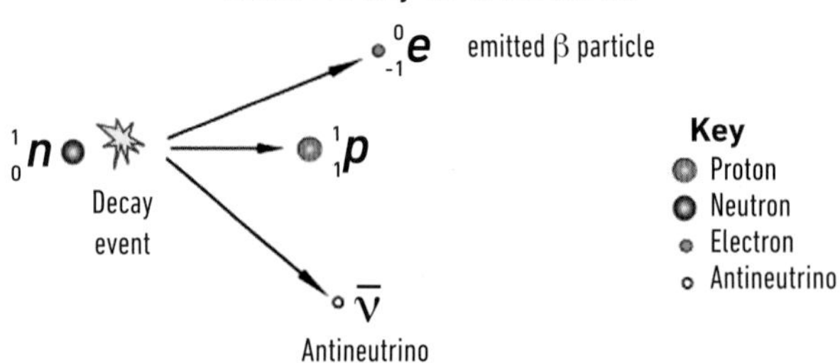

This schematic shows the beta decay of a neutron. Neutrons that are not bound in a nucleus with protons decay relatively quickly to form a proton, an electron, and an antineutrino.

Neutrinos and the "Cosmic Neutrino Background"

Neutrinos were involved in many interactions with protons and neutrons in the early universe. But once the universe was a few seconds old (as nucleosynthesis began), the universe had become transparent to neutrinos. In the same way that neutrinos come streaming out of the dense core of the Sun unimpeded, they were "decoupled" from ordinary matter.

These neutrinos and antineutrinos released at this early time should fill the universe just like the cosmic microwave background that appeared at recombination (see "Recombination"). Therefore, there is much interest in the "cosmic neutrino background," as it comes from a time when the universe was only a few seconds old. As you have read, though, neutrinos are very difficult to detect, and the cosmic neutrino background has not yet been observed.

While the direct detection of the cosmic neutrino background will be a difficult task, the indirect effects of the cosmic neutrino background can be measured by instruments like the Planck spacecraft, which is measuring tiny fluctuations in the cosmic microwave background. The presence of these neutrinos also affects the growth of large-scale structures in the early universe.

An all-sky view of the cosmic microwave background as imaged by the ESA's Planck mission. The tiny fluctuations in the temperature of the early universe are thought to have led to the formation of the first generation of stars and galaxies. These data allow astronomers to look for indirect evidence of the cosmic neutrino background.

Recombination

So far, you have followed the expanding universe from its earliest moments to the release of the first neutrinos to decouple from matter and move out into space. At this early time, the universe was still devoid of stars and galaxies but was filled with familiar atoms—mostly hydrogen and helium, with a tiny admixture of lithium and beryllium from Big Bang nucleosynthesis. In addition, there were electrons, neutrinos, and photons.

From the time the universe was about 15 minutes old (the end of Big Bang nucleosynthesis) until it was about 380,000 years old, the universe was so hot and particles were moving around so quickly that every time an electron and a proton came together (pulled close by their opposite charges), they were knocked apart. The entire universe was an ionized plasma, with atomic nuclei and electrons unable to come together. However, the universe continued to expand and cool over those first few hundred thousand years.

Things began to change at about 380,000 years after the Big Bang, when the temperature of the universe (on average) was about 3,000 K. At this temperature, for the first time, photons did not have enough energy to knock an electron out of its orbit around a proton. So as the universe cooled past this temperature, hydrogen atoms (one proton and one electron) could survive for the first time. Physicists refer to the capture of an electron by a proton as *recombination;* therefore, this epoch in the early universe is known as the *Epoch of Recombination.*

Recombination had a big effect. If you recall, photons have a hard time moving through a sea of free electrons (see "The Solar Interior"). Fusion creates gamma ray photons in the core of the Sun, but those photons have a hard time getting out of the plasma of charged particles that fill the Sun's interior. In the same way, photons in the early universe were "trapped," continually being scattered by free protons and electrons. But as the universe cooled past 3,000 K, there were suddenly no more free electrons. It was as if a haze had cleared, and the photons were free to move across the universe (known as *photon decoupling*). These photons that "got out" at this time are what observers can now see as the cosmic microwave background.

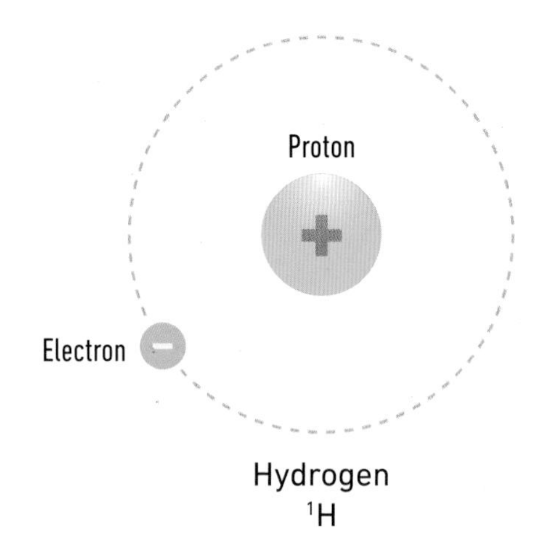

For the first time in the Epoch of Recombination, electrons and protons combined to form hydrogen atoms, pulled together by electromagnetic forces. This image shows the makeup of a hydrogen atom.

Surface of Last Scatter

The nature of the cosmic microwave background becomes clearer when you think of a cloud. When you look up at the sky and see the bottom of a cloud, you are seeing what is called a *surface of last scatter.* What that means is that the photons in the atmosphere scattered off material in the lower layer of the cloud and then hit your eye. So you can't see anything behind the cloud, just its lower surface. In the same way you can't see blue sky behind a cloud, you can't see the universe beyond the cosmic microwave background. Instead, you see the surface of last scatter for the early universe, which is the earliest picture of the universe you can see.

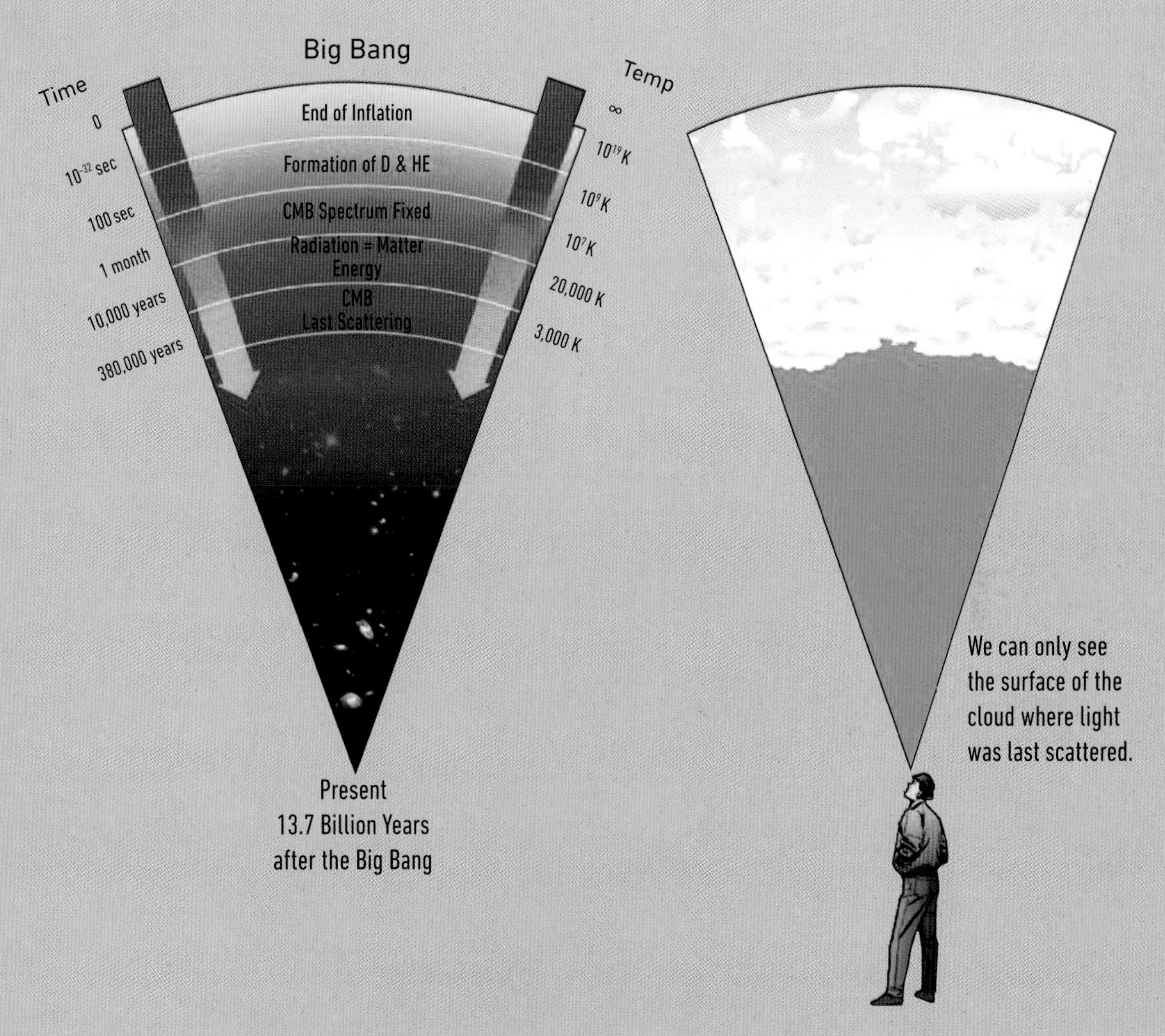

The cosmic microwave background can be thought of as a surface of last scatter. You can't see the events on the "far side" of the cosmic microwave background, because photons could not yet escape. In a similar way, you only see the near, lower surface of a cloud and nothing beyond it.

Matter and Energy

The universe is filled with matter and energy, which are related by Einstein's famous equation, $E = mc^2$ (energy equals mass times the speed of light squared). So even though radiation (a form of energy) has no mass, it can be thought of contributing some part of the entire mass density of the universe.

In the current universe, there are many more photons per cubic meter of the universe than there are atoms (by a factor of about 1 billion to 1). But most of those photons are from the cosmic microwave background, and each photon carries very little energy.

Mass Density in the Past

The photons visible in the cosmic microwave background make up most of the universe in terms of a head count of particles. Because there are currently many more photons per cubic meter than atoms, that means the universe must have been dominated by the mass density in radiation (rho_r) in the past, when each photon carried much more energy.

The transition from a universe that was dominated by radiation to one dominated by matter occurred about 24,000 years after the Big Bang. At that time, the black-body temperature of the entire universe was about 14,000 K, or hotter than the surface of the Sun. This means a typical cosmic microwave background photon would have been in the ultraviolet part of the spectrum, carrying much more energy than it does now in the microwave.

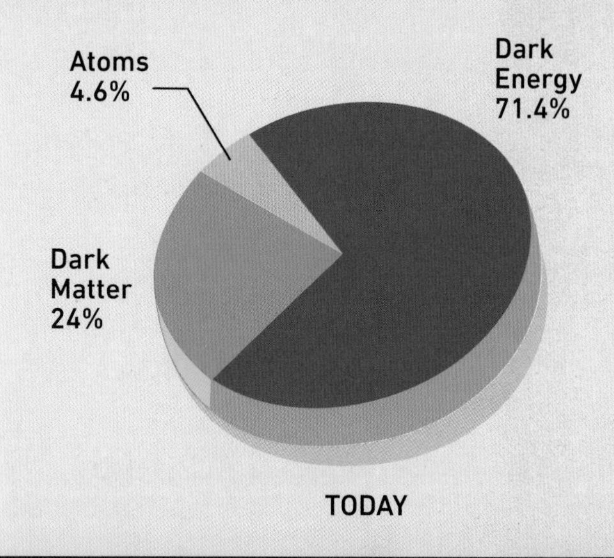

The current mass density of the universe is made up of about 71 percent dark energy, 24 percent dark matter, and less than 5 percent "ordinary" matter. Most of what people observe in the cosmos makes up only a tiny fraction of the mass density of the universe.

What Is Mass Density?

Mass density is the amount of matter or energy contained in a volume of the universe. For example, in every cubic meter of the universe, astronomers can determine the number of protons and the number of photons. Dividing the number of protons by the volume gives the *mass density in matter (rho_m)*. Dividing the mass equivalent ($E = mc^2$) of all the photons in that same cubic meter by the volume gives the *mass density in radiation (rho_r)*.

The galaxy cluster Abell 2744 is 3.5 billion light-years from Earth and contains a collision between four or more galaxy clusters. In this image, X-ray emission is in red, optical data are yellow/white, and dark matter (as revealed by gravitational lensing) is in blue. Matter in the universe is mainly comprised of dark matter.

What Is the Universe Made Of?

In terms of matter, the universe is made up of both luminous matter and dark matter. *Luminous matter* is everything you can see (planets, stars, and galaxies), while *dark matter* is everything you can only detect from its gravitational effects (like in the rotation curves of galaxies and galaxy clusters). Surprisingly, dark matter comprises most of the matter in the universe; however, its composition remains a mystery (see "What Is Dark Matter?").

Radiation is one form of energy in the universe, and photons of light carry radiation throughout the universe. But recent observations of the accelerating expansion of the universe (see "The Runaway Universe") indicate that there is another form of energy in the universe, dubbed "dark energy." Like dark matter, the nature of dark energy is not yet understood by astronomers, though it is detected by its effects on the expansion of the universe.

Astronomer's best measurements show that the universe is about 71 percent dark energy, 24 percent dark matter, and 5 percent luminous matter. So as much as astronomers know about the universe, they still do not know what 95 percent of it is.

The First Stars and the Dark Ages

If the universe had been perfectly uniform (homogeneous), it would still be so today, and the universe would be evenly filled with a few atoms per cubic meter. Instead, the current universe is very inhomogeneous, filled with dense concentrations of matter (like stars) that are separated by huge distances where there is very little matter at all.

When the first maps of the cosmic microwave background were made, it appeared that the radiation indicated the early universe was very smooth, so it was a puzzle how this smooth early universe evolved into the clumpy universe Earth inhabits today.

As you have seen, detailed observations have shown that the cosmic microwave background is not perfectly smooth, but has small "anisotropies" (meaning it is not the same in all directions). These small temperature fluctuations correspond to similarly small density fluctuations, which are amplified by gravity. Over time, gravitational forces gathered matter to sufficient densities and temperatures to form the first stars.

The Cosmic Microwave Background as seen by Planck and WMAP

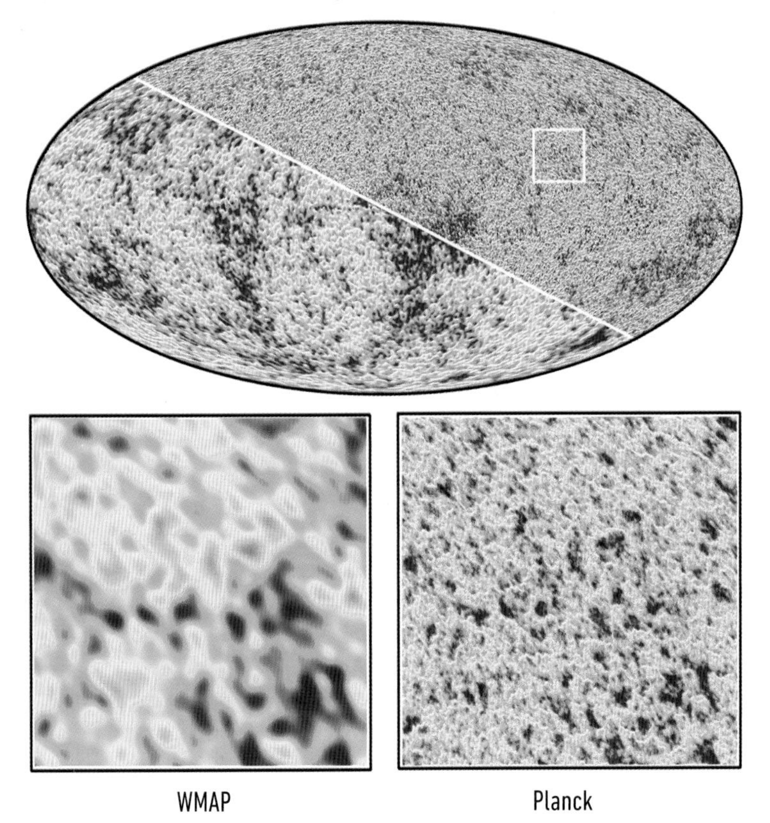

This image shows a detail of the temperature fluctuations detected in the cosmic microwave background in the data from the WMAP and Planck missions.

Globular Clusters and Population III Stars

The earliest structures to form in the universe probably were the size of globular clusters (see "Galaxy Detail: Globular Clusters"). Globular clusters contain some of the oldest stars you can observe. In fact, the age of stars in globular clusters has been used as an independent check on the age of the universe.

But the universe must have had a generation of stars even older than the Population II stars you see in globular clusters. These earlier stars, known as *Population III stars,* would have had very low metallicity, since they formed from the material present after Big Bang nucleosynthesis—hydrogen, helium, and a tiny amount of lithium and beryllium (see "Big Bang Nucleosynthesis"). These stars would have been behemoths, with masses up to 1,000 solar masses (a size that higher-metallicity stars can't achieve).

Globular clusters contain some of the oldest observable stars. This globular cluster, NGC 2808, is thought to be about 12.5 billion years old.

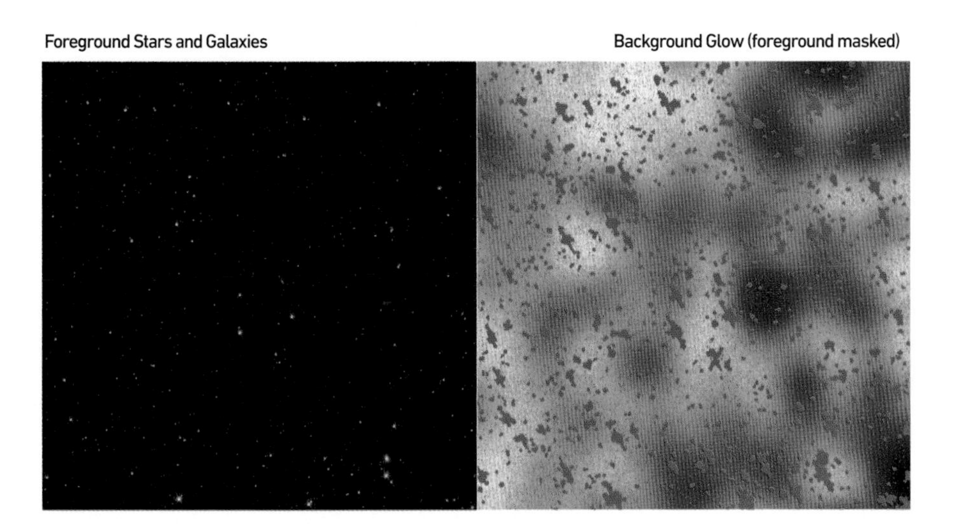

Spitzer Space Telescope may have detected the faint glow from a first generation of stars. On the left is an infrared view of a part of the night sky called the *Boötes field.* On the right is an image of the faint infrared glow coming from objects beyond the foreground stars and galaxies. The grey areas show where the light from foreground objects has been subtracted.

The Dark Age in the Universe

The formation of Population III stars would have produced enormous amounts of high-energy photons that would once again have been able to strip electrons from protons, a process known as *ionization.* Therefore, the formation of these massive stars would have coincided with the reionization of hydrogen nuclei in the universe.

Reionization and the formation of these stars in the universe is thought to have occurred about 400 million years ago. The explosions of these massive stars would have enriched the universe with heavier elements and allowed for the formation of smaller stars. This time in the universe, from recombination (380,000 years) to reionization (400 million years), is sometimes called the *Dark Ages.*

The First Galaxies and Large-Scale Structure

As you have seen, most of the matter in the universe—when you look at things the size of galaxies and galaxy clusters—is dark matter. This means early on, the tiny fluctuations in the density of the early universe you see in the cosmic microwave background must have been amplified by the inflation that occurred at about 10^{-35} seconds after the Big Bang. These minor variations in density allowed gravity to begin to pull dark and luminous matter into structures you see today.

The earliest galaxies were metal-poor dwarf galaxies, smaller than the galaxies observed nearer to the Milky Way. These galaxies emitted mostly ultraviolet light as they formed stars at 10 times the rate of normal galaxies today; that light has been shifted to the infrared part of the spectrum by the expansion of the universe.

Distribution of Galaxies and Dark Matter

As astronomers have mapped the universe, they have found that superclusters of galaxies are not randomly distributed throughout space, but are arranged in sheets and filaments, with large voids in dark and luminous matter between them (see "Very-Large-Scale Structures"). These observations are consistent with computer models that follow the evolution of a universe filled with mostly dark matter, of which both MACHOs and WIMPs are leading candidates (see "What Is Dark Matter?")

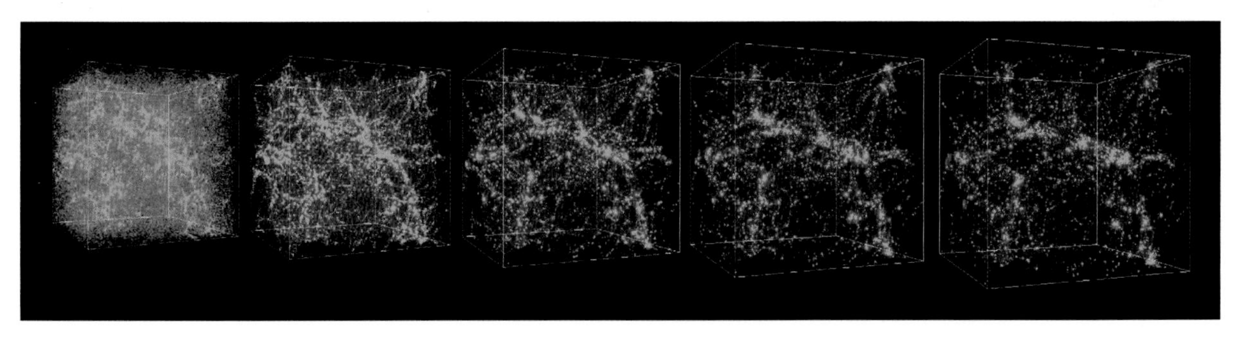

Computer models allow astronomers to model the evolution of the early universe from a smooth mixture of dark and luminous matter and dark energy. The box shown is 140 million light-years across, and the model follows the evolution of the universe from a redshift of z = 30 (100 million years old) to z = 0 (current day).

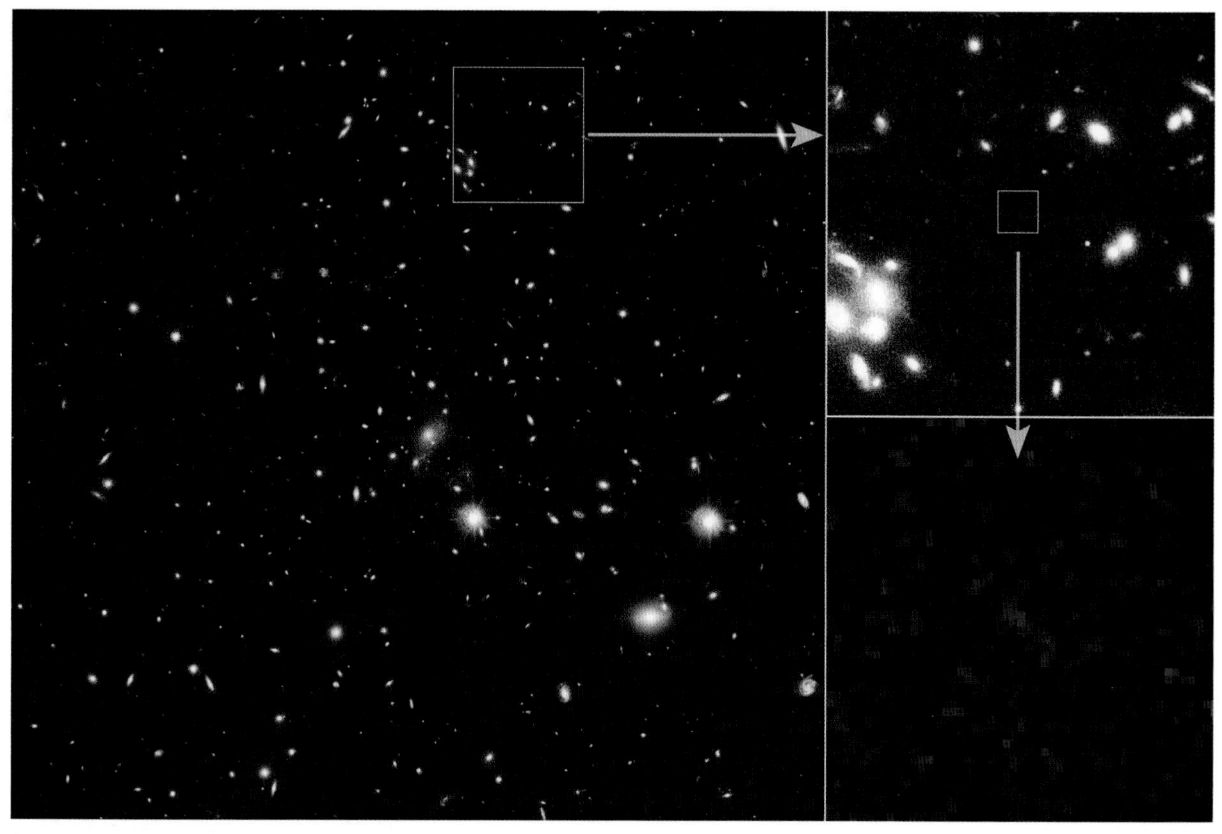

Gravitational lensing by dark matter in the massive cluster of galaxies called MACS J1149+2223 is making a very-distant source about 15 times brighter, which in turn makes it visible. The two boxes to the right show zoomed-in details of the galaxy. The original ultraviolet radiation from the galaxy has been shifted into the infrared.

Viewing Early Galaxies

One way astronomers can see even-more-distant galaxies is with the help of gravitational lenses. These can be thought of as telescopes made out of dark matter that make galaxies from the very early universe visible within the first few hundred million years (see "Gravitational Lenses").

Images like the Hubble Deep Field and the Hubble Ultra Deep Field are also perfect places to look for the youngest, most-distant galaxies you can find. In these fields, crowded with nearby galaxies, there are faint red smudges of light that come from very distant galaxies.

In the Hubble Ultra Deep Field, over 500 such distant galaxies have been discovered, from a time when galaxies were just beginning to form. Observations like these allow astronomers to understand how the universe changed from a smooth, featureless sea of particles to the stars and galaxies you see today.

The Geometry of the Universe

After the detection of the recession of galaxies in the 1920s, most astronomers were of the opinion the expansion of the universe would eventually slow. Some even thought that the expansion of the universe would halt and reverse, leading to what was called a *Big Crunch.* Whether the universe expands forever or eventually reverses its expansion depends on one important factor—the mass density of the universe (see "Matter and Energy"). That is, how much stuff is there in every cubic meter of the universe—not just luminous matter, stars, and galaxies, but also dark matter and dark energy?

Expansion and Its Relation to Geometry

A Big Crunch was ruled out early on (there isn't enough mass density for the universe to collapse again). In recent decades, observations of galaxies and clusters have made it clear that most of the matter in the universe was dark matter. The problem was, even taking into account all of the luminous matter apparent in galaxies and dark matter inferred from its gravitational effects, there was not nearly enough matter in the universe to halt its expansion. Earth appeared to live in an open universe that would expand forever.

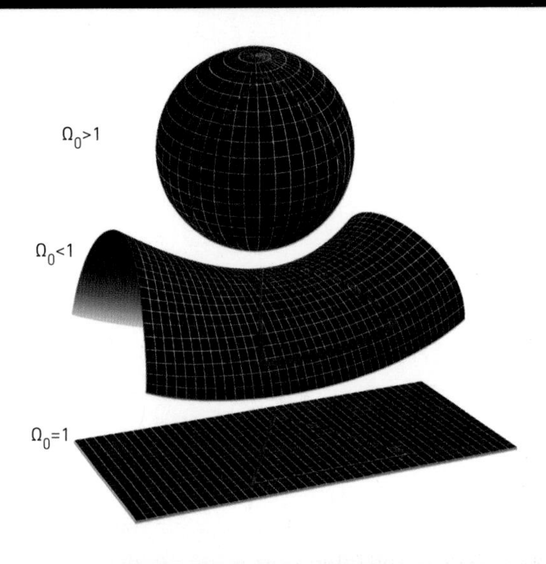

In these images, the universe is represented by a two-dimensional sheet. The mass density of the universe determines the geometry of the universe. In a universe that is flat, the density parameter (Ω_0) is exactly 1.

Mass Density, Critical Density, and the Density Parameter

The mass density of the universe (ρ) refers to how much mass there is in each cubic meter of the universe. People refer to the current observed mass density of the universe as ρ_0 *("rho-naught").* The amount of mass density required to have a flat universe is called the *critical density* (ρ_c).

The Greek letter Ω is called the *density parameter,* and it is the ratio of the observed mass density to the critical mass (ρ_0/ρ_c) density. In a universe with a flat geometry, the density parameter is 1. If the universe has more than critical density, Ω is greater than 1. If the universe has less than critical density, Ω is less than 1.

A Flat Universe

One of the best ways to determine the geometry of the universe is to carefully study the fluctuations apparent in the cosmic microwave background. The size scale of the fluctuations tells you the geometry of universe that you live in.

In analyzing cosmic microwave background images, astronomers carefully look at the size of the "spots" in the image. These spots change in size depending on whether the universe is open, flat, or closed.

Results from the Wilkinson Microwave Anisotropy (WMAP) probe of the cosmic microwave background indicate a flat universe with a density parameter (Ω_0) of 1.

The remaining question is what gives the universe its critical density, since there is not enough matter. It turns out that the remaining mass density of the universe comes not from mass, but from dark energy (see "Mass and Energy").

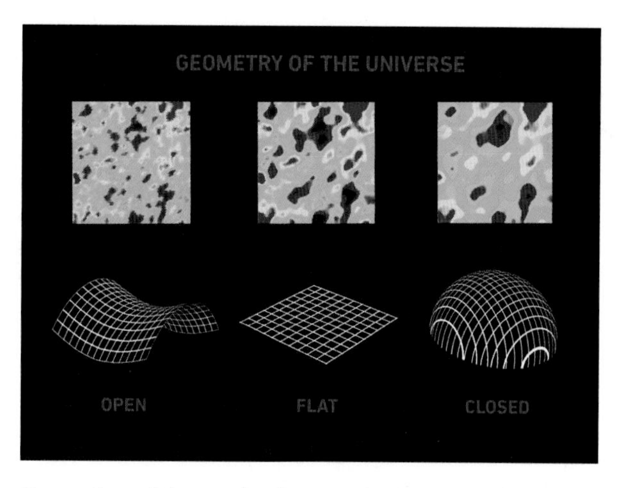

Observations of the cosmic microwave background can tell you the sort of universe you live in. The size scale of the temperature fluctuations will change depending on whether you live in an open, flat, or closed universe. Data from the WMAP indicate that this is a flat universe.

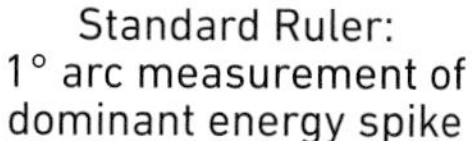

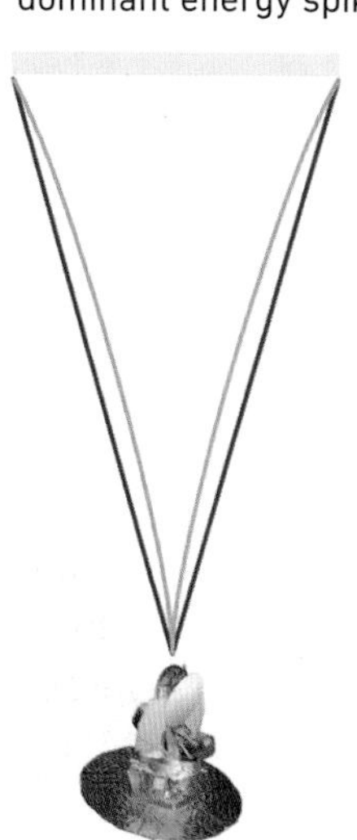

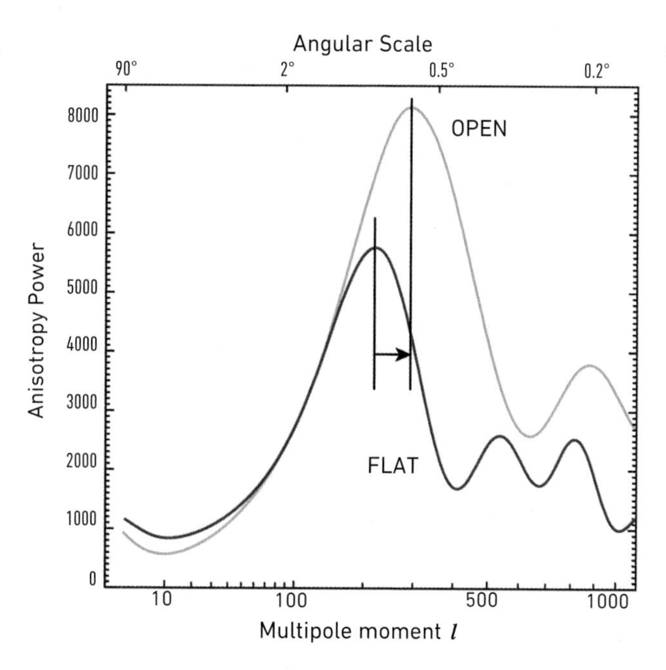

In this image, the red lines (and red data) show the size scale of cosmic microwave background fluctuations that would be detected if the universe is flat—about 1 degree. The grey lines (and grey data) show the smaller size scale expected in an open universe—about 0.5 degree. The data from WMAP is consistent with a flat universe.

Critical Density of the Universe

Observations of the cosmic microwave background by WMAP have determined that the universe is flat. Ordinary and dark matter together contribute only about 28 percent of the critical density. So what makes up the bulk of the critical density of the universe? The largest contributor (at 73 percent) is dark energy (see "Matter and Energy").

The presence of dark energy in the universe can explain two phenomena—the accelerating expansion of the universe (see "The Runaway Universe") and the effective mass density needed for a flat universe.

Density Parameter

Now that I have defined the density parameter (see "The Geometry of the Universe"), I'd like to pull that density parameter into its component parts.

The density parameter (Ω) is the ratio of the mass density of the universe to the critical mass density required for a flat universe (ρ_0/ρ_c). Ω is made up of three components:

- The mass density of ordinary matter (baryons) plus dark matter, or Ω_m
- The equivalent mass density of relativistic particles (photons and neutrinos), or Ω_{rel}
- The effective mass density of dark energy, or Ω_λ

This is written as follows:

$$\Omega = \Omega_m + \Omega_{rel} + \Omega_\lambda$$

In a flat universe, Ω is 1.

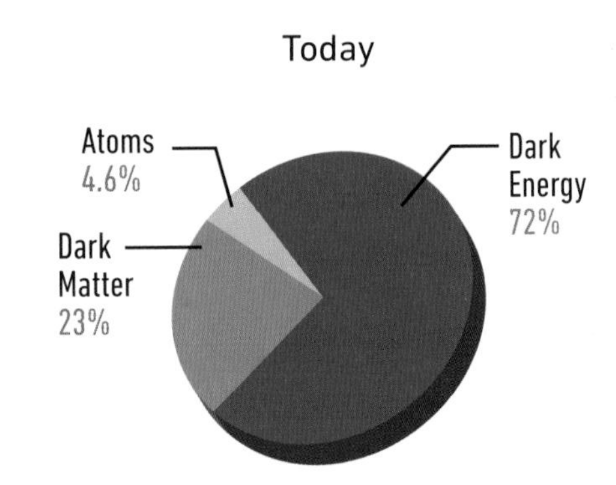

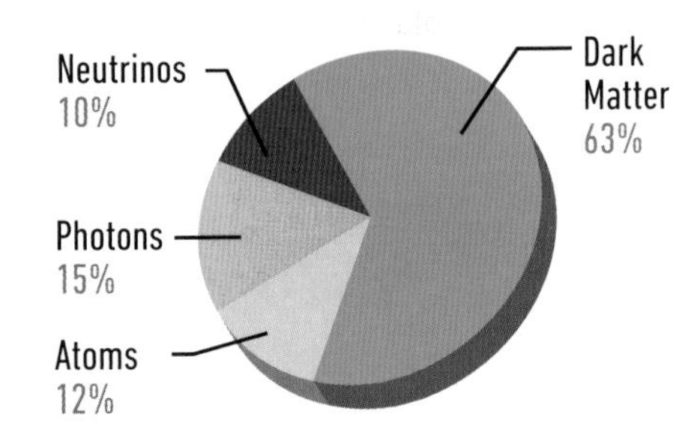

This image shows the components of the mass density of the universe now (top) and when the universe was young (bottom).

Astronomers refer to the current value of the density parameter (Ω) as Ω_0, which is the ratio of the observed mass density of the universe to its critical mass density.

Even before astronomers had proposed that dark energy could provide the remaining critical density, the fact that the density parameter was relatively close to 1 suggested that it must be exactly 1. In cosmological models, a value of the density parameter (Ω) that was not almost exactly 1 in the early universe would have rapidly diverged from 1 and been either a very large or a very small number.

Inflation and Flatness

You have seen that the theory of inflation (see "The Very Early Universe and Inflation") can explain a number of cosmological observations, including the isotropy (looking the same in all directions) of the cosmic microwave background.

Inflation may also explain why the universe appears to be flat. Inflation, which suddenly changed the scale of the universe by a factor of 10^{50}, could have made the universe appear locally flat. In your daily life, you experience that the planet (locally) is pretty flat, even though you know that Earth has a curved surface. In the same way, the part of the universe you observe (out to the cosmic microwave background) may appear flat, even though the universe on larger, unobservable scales may be curved.

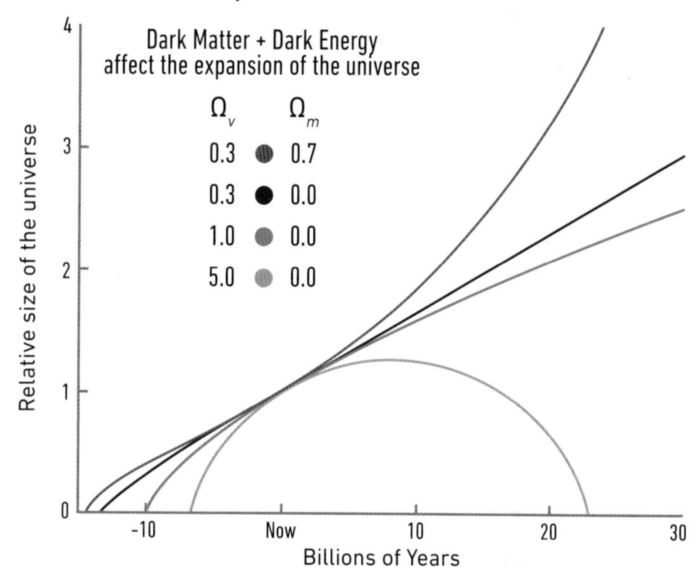

The relative size of the universe is shown as a function of time. The possible universes shown here are closed (orange), flat (green), open (blue), and flat but accelerating (red). The universe appears to be expanding in the way plotted in the red curve, showing that the universe must contain a large amount of dark energy.

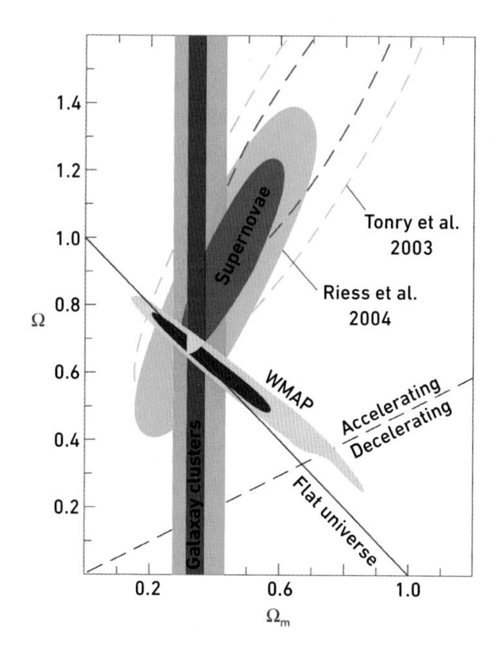

A summary plot showing the independent observational constraints on the density parameter of the universe (Ω_0), with the *x*- and *y*-axes showing contributions from matter and dark energy respectively. The data indicates that the universe is flat and accelerating in its expansion.

The Runaway Universe

Astronomers have known that the universe is expanding since the 1920s, and improvements in telescope technology and the discovery of standard candles have helped them to measure the redshifts of galaxies at greater and greater distances.

By the 1990s, observations of luminous and dark matter in the universe indicated that the universe was, at best, open, and perhaps flat—that is, there was not enough matter in the universe to reverse its expansion. But most astronomers assumed that the density parameter was 1, meaning they should expect to see a slowing expansion when observing the universe.

Discovering the Runaway Universe

When a group of astronomers started to use Type Ia supernovae (which are excellent standard candles for measuring large distances; see "How Distant Are Galaxies?") to measure the rate at which the expansion of the universe was slowing down, they were shocked to find that the expansion of the universe was not slowing down—it was accelerating. The universe was expanding more quickly now than it had in the past, leading some to refer to this phenomenon as the "runaway universe."

The accelerating expansion of the universe has to draw its energy from somewhere. Since the discovery of the accelerating expansion, this energy has been referred to with the catch-all name "dark energy" (see "Matter and Energy").

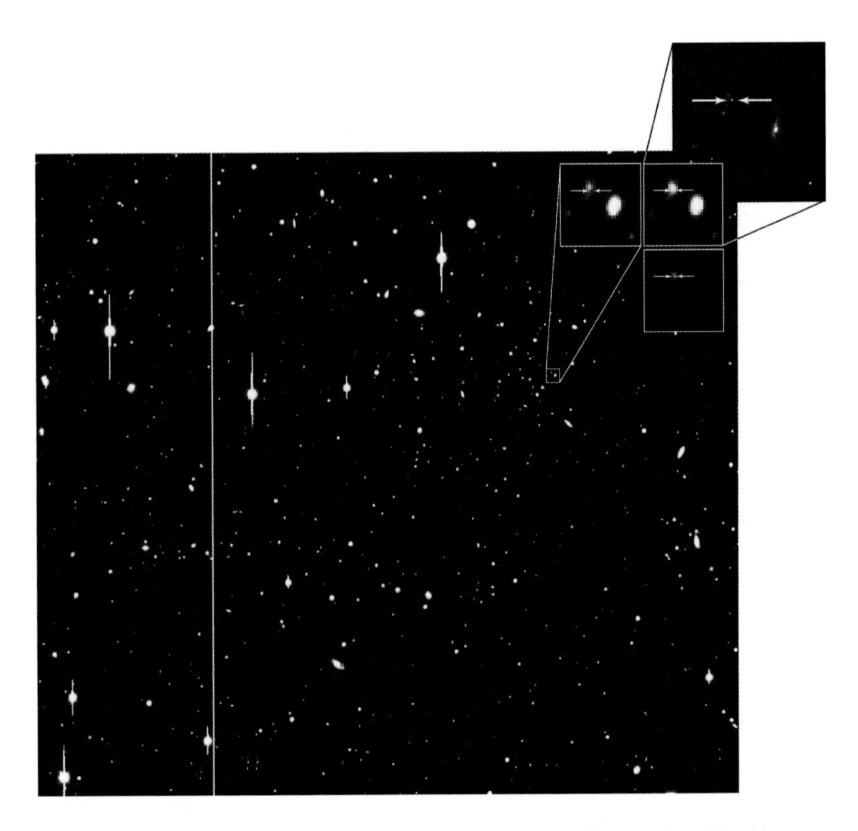

This image shows the detection of a high-redshift supernova. Observations like this one helped to determine that the expansion of the universe is accelerating. The discovery of this acceleration led to the 2011 Nobel Prize in Physics being awarded to Saul Perlmutter, Adam Reiss, and Brian Schmidt.

Dark Energy: The Cosmological Constant?

You may recall that in 1917, Einstein thought that there must be a force that kept the universe from collapsing on itself gravitationally (see "Einstein and the Steady State Universe"). He introduced a term called the *cosmological constant* (λ), which provided the energy needed to keep the universe from collapsing. When Hubble later discovered that galaxies were rushing away from one another, Einstein removed the constant as unnecessary. But the discovery of the accelerating expansion of the universe has renewed interest in Einstein's cosmological constant.

What is clear from the observations is that most of the mass density of the universe (about 70 percent) must come from dark energy. Where this dark energy comes from is a complete mystery. Even Einstein's constant, while mathematically intriguing, does not explain the origin of dark energy.

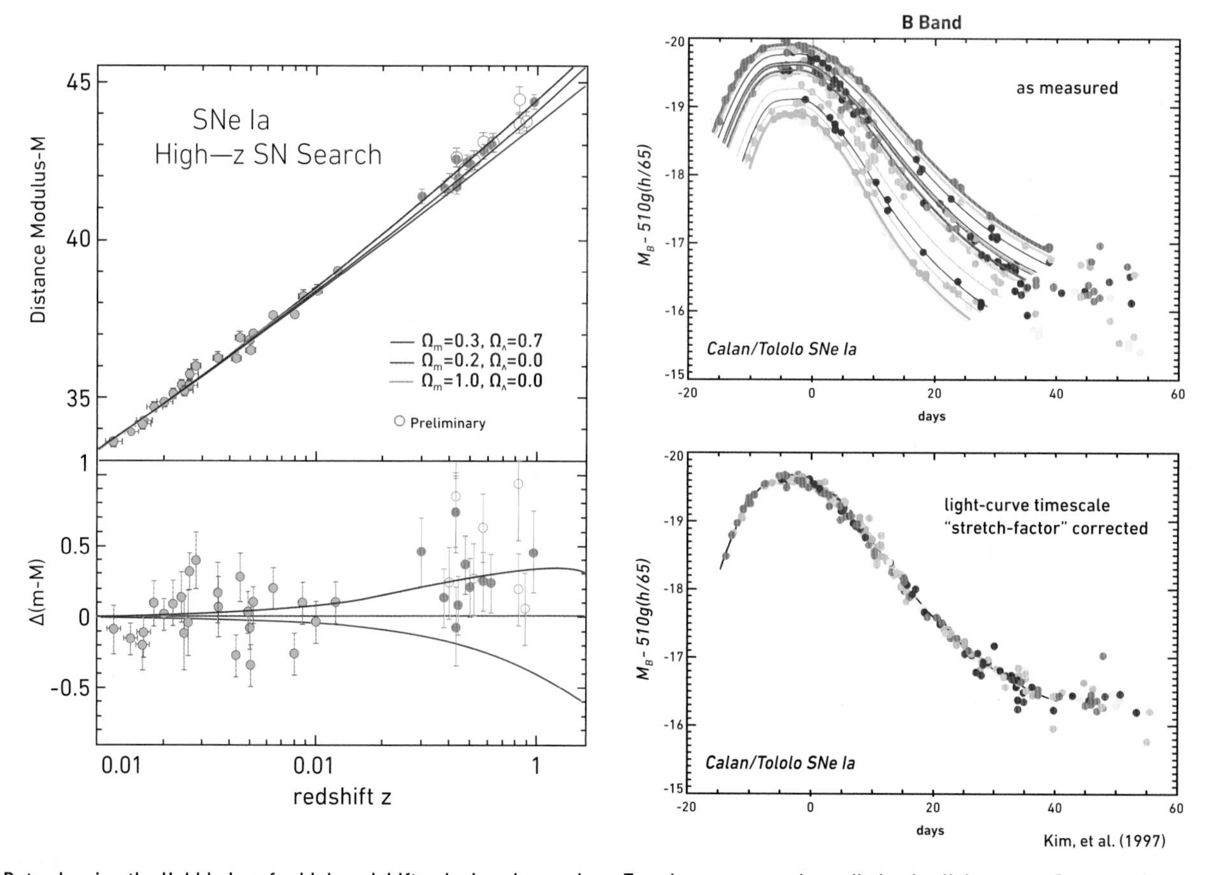

Data showing the Hubble Law for high-redshift galaxies observed with Type Ia supernovae. The high-redshift (red) data points are most consistent with a universe that has about 30 percent mass density in matter and 70 percent in dark energy and inconsistent with a universe that has no dark energy.

Type Ia supernovae have distinctive light curves. By correcting for the distance to a particular supernova, it is clear these events have a distinctive rise and fall, and their peak brightness can be used as a standard candle. These supernova are so bright that they allow the measurement of very large distances.

The Anthropic Principle

You live in a universe that is perfectly balanced for atoms, stars, and carbon-based life to exist. This near-infrared image shows stars, gas, and dust near the Galactic Center.

The anthropic principle is perhaps more philosophical than astronomical, but it does provide some guidance as people consider the structure and evolution of the universe.

What Is the Anthropic Principle?

Fundamentally, the anthropic principle states something fairly obvious: if the universe were not exactly the way it is in very fundamental ways, you would not be here to observe it. Some of the fundamental properties of the universe that I have discussed—things like the ratio of the mass of the proton to the mass of the electron or the relative strengths of the four fundamental forces—had to have their exact measured values, or the universe would not have been able to support the formation of atoms, stars, and galaxies, much less life.

You could also think of it as living in a "Goldilocks" universe; the conditions in the early universe were just right for intelligent life to exist. If the universe had expanded more quickly, stars might never have formed; if it had expanded more slowly, the universe could have collapsed in a Big Crunch (see "The Geometry of the Universe") before life ever evolved. If the strength of the strong nuclear force had been slightly different, the fusion of hydrogen would have proceeded more quickly, shortening the lifetimes of stars. However, this doesn't necessarily mean that humans live at a privileged time or place in the universe; instead, it may just mean that if the correct conditions to bring forth life did not exist, then people would not be here to discuss astronomy or anything else.

Perhaps the most interesting aspect of the anthropic principle is the way in which it limits the values of fundamental constants in the universe. That is, fundamental parameters like the gravitational constant and the relative strengths of the fundamental forces can't have just any value; they must have values that are compatible with the origin and evolution of carbon-based life. In this way, the anthropic principle points out that the universe had to fall within a narrow set of parameters in order for humans to exist at this point in space and time.

This 2003 view from the International Space Station shows life-bearing planet Earth and the Sun.

What Does It All Mean?

Where discussions of the anthropic principle diverge wildly is when it comes to interpretation. What does it all mean?

One way out of the conundrum is the multiple universes theory. Simply stated, there are an infinite number of universes that exist, each of which can have a variety of values for the fundamental constants—for example, some may allow the formation of atoms, while some may not. Humans simply live in one of those universes in which the conditions necessary for carbon-based life do exist.

Alternatively, progress toward one of the great unsolved questions—a Theory of Everything (ToE)—might explain why for some as-yet-unknown reason the physical constants of the universe must have their particular values and ratios.

The anthropic principle is one of those topics that leads to interesting late-night discussions, even if it leaves people with more questions than answers.

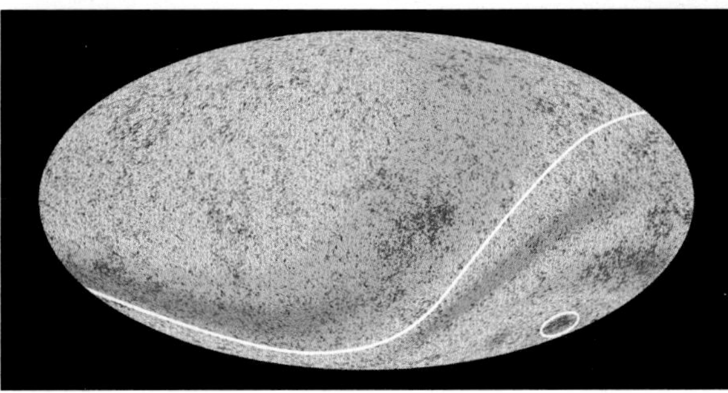
Ever-more-detailed views of the cosmic microwave background might give astronomers observational clues about the presence of multiple universes.

Unsolved Questions in the Cosmos

Because astronomy is a continually changing discipline, any list of unsolved questions in astronomy is doomed to be quickly outdated. But there are still some big unanswered questions, which I've listed here in no particular order. Part of the excitement of astronomy is that many of these questions may ultimately be answered in the coming decades.

Are We Alone in the Universe?

As you read in "Searching for Life in the Milky Way," one fundamental question is whether humans have company in the universe. Unless astronomers detect signs of life elsewhere in the solar system or around other stars—providing a definitive answer to the question—we may never know. However, the absence of evidence is not evidence of absence.

Is Life Inevitable?

This is more a question for the biologists and chemists, but the fundamental laws of physics are set up in such a way that life is certainly possible (see "The Anthropic Principle"). But possible is different than inevitable. Scientists have not yet been able to show how life emerged from a sea of matter and energy.

What Exactly Is Dark Matter?

The stars and galaxies that fill this book make up only about 17 percent of the matter in the universe and only about 4 percent of all the matter and energy in the universe (see "Critical Density of the Universe"). The rest of the mass of the universe, apparent in its gravitational effects, is simply called dark matter (see "What Is Dark Matter?"). Astronomers have clues as to what dark matter might or might not be, but it is humbling to realize that so much of the matter in the universe remains a mystery.

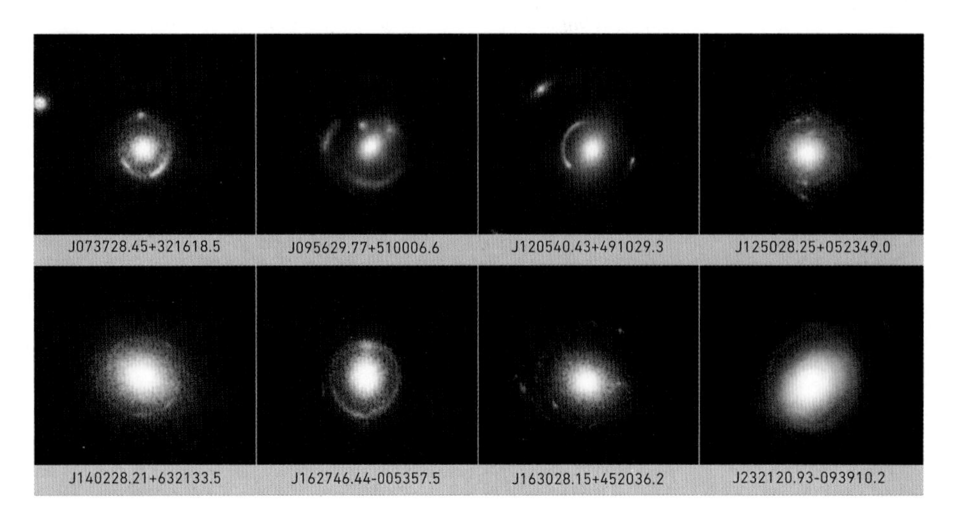

This image shows a variety of gravitational lenses called *Einstein Rings*. In these sources, foreground dark matter has distorted the light from a background galaxy. Astronomers know dark matter is there and how it's distributed; however, they don't yet know what it is.

What Was There Before the Universe?

The universe began with the Big Bang, and the evolution of it can be followed from the time it was 10^{-43} seconds old. But what there was before the universe is unanswered, and perhaps unanswerable.

Why Was There More Matter than Antimatter?

Early in the history of the universe, matter and antimatter were generated from the vacuum as the universe expanded suddenly in the era of inflation. Most of the matter and antimatter annihilated, but there was slightly more (1 part in 1 billion) more matter than antimatter. Why is that? What caused this imbalance? Without this slight asymmetry, there would be no matter in the observable universe, just light.

What Is Dark Energy?

About 70 percent of the critical density of the universe is in a form called *dark energy.* There is no theory that comes close to describing what dark energy might be. This is a big one; surely a Nobel Prize awaits the young minds who figure this one out.

Why Are There Three Generations of Matter?

Astronomers can observe that leptons and quarks come in three generations (see "Fundamental Particles"). But why there are three generations (for example, the electron, muon, and tau neutrino) is a fundamental question for cosmologists and particle physicists.

What Is the Fate of the Universe?

The universe is accelerating in its expansion (see "The Runaway Universe"), but what will that lead to in the future? Will there be a Big Rip, in which the distances between objects become infinite? Or will the universe slowly fade into a cold, dark state?

Is There a Theory of Everything (ToE)?

Cosmologists and physicists hope one day to arrive at a theory that combines quantum mechanics and general relativity into a single theoretical basis. This theory might also explain why the fundamental constants in the universe have the values that they do (see "The Anthropic Principle").

The recent announcement of gravitational signatures in the cosmic microwave background are a promising development in this search.

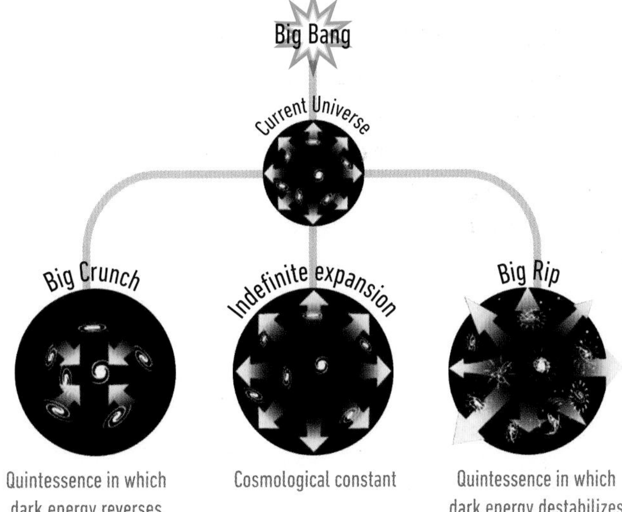

What is the ultimate fate of the universe? Some possibilities, from a Big Crunch to a Big Rip, are shown here.

Index

D

E

F

G

N

O

P

Q

R

S

T

U

V

W

X–Y–Z

Photo Credits:

iv, 170 ESO/Y. Beletsky
4 Gemini Observatory/Aura/NSF/P. Michaud
6, 45, 174 NASA/JPL
9 NASA/ISS
11 U.S. Naval Observatory
12, 18-21, 23, 25, 27-28, 30-31, 34-39, 42, 45, 55, 69, 86-87, 91-92, 104, 110, 117, 130-131, 145, 152-153, 179, 188, 191-192, 211, 233, 235, 240-241, 256-258, 264-267 NASA
14 NASA/MSFC
15 Hinode JAXA/NASA/PPARC
21 NASA/SOHO
24 Jesse Allen, Earth Observatory/NASA/GSFC
25, 32, 33 NOAA
26-27, 96 USGS
29 LANL/U.S. Dept. of Energy
33 NSIDC
35 NASA/GRAIL
39 T. A. Rector, I. P. Dell'Antonio/NOAO/AURA/NSF #
43 HST/NASA
48, 132, 149, 165, 221, 261, 270 NASA/JPL-Caltech
50 Robert Harry Van Gent/University of Utrecht
52 NASA/JHU Applied Physics Lab/Carnegie Institution of Washington
53 National Science Foundation
54 NASA/JPL/Mosaic by Mattias Malmer
55 NASA/SDO
55 Instituto di Linguistica Computazionale
57 ESA/NASA/JPL
57 ESA/AOES
57 NASA/JPL/RPIF/DLR
58-59 NSSDC Photo Gallery
58 NASA/JPL/USGS
59, 61, 63, 66-67, 69, 73- 75, 77-81, 84, 88, 91, 174 NASA/JPL
60 NASA/JPL/MSSS
60 NASA/LPI
61 NASA/JPL/USGS
62 MSSS/MGS/JPL/NASA
63, 163 NASA/STSCI
64 NASA/JPL/GSFC
64 NASA/JPL/Arizona State University
65 NASA/MGS
65 NASA/JPL-Caltech/University of Arizona
68, 72-73 NASA/JPL/University of Arizona
69 NASA/JPL/STSCI
70 R. Evans, J. Trauger, H. Hammel, and the HST Comet Science Team and NASA
71 JPL/NASA/STSCI
72, 74 NASA/JPL/DLR
76 NASA/JPL-Caltech/SSI
77 Lunar and Planetary Institute
81, 233 NASA/ESA/STSCI
82 NASA/JPL/SSI
83 NASA/JPL/ESA/University of Arizona
83 NASA/JPL-Caltech/USGS
85 Lawrence Sromovsky, University of Wisconsin-Madison/ W. M. Keck Observatory
86 Mike Brown/Caltech
87 Imke de Pater/Keck
89 USGS/NASA/JPL
89 NASA/JPL/LPI
90, 114 NASA/STSCI
90 NASA/ESA/Caltech
93 NASA/JHUAPL
93 NASA/ESA/SWRI/Cornell University/University of Maryland/STSCI
93 NASA/JPL-Caltech/UCAL/MPS/DLR/IDA
94 Halley Multicolor Camera Team, Giotto Project, ESA
95 NASA/JPL-Caltech/UMD
97 NASA/Don Davis
97, 103, 109 NASA/HST
101 NOAO
101 ESA/AOES Medialab
102 CTIO/NOAO/AURA/NSF
104-105, 112, 128, 159, 171, 173, 183, 225, 267 European Southern Observatory
106 NASA/HST/ESO
107 NASA/HST/Spitzer
108 NASA/CXC/Spitzer
109 HST/Wide Field Planetary Camera 2
111 NASA/ESA/STSCI
111 NASA/ESA/P. Hartigan (Rice University)
113, 162, 222, 253, 255 NASA/ESA
115 NASA/CXC/SAO
116 NASA, ESA, and the Hubble SM4 ERO Team
118 ESO/L. Calçada
119 Harvard-Smithsonian CFA/Nanjing University/K. Qiu
119, 157, 161, 199 NRAO
120 Submillimetre: ESO/APEX/T. Preibisch et al.; Optical: N. Smith, University of Minnesota/NOAO/AURA/NSF
120-121 NRAO/AUI/NSF
121 A. Ginsburg (University of Colorado—Boulder) et al., BGPS Team, Glimpse II Team]
122 NSF
123 ESA/HERSCHEL/PACS/L. Decin et al.
125 ORNL
125 NASA/Goddard Space Flight Center
126 NRAO/NASA/CXC
127 NASA/CXC
129 NASA/CXRC
129 Hubble Heritage Team (STSCI/AURA)
131 NASA/CXC/PSU/G. Pavlov et al.
133 ESA/XMM-NEWTON/M. Sasaki et al.
133 NASA/Goddard Space Flight Center
134 U.S. Nuclear Regulatory Commission
135 Sudbury Neutrino Observatory
136-137 NASA/CXC/M. Weiss
138 NASA/JPL/Gemini Observatory/AURA/NSF
139 Light Curve: NASA/CXC/GSFC/T. Strohmayer; Illustration: GSFC/D. Berry
142 NSF/GEMINI
143 Bruno Gilli/ESO
143 STELLARIUM.ORG
145, 157 NASA/GSFC
146 2MASS/G. Kopan, and R. Hurt
147 NASA/JPL-Caltech/EDA/CXC/STSCI
147 NRAO/AUI/NSF F. Yusef-Zadeh
148 ESA
149 NRAO/VLA/NASA/CXC/UCLA/Z. Li
149 NASA/DOE/FERMI LAT/D. Finkbeiner
150-151 Keck/UCLA Galactic Center Group
151 ESO/MPE/Marc Schartmann
151 ESO/R. Schoedel
154 NASA/JPL-Caltech/M. Povich (Penn State University)
155 NASA/Jim Ross
155 NASA, ESA/JPL-Caltech/R. Sankrit and W. Blair (Johns Hopkins University)
156 ESA/HERSCEL/SPIRE
158 Upper panel: ESO/Digitized Sky Survey 2; Lower panel: ESA/PACS/SPIRE/Tracey Hill, and Frédérique Motte, Laboratoire aim Paris-Saclay, CEA/IRFU—CNRS/INSU—UNIV. Paris Diderot, France
159 T. Dame (CFA)/COLUMBIA
159 NASA/JPL-Caltech/T. Bourke (Harvard-Smithsonian CFA)
160 X-ray: NASA/CXC/RUTGERS/G. Cassam-Chenaï, J. Hughes et al.; Radio: NRAO/AUI/NSF/GBT/VLA/Dyer, Maddalena, and Cornwell; Optical: Middlebury College/F. Winkler, NOAO/AURA/NSF/CTIO SCHMIDT and DSS)
161 INT/A. Zijlstra/University of Manchester/J. Irwin
162 NASA/JPL-Caltech/H. Kobulnicky (University of Wyoming)
163, 177 NASA/ESA/STSCI/AURA
164, 206, 216 ESA/Hubble & NASA
165 J. S. Urquhart/Robert Hurt, SSC/R. Benjamin
166 VLA/NRAO/C. De Pree
167 Near-infrared image: NASA; K. L. Luhman (Harvard-Smithsonian Center for Astrophysics, Cambridge, MA); and G. Schneider, E. Young, G. Rieke, A. Cotera, H. Chen, M. Rieke, R. Thompson (Steward Observatory, University of Arizona, Tucson, AR). Visible-light picture: NASA, C.R. O'Dell, and S.K. Wong (Rice University)
167 ESO/APEX/DSS2/SUPERCOSMOS/DEHARVENG (LAM)/ZAVAGNO (LAM)
168 MPIFR
169 Max Planck Institute for Astrophysics
169 Radio image: Ettore Carretti, CSIRO; Radio data: S-Pass Survey Team; Optical image: Axel Mellinger, Central Michigan University; Composition: Eli Bressert, CSIRO
171 David L. Nidever, et al., NRAO/AUI/NSF and Mellinger, Lab Survey, Parkes Observatory, Westerbork Observatory, and ARECIBO Observatory
171, 196 ESA/NASA/JPL-Caltech/STSCI
173 2MASS/NASA/NSF/J. Read and O. Agertz
175 OGLE/J. Skowron
175 LUX/D. Taylor, J. White, and C. Faham
177 NASA/JPL-Caltech/Carnegie
178 NASA/JPL/University of Arizona
179 SETI Institute
180 Wikipedia/A. Nordmann
181 NASA/AMES/JPL-Caltech
182 NASA, ESA, and P. Kalas (UC Berkeley)
183 NASA AMES
184-185 NASA/Kepler
188 IAU and Sky and Telescope Magazine (Roger Sinnott and Rick Fienberg)
189 Infrared: ESA/HERSCHEL/PACS/SPIRE/J. Fritz, U. Gent; X-ray: ESA/XMM-NEWTON/EPIC/W. Pietsch, MPE; Optical: R. Gendler
189 HSC Project/NAOJ
190 Hélène M. Courtois et al. 2013. The Astronomical Journal,146: 69
190 OVRA/Caltech/NOAO/KPNO
191 NAOJ/R. Gendler
193 ESO/P. Grosbel
193 NASA/JPL-Caltech/UCLA
194 M. West (ESO, CHILE), NASA, ESA and the Hubble Heritage (STSCI/AURA)-ESA/Hubble Collaboration
195 NASA, ESA, R. M. Crockett (University of Oxford, UK), S. Kaviraj (Imperial College London and University of Oxford, UK), J. Silk (University of Oxford), M. Mutchler (Space Telescope Science Institute, Baltimore), R. O'Connell (University of Virginia, Charlottesville), and the WFC3 SCientific Oversight Committee
195 Roeland P. Van Der Marel (STSCI), Frank C. Van Den Bosch (University of Washington), ANDNASA/ESA
196 NASA, ESA, and the Hubble Heritage Team (STSCI/AURA)
197 NAOJ/N. Arimoto/V. Vansevicius
197 X-ray: NASA/CXC/Virginia/A. Reines et al.; Radio: NRAO/AUI/NSF; Optical: NASA/STSCI
198 University of London Observatory/Steve Fossey/Ben Cooke/Guy Pollack/Matthew Wilde/Thomas Wright
199 NASA/SWIFT/P. Brown, TAMU
200 CLUES/Gottlöber, Hoffman, and Yepes
201 Optical: SDSS; Infrared: Spitzer Space Telescope
201 ESA/HST/NASA/J. Schmidt
201 Illustration credit: NASA, ESA, Z. Levay, and L. Frattare (STSCI); Image credit: A. Karick and M. Gregg (LLNL/UC, D) using the Michigan Curtis Schmidt Telescope at CTIO: Digital color composite/Mosaic in B, V and I Bands; Image courtesy: M. Drinkwater (University of Queensland)
202 JPL/Caltech/Planck
203, 238, 269 PNAS
203, 268 SDSS
204 NASA, ESA, A. Evans (University of Virginia, Charlottesville/NRAO/Stony Brook University), and the Hubble Heritage Team (STSCI/AURA)-ESA/Hubble Collaboration
204 NASA/ESA/M. Livio/STSCI/AURA
205 Credit: (NRAO/AUI/NSF); ALMA (ESO/NAOJ/NRAO); HST (NASA, ESA, AND B. Whitmore (STSCI); J. Hibbard, (NRAO/AUI/NSF); NOAO/AURA/NSF
205 Palomar Sky Survey/VLA/M. S. Yun
206 ESA/NASA/JPL-Caltech Keck/SMA
207 Andrew Fruchter (STSCI) et al., WFPC2, HST, NASA
208 NASA, ESA, C. Heymans (University of British Columbia, Vancouver), M. Gray (University of Nottingham, UK), M. Barden (Innsbruck), and the Stages Collaboration
209 NASA, ESA, M. Jee, and H. Ford (Johns Hopkins University)
209 X-ray: NASA/CXC/CFA/M. Markevitch; Optical and lensing map: NASA/STSCI, Magellan/University of Arizona/D. Clowe; Lensing map: ESO WFI
210 T. Weinzirl, S. Jogee (University of Texas)/STSCI/NASA/SDSS
211 NASA, H. Ford (JHU), G. Illingworth (UCSC/LO), M. Clampin (STSCI), G. Hartig (STSCI), the ACS Science Team, and ESA
212 ESO/M. Kornmesser
213 NRAO/AUI
214 ASA, ESA, S. Baum and C. O'Dea (RIT), R. Perley, and W. Cotton (NRAO/AUI/NSF), and the Hubble Heritage Team (STSCI/AURA)
215 X-Ray: NASA/CXC/CFA/W. Forman et al.; Radio: NRAO/AUI/NSF/W. cotton; Optical: NASA/ESA/Hubble Heritage Team (STSCI/AURA), and R. Gendler
215 NRAO/NASA/ESA
216-217 NASA/DOE/FERMI LAT
217 NASA/STSCI
218 ESO, ALMA (ESO/NAOJ/NRAO), T. Izumi
218 OSU/S. Kazantzitis
219 NASA/SWIFT/NOAO/Michael Koss and Richard Mushotzky (University of Maryland)
220 Optical: Capella Observatory; Radio data: Ilana Feain, Tim Cornwell, and Ron Ekers (CSIRO/ATNF), R. Morganti (ASTRON), and N. Junkes (MPIFR), NASA/TANAMI/Müller et al. (right image)
221 NASA'S Goddard Space Flight Center/J. Schnittman, J. Krolik (JHU) and S. Noble (RIT)
222 STSCI/NASA
223 NASA/ESA/Z. Levay, F. Summers
224 2MRS TEAM
225 2MASS/T. H. Jarrett, J. Carpenter, and R. Hurt
226 AAO/6DFGS
227 AAO/A. Jones, IPAC/Caltech/T. Jarrett, 2MASS
227 MPIFA/V. Springel
230 FAMSI.ORG
232 A. Eddington
237 University of Oregon/J. Imamura
239 NASA/JPL-Caltech/W. Freedman (Carnegie)
239 NASA/HST/ESA
241, 243, 254 NASA/WMAP Science Team
242 NASA/COBE
243 NASA/ESA
244 NASA/JPL-Caltech/A. Kashlinsky (Goddard Space Flight Center)
245 ESA/C. Carreau
246 AAAS
247 CERN/M. Brice
250 Particle Data Group/LBNL/DOE/NSF
251 CERN
254 ATNF/CSIRO
259 X-ray: NASA/CXC/ITA/INAF/J. Merten et al, Lensing: NASA/STSCI; NAOJ/SUBARU; ESO/VLT, Optical: NASA/STSCI/R. Dupke
260 ESA and the Planck Collaboration; NASA/WMAP Science Team
261 NASA, ESA, G. Piotto (University OF Padua), and A. Sarajedini (University of Florida)
262 University of Chicago/CCP
263 NASA/ESA/STSCI/JHU
269 LBNL/SCP
271 ISS/EOL/NASA
271 ESA/Planck Collaboration
272 NASA/ESA/A. Bolton (CFA)/SLACS Team
273 NASA/STSCI/A. Field